A·N·N·U·A·L EDITIONS

ENVIRONMENT 01/02
Twentieth Edition

EDITOR

John L. Allen
University of Wyoming

John L. Allen is professor of geography at the University of Wyoming. He received his bachelor's degree in 1963 and his M.A. in 1964 from the University of Wyoming, and in 1969 he received his Ph.D. from Clark University. His special area of interest is the impact of contemporary human societies on environmental systems.

McGraw-Hill/Dushkin
530 Old Whitfield Street, Guilford, Connecticut 06437

Visit us on the Internet
http://www.dushkin.com

Credits

1. The Global Environment: An Emerging World View
Unit photo—Courtesy of NASA.
2. The World's Population: People and Hunger
Unit photo—United Nations photo by Ray Witlin.
3. Energy: Present and Future Problems
Unit photo—© Sweet By & By/Cindy Brown. All rights reserved.
4. Biosphere: Endangered Species
Unit photo—United Nations photo by M. Gonzalez.
5. Resources: Land, Water, and Air
Unit photo—United Nations photo.
6. Pollution: The Hazards of Growth
Unit photo—McGraw-Hill/Dushkin photo courtesy of Jean Bailey.

Copyright

Cataloging in Publication Data
Main entry under title: Annual Editions: Environment. 2001/2002.
   1. Environment—Periodicals. 2. Ecology—Periodicals. I. Allen, John L., comp.
II. Title: Environment.
ISBN 0-07-243359-0   301.31'05   79-644216   ISSN 0272-9008

© 2001 by McGraw-Hill/Dushkin, Guilford, CT 06437, A Division of The McGraw-Hill Companies.

Copyright law prohibits the reproduction, storage, or transmission in any form by any means of any portion of this publication without the express written permission of McGraw-Hill/Dushkin, and of the copyright holder (if different) of the part of the publication to be reproduced. The Guidelines for Classroom Copying endorsed by Congress explicitly state that unauthorized copying may not be used to create, to replace, or to substitute for anthologies, compilations, or collective works.

Annual Editions® is a Registered Trademark of McGraw-Hill/Dushkin, A Division of The McGraw-Hill Companies.

Twentieth Edition

Cover image © 2001 by PhotoDisc, Inc.

Printed in the United States of America   1234567890BAHBAH54321   Printed on Recycled Paper

Editors/Advisory Board

Members of the Advisory Board are instrumental in the final selection of articles for each edition of ANNUAL EDITIONS. Their review of articles for content, level, currentness, and appropriateness provides critical direction to the editor and staff. We think that you will find their careful consideration well reflected in this volume.

EDITORS

John L. Allen
University of Wyoming

ADVISORY BOARD

Daniel Agley
Towson University

Matthew R. Auer
Indiana University

Robert V. Bartlett
Purdue University

Susan W. Beatty
University of Colorado
Boulder

William P. Cunningham
University of Minnesota
St. Paul

Lisa Danko
Mercyhurst College

Dianne Draper
University of Calgary

Juris Dreifelds
Brock University

Debbie R. Folkerts
Auburn University–Main

Theodore D. Goldfarb
SUNY at Stony Brook

Gian Gupta
University of Maryland

George N. Huppert
University of Wisconsin
La Crosse

Jeffrey W. Jacobs
National Research Council

Vishnu R. Khade
Eastern Connecticut
State University

Adil Najam
Boston University

David J. Nemeth
University of Toledo

Shannon O'Lear
Illinois State University

David Padgett
Tennessee State University

John H. Parker
Florida International University

Charles R. Peebles
Michigan State University

Joseph David Shorthouse
Laurentian University

Bradley F. Smith
Western Washington University

Nicholas J. Smith-Sebasto
Montclair State University

Ben E. Wodi
SUNY at Cortland

Staff

EDITORIAL STAFF

Ian A. Nielsen, Publisher
Roberta Monaco, Senior Developmental Editor
Dorothy Fink, Associate Developmental Editor
Addie Raucci, Senior Administrative Editor
Robin Zarnetske, Permissions Editor
Joseph Offredi, Permissions Assistant
Diane Barker, Proofreader
Lisa Holmes-Doebrick, Senior Program Coordinator

TECHNOLOGY STAFF

Richard Tietjen, Senior Publishing Technologist
Jonathan Stowe, Director of Technology
Janice Ward, Software Support Analyst
Ciro Parente, Editorial Assistant

PRODUCTION STAFF

Brenda S. Filley, Director of Production
Charles Vitelli, Designer
Laura Levine, Graphics
Mike Campbell, Graphics
Tom Goddard, Graphics
Eldis Lima, Graphics
Nancy Norton, Graphics
Juliana Arbo, Typesetting Supervisor
Marie Lazauskas, Typesetter
Karen Roberts, Typesetter
Jocelyn Proto, Typesetter
Larry Killian, Copier Coordinator

To the Reader

In publishing ANNUAL EDITIONS we recognize the enormous role played by the magazines, newspapers, and journals of the public press in providing current, first-rate educational information in a broad spectrum of interest areas. Many of these articles are appropriate for students, researchers, and professionals seeking accurate, current material to help bridge the gap between principles and theories and the real world. These articles, however, become more useful for study when those of lasting value are carefully collected, organized, indexed, and reproduced in a low-cost format, which provides easy and permanent access when the material is needed. That is the role played by ANNUAL EDITIONS.

As a new millennium begins, environmental dilemmas long foreseen by natural and social scientists have begun to emerge in a number of guises: population/food imbalances, problems of energy scarcity, acid rain, toxic and hazardous wastes, ozone depletion, water shortages, massive soil erosion, global atmospheric pollution and possible climate change, forest dieback and tropical deforestation, and the highest rates of plant and animal extinction the world has known in 65 million years.

These and other problems have worsened in spite of an increasing amount of national and international attention to environmental issues and increased environmental awareness and legislation. The problems have resulted from centuries of exploitation and unwise use of resources, accelerated recently by the short-sighted public policies that have favored the short-term, expedient approach to problem solving over longer-term economic and ecological good sense. In Africa, for example, the drive to produce enough food to support a growing population has caused the use of increasingly fragile and marginal resources, resulting in the dryland deterioration that brings famine to that troubled continent. Similar social and economic problems have contributed to massive deforestation in Middle and South America and Southeast Asia.

During the decade of the eighties, economic problems generated by resource scarcity caused the relaxation of environmental quality standards and contributed to the legislature's refusal to enact environmentally sound protective measures, which were viewed as too costly. More recently, in the nineties, governments adopted environmental protection policies that were often cosmetic, designed for good press but little else. Even with these public relations policies, governments often lacked either the will or the means to implement them properly. The absence of effective environmental policy has been particularly apparent in those countries that are striving to become economically developed. But even in the more highly developed nations, economic concerns tend to favor a loosening of environmental controls. In the interests of maintaining jobs for the timber industry, for example, many of the last areas of old-growth forests in the United States are imperiled, and in the interests of maintaining agricultural productivity at all costs, destructive and toxic chemicals continue to be used on the nation's farmlands. In addition, concerns over energy availability have created the need for foreign policy and military action to protect the developed nations' access to cheap oil and have prompted increasing reliance on technological quick fixes, as well as the development of environmentally-sensitive areas to new energy resource exploration and exploitation.

There is some reason to hope that, globally, a new environmental consciousness is awakening at the dawning of a new millennium. The dissolution of the Soviet Union lifted the Iron Curtain, and the environmental horror stories that emerged from Eastern Europe and the newly independent states that made up the former USSR gave new incentives to international cooperation. International conferences have been held on global warming and other environmental issues, and, in spite of the recalcitrance of some of the world's most important countries, there is some evidence of an increased international desire to do something about environmental quality before it is too late.

The articles contained in *Annual Editions: Environment 01/02* have been selected for the light they shed on these and other problems and issues. The selection process was aimed at including material that will be readily assimilated by the general reader. Additionally, every effort has been made to choose articles that encourage an understanding of the nature of the environmental problems that beset us and how, with wisdom and knowledge and the proper perspective, they can be solved or at least mitigated. Accordingly, the selections in this book have been chosen more for their intellectual content than for their emotional tone. They have been arranged into an order of topics—the global environment, population and food, energy, the biosphere, resources, and pollution—that lends itself to a progressive understanding of the causes and effects of human modifications of Earth's environmental systems. We will not be protected against the ecological consequences of human actions by remaining ignorant of them. Although the knowledge gained through the use of this book may not allow any of us to escape the environmental predicament, it should ensure that we do not continue to act and react in ways that will make that predicament worse.

The *World Wide Web* sites in this edition can be used to further explore the topics. These sites will be cross-referenced by number in the *topic guide*. In addition, this edition contains both a newly refreshed *Environmental Information Retrieval* guide and *glossary*.

Readers can have input into the next edition of *Annual Editions: Environment* by completing and returning the postpaid *article rating form* at the back of the book.

John L. Allen
Editor

Contents

To the Reader   iv
Topic Guide   2
◉ Selected World Wide Web Sites   4

Overview   6

1. **Environmental Surprises: Planning for the Unexpected,** Chris Bright, *The Futurist,* July/August 2000.   8

   Those scientists who investigate the different possibilities of the future have come up with some interesting ways to anticipate the consequences of the complex processes at work in today's world. Two of their major findings are that ***ecological change*** is nearly always irreversible and that ***human systems*** need to find a way to work with, rather than against, ***natural systems.***

2. **The Nemesis Effect,** Chris Bright, *World Watch,* May/June 1999.   14

   The complexity of ***environmental systems*** and human interactions with them suggests that environmental pressures produced by human activities will begin to converge in ways that will produce a number of unexpected ***environmental crises.*** The only way to avoid the "nemesis" of unpredicted environmental crisis is to do a better job of managing the human/environmental systems.

3. **Harnessing Corporate Power to Heal the Planet,** L. Hunter Lovins and Amory B. Lovins, *The World & I,* April 2000.   24

   One of the great intellectual shifts of the late twentieth century is the emergence of a new form of economics called ***"natural capitalism."*** The basis of the new corporate thinking is to enhance ***resource productivity*** through recycling and other efforts to eliminate the concept of waste. The primary goal of natural capitalism is to reverse the global trend of ecosystem destruction by using nature as a mentor and model in manufacturing.

4. **Crossing the Threshold: Early Signs of an Environmental Awakening,** Lester R. Brown, *World Watch,* March/April 1999.   30

   An ***environmental threshold*** is a critical parameter beyond which significant ***environmental change*** becomes inevitable. Globally, we are approaching some important environmental thresholds (the number of endangered species, for example). But we may also be approaching some key breakthroughs in social concepts and organization that will enable us to restructure the ***global economy*** before environmental deterioration can lead to irreversible economic decline.

UNIT 1

The Global Environment: An Emerging World View

Four selections provide information on the current state of Earth and the changes we will face.

The concepts in bold italics are developed in the article. For further expansion please refer to the Topic Guide, the Glossary, and the Index.

UNIT 2

The World's Population: People and Hunger

Four selections examine the problems the world will have in feeding its ever-increasing population.

Overview   **40**

5. **The Population Surprise,** Max Singer, *The Atlantic Monthly,* August 1999.   **42**

   For years, population experts have been predicting that the world's population will continue to grow well into the twenty-first century before stabilizing sometime after 2050. Most of these predictions have been based on the ***demographic transition,*** a pattern of ***population growth*** experienced in the industrialized nations in which population growth eventually approaches zero. Recent evidence suggests that rather than leveling off, the world's population will actually enter a period of decline.

6. **Population and Consumption: What We Know, What We Need to Know,** Robert W. Kates, *Environment,* April 2000.   **44**

   A general consensus exists among scientists that the roots of the current ***environmental crisis*** are to be found in a combination of ***population growth,*** affluence, and increasing ***technology.*** No such consensus exists, however, about the ultimate causes of either population growth or the desire to consume. Notwithstanding this lack of agreement, society needs to sublimate the desire to acquire things for the good of the ***global commons.***

7. **Food for All in the 21st Century,** Gordon Conway, *Environment,* January/February 2000.   **52**

   For people in the world's industrialized countries, there is little awareness of the depth of the global ***food problem.*** In order for ***agricultural systems*** to provide enough food to feed the world's population in the twenty-first century, that awareness needs to develop, as does agricultural and natural resource production aimed at not just equitable ***food production*** but at ***sustainability.***

8. **Escaping Hunger, Escaping Excess,** Gary Gardner and Brian Halweil, *World Watch,* July/August 2000.   **61**

   Malnutrition is a growing global epidemic. Malnourishment includes both the underfed of the poorer countries and the overfed of the richer ones, and because misunderstanding exists as to the nature of malnutrition, policy responses to it have been inadequate. When taken in total, the world's ***food problem*** often appears to be more a problem of allocation than a problem of supply.

The concepts in bold italics are developed in the article. For further expansion please refer to the Topic Guide, the Glossary, and the Index.

Overview 70

9. **King Coal's Weakening Grip on Power,** Seth Dunn, *World Watch,* September/October 1999. 72

   Although the fuel of choice today is petroleum, for most of the last thousand years coal has been the **fossil fuel** most in demand for uses ranging from the village blacksmith to the modern electrical generation power plant. The use of coal has left a legacy of human and **environmental costs** that we have only now begun to assess. Initiatives to replace coal as a primary fuel in all societies suggest that the reign of this destructive **energy source** is nearing an end.

10. **Oil, Profit$, and the Question of Alternative Energy,** Richard Rosentreter, *The Humanist,* September/October 2000. 79

    Even though the costs of **fossil fuels,** particularly oil and natural gas, continue to rise, there has been public policy encouragement for the development of **alternative energy** sources, such as solar power, wind power, and others. It could be argued that the failure of alternative energy to capture either public attention or public money is the result of the political influence wielded by large corporations wedded to nonrenewable energy.

11. **Here Comes the Sun: Whatever Happened to Solar Energy?** Eric Weltman, *In These Times,* February 7, 2000. 83

    Amidst the oil embargos and nuclear accidents of the seventies, the future seemed to belong to **solar energy,** but federal dollars for research funding for solar power has decreased more than 600 percent since 1980, and **renewable energy** is expected to supply only about 3 percent of the needs of the United States in 2020. The primary reason for not implementing these alternative energy sources is the **political environment** surrounding **renewable energy** issues.

12. **The Hydrogen Experiment,** Seth Dunn, *World Watch,* November/December 2000. 87

    Iceland is an isolated volcanic island long adjusted to using **alternative energy** sources, particularly **geothermal energy** from the country's many hot springs and thermal vents. Now a new alternative energy experiment has been set into motion in Iceland's capital city—**hydrogen power.**

13. **Power Play,** *Business Week,* February 12, 2001. 97

    The energy crisis in California is a prime example of overextended expectations on environmental issues. This aticle focuses on the manipulation and mismanagement of generating electrical energy and how these actions can impact on the average citizen. Implications are that the deregulation of energy sources could affect the entire nation.

UNIT 3

Energy: Present and Future Problems

Six articles consider the problems of meeting present and future energy needs. Alternative energy sources are also examined.

14. **Bull Market in Wind Energy,** Christopher Flavin, *World Watch,* March/April 1999. 103

   Wind power is one of the world's most rapidly expanding industries, and both industrialized and developing countries are discovering that **electrical energy** from wind-driven turbines is not only cheap and environmentally protective but also technologically accessible. In some rapidly developing countries like China, the potential of wind energy exceeds the current demand for electricity.

Overview 106

15. **Planet of Weeds,** David Quammen, *Harper's,* October 1998. 108

   Earth has undergone periods of major **biological extinction** requiring millions of years of recovery time. Biologists believe that we are entering another such period: a significant reduction in **biodiversity** brought about not by natural forces but by human action. Over the next century huge percentages of Earth's plants and animals will disappear, leaving behind impoverished ecosystems dominated by "weeds," the hardiest, most adaptable plants and animals, including the consumate weed: ***Homo sapiens.***

A. PLANTS

16. **Invasive Species: Pathogens of Globalization,** Christopher Bright, *Foreign Policy,* Fall 1999. 116

   One of the least expected and least visible consequences of economic globalization has been the spread of **invasive species**—plants and animals that are hitchhiking through the global trading network and finding niches where they can survive better than native species. This **bioinvasion** is difficult to control because it means altering the nature of the **global trading economy** that released invasive species in the first place.

B. ANIMALS

17. **Mass Extinction,** David Hosansky, *CQ Researcher,* September 15, 2000. 123

   Biologists cannot be precise about the number of species that are becoming extinct because they don't know how many species exist. What they can say, however, is that the rate of **biological extinction** over the past few centuries is 100 to 10,000 times the normal rate. The Earth has lost over 30 percent of its species since 1970 and it may be that half of all existing **animal species** in 2000 will be gone by the end of the twenty-first century.

UNIT 4

Biosphere: Endangered Species

Four articles examine the problems in the world's biosphere. Not only are plants and animals endangered, but many human groups are also disastrously affected by deforestation and primitive agricultural policies.

18. **Watching vs. Taking,** Howard Youth, *World Watch,* 130
May/June 2000.
In many parts of the world, people have turned to ***wildlife watching*** for recreation and business instead of for hunting and poaching. The reasons are largely economic: a live lion in a Kenyan game reserve may generate as much as $575,000 in tourist dollars. This is ***ecotourism*** at its most productive. Certainly the growing awareness of global ***biodiversity*** loss is an issue, as the public becomes more aware that watching rare and endangered species can be a valuable experience—and one not easily replicated.

Overview 140

19. **The Tragedy of the Commons: 30 Years Later,** 142
Joanna Burger and Michael Gochfeld, *Environment,*
December 1998.
In 1968, the pioneering human ecologist Garrett Hardin argued in an article entitled "***The Tragedy of the Commons***" that increasing human ***population*** would create such pressure on finite resources at both local and global levels that the inevitable consequence would be overexploitation and ***environmental crisis.*** Hardin's work spawned new approaches to ***resource management,*** but, 30 years later, the problem of the commons still exists.

A. LAND

20. **Where Have All the Farmers Gone?** Brian Halweil, 150
World Watch, September/October 2000.
The movement toward a ***global economy*** has meant a standardization in the management of much of the world's land. These new standardized ***land management*** practices have, in turn, led to a decrease in the number of farmers. As ***agribusiness,*** in the form of large corporations, takes over more and more of the world's farmland, not only is a way of life lost but also crop diversity, ecosystems, and cultures are threatened.

B. WATER

21. **When the World's Wells Run Dry,** Sandra Postel, 162
World Watch, September/October 1999.
Most of the water used in ***irrigation agriculture*** is ***groundwater*** rather than water drawn from surface reservoirs. And because groundwater is being extracted or withdrawn at rates far in excess of its renewal or recharge, the world is quickly running short of one of its most precious resources. The only solution is to develop plans to reduce overconsumption of groundwater and to ensure sustainable groundwater use.

UNIT 5

Resources: Land, Water, and Air

Six selections discuss the environmental problems affecting our land, water, and air resources.

22. Oceans Are on the Critical List, Anne Platt McGinn, *USA Today Magazine (Society for the Advancement of Education),* January 2000.   170

The world's oceans are both central to the **global economy** and to human and planetary health. Yet these critical areas are being threatened by overfishing, **habitat degradation,** pollution, introduction of alien species, and **climate change.** Unfortunately, protection efforts are being hampered and the human impact on oceans is in danger of disrupting life on the planet.

C. AIR

23. The Human Impact on Climate, Thomas R. Karl and Kevin E. Trenberth, *Scientific American,* December 1999.   174

Scientists are in general agreement that the **global climate** is undergoing a warming trend. There is even substantial agreement that much of this temperature increase is human-induced. What is not known is exactly how much of the **global warming** can be attributed to natural processes or to human ones. We can have accurate **climate monitoring** systems in place by the middle of the century, but to wait until those systems are in place before taking action to halt the warming trend would be foolish.

24. Warming Up: The Real Evidence for the Greenhouse Effect, Gregg Easterbrook, *The New Republic,* November 8, 1999.   180

Despite the scientific and nonscientific rhetoric and wrangling over **global warming,** it is clear that the planet is becoming warmer and that human activities are playing some role in that process. Whoever develops a **clean energy** system will not only aid in the global warming problem but also will have a significant competitive advantage in twenty-first-century economics.

Overview   186

25. Making Things Last: Reinventing Our Material Culture, Gary Gardner and Payal Sampat, *The Futurist,* May 1999.   188

Consumption of industrial products in countries like the United States have increased nearly twentyfold in this century and manufacturing has converted unprecedented amounts of **raw materials** to usable products that then end up as **solid waste.** The waste that characterizes the industrialized countries of the world has produced enormous damage to both human and **environmental health.**

UNIT 6

Pollution: The Hazards of Growth

Four selections weigh the environmental impacts of the growth of human population.

The concepts in bold italics are developed in the article. For further expansion please refer to the Topic Guide, the Glossary, and the Index.

26. **Groundwater Shock: The Polluting of the World's Major Freshwater Stores,** Payal Sampat, *World Watch,* January/February 2000. 193

Most of the planet's freshwater—97 percent to be exact—is stored in vast underground **aquifers,** which supply nearly 40 percent of the world's population with drinking water and over 90 percent of the world's irrigated agriculture with the water necessary to sustain it. While the pollution of surface water is easily recognizable and has more readily understood sources, **groundwater pollution** is not only more difficult to identify but also tends to have sources that are less visible.

27. **POPs Culture,** Anne Platt McGinn, *World Watch,* March/April 2000. 204

While industrial innovation is usually viewed as a good thing, at least economically, one form of innovation that the world could do without is that which produces **persistent organic pollutants** or "POPs." Many of these substances, widely used in both agricultural and industry, are so toxic and so durable that they may be creating **public health** problems 1,000 years from now.

28. **It's a Breath of Fresh Air,** David Whitman, *U.S. News & World Report,* April 17, 2000. 212

When the world celebrated the 30th anniversary of the original **Earth Day** in April 2000, the United States was pleased with the progress made in **environmental cleanup.** While most Americans surveyed in 2000 believed that the environment had improved only slightly since 1970, the fact is that the environmental trend in the United States has produced substantial reductions in air and water **pollution,** without forestalling **economic growth.**

Environmental Information Retrieval	215
Glossary	221
Index	225
Test Your Knowledge Form	228
Article Rating Form	229

The concepts in bold italics are developed in the article. For further expansion please refer to the Topic Guide, the Glossary, and the Index.

Topic Guide

This topic guide suggests how the selections in this book relate to the subjects covered in your course.

The Web icon (☉) under the topic articles easily identifies the relevant Web sites, which are numbered and annotated on the next two pages. By linking the articles and the Web sites by topic, this ANNUAL EDITIONS reader becomes a powerful learning and research tool.

TOPIC AREA	TREATED IN	TOPIC AREA	TREATED IN
Agribusiness	20. Where Have All the Farmers Gone? ☉ **11, 12, 13, 14**	Environmental Change	4. Crossing the Threshold: Early Signs of an Environmental Awakening ☉ **2, 4, 10, 17, 34**
Agricultural Systems	7. Food for All in the 21st Century ☉ **11, 12, 13, 14**	Environmental Cleanup	28. It's a Breath of Fresh Air ☉ **2, 10, 34**
Alternative Energy	10. Oil, Profit$, and the Question of Alternative Energy 11. Here Comes the Sun: Whatever Happened to Solar Energy? ☉ **15, 16, 17, 18, 19**	Environmental Costs	9. King Coal's Weakening Grip on Power ☉ **4, 10, 34**
		Environmental Crisis	2. Nemesis Effect 6. Population and Consumption: What We Know, What We Need to Know 19. Tragedy of the Commons: 30 Years Later ☉ **2, 4, 10, 34**
Animal Species	17. Mass Extinction ☉ **1, 20, 21, 22, 23, 24**		
Aquifers	26. Groundwater Shock: The Polluting of the World's Major Freshwater Stores ☉ **28, 29**		
		Environmental Health	25. Making Things Last: Reinventing Our Material Culture
Biodiversity	15. Planet of Weeds ☉ **20, 21, 22, 23, 24**	Environmental Systems	2. Nemesis Effect ☉ **2, 4, 10, 34**
Bioinvasion	16. Invasive Species: Pathogens of Globalization ☉ **2, 3, 7**	Environmental Threshold	4. Crossing the Threshold: Early Signs of an Environmental Awakening ☉ **2, 4, 10, 17, 34**
Biological Extinction	15. Planet of Weeds 17. Mass Extinction ☉ **30, 33**	Food Problem	7. Food for All in the 21st Century 8. Escaping Hunger, Escaping Excess ☉ **11, 12, 13, 14**
Clean Energy	24. Warming Up: The Real Evidence for the Greenhouse Effect ☉ **15, 16, 17, 18, 19**	Food Production	7. Food for All in the 21st Century ☉ **11, 12, 13, 14**
Climate Change	22. Oceans Are on the Critical List ☉ **25, 26, 28, 29**	Fossil Fuels	9. King Coal's Weakening Grip on Power 10. Oil, Profit$, and the Question of Alternative Energy ☉ **4, 10, 34**
Climate Monitoring	23. Human Impact on Climate ☉ **25, 26, 28, 29**		
Demographic Transition	5. Population Surprise ☉ **11, 12, 13, 14**	Geothermal Energy	12. Hydrogen Experiment ☉ **5, 15, 16, 18, 19**
Ecological Change	1. Environmental Surprises: Planning for the Unexpected ☉ **2, 4, 34**	Global Climate	23. Human Impact on Climate ☉ **6, 7, 8, 25**
Economic Growth	28. It's a Breath of Fresh Air ☉ **16, 34**	Global Commons	6. Population and Consumption: What We Know, What We Need to Know ☉ **6, 7, 8, 25, 26, 29**
Ecotourism	18. Watching vs. Taking ☉ **21**		
Electrical Energy	14. Bull Market in Wind Energy ☉ **15, 17, 18, 19**	Global Economy	4. Crossing the Threshold: Early Signs of an Environmental Awakening 20. Where Have All the Farmers Gone? 22. Oceans Are on the Critical List 26. Groundwater Shock: The Polluting of the World's Major Freshwater Stores ☉ **6, 7, 8, 11, 12, 13, 14**
Energy Source	9. King Coal's Weakening Grip on Power ☉ **15, 16, 17, 18, 19**		

TOPIC AREA	TREATED IN	TOPIC AREA	TREATED IN
Global Warming	23. Human Impact on Climate 24. Warming Up: The Real Evidence for the Greenhouse Effect ● **2, 6, 7, 8, 20, 25, 28, 29**	**Population**	5. Population Surprise 6. Population and Consumption: What We Know, What We Need to Know 7. Food for All in the 21st Century 19. Tragedy of the Commons: 30 Years Later ● **9, 10, 12, 14**
Groundwater	21. When the World's Wells Run Dry ● **28, 29**	**Population Growth**	5. Population Surprise 6. Population and Consumption: What We Know, What We Need to Know ● **9, 10, 12, 14**
Groundwater Pollution	26. Groundwater Shock: The Polluting of the World's Major Freshwater Stores ● **28, 29**	**Public Health**	27. POPs Culture ● **32, 34**
Habitat Degradation	22. Oceans Are on the Critical List ● **2, 7, 9, 10, 26, 34**	**Raw Materials**	25. Making Things Last: Reinventing Our Material Culture ● **1, 4**
Homo sapiens	15. Planet of Weeds 17. Mass Extinction ● **34**	**Renewable Energy**	10. Oil, Profit$, and the Question of Alternative Energy ● **15, 16, 17, 18, 19**
Human Systems	1. Environmental Surprises: Planning for the Unexpected ● **34**	**Resource Management**	19. Tragedy of the Commons: 30 Years Later ● **9, 10, 12, 14**
Hydrogen Power	12. Hydrogen Experiment ● **15, 16, 19**	**Resource Productivity**	3. Harnessing Corporate Power to Heal the Planet ● **9, 10, 12, 27, 28, 29**
Invasive Species	16. Invasive Species: Pathogens of Globalization ● **2, 3, 7**	**Solar Energy**	11. Here Comes the Sun: Whatever Happened to Solar Energy? ● **25, 29**
Irrigation Agriculture	21. When the World's Wells Run Dry ● **26**	**Solid Waste**	25. Making Things Last: Reinventing Our Material Culture ● **1, 4**
Malnutrition	8. Escaping Hunger, Escaping Excess ● **11, 12, 13**	**Sustainability**	7. Food for All in the 21st Century ● **2, 6, 8**
Mass Extinction	17. Mass Extinction ● **16, 30, 33, 34**	**The Tragedy of the Commons**	19. Tragedy of the Commons: 30 Years Later ● **9, 10, 12, 14**
Natural Capitalism	3. Harnessing Corporate Power to Heal the Planet	**Wildlife Watching**	18. Watching vs. Taking ● **21**
Natural Systems	1. Environmental Surprises: Planning for the Unexpected	**Wind Power**	10. Oil, Profit$, and the Question of Alternative Energy 14. Bull Market in Wind Energy ● **25, 26, 29**
Persistent Organic Pollutants	27. POPs Culture ● **31**		
Political Environment	11. Here Comes the Sun: Whatever Happened to Solar Energy? ● **1, 2, 16, 34**		
Pollution	22. Oceans Are on the Critical List 26. Groundwater Shock: The Polluting of the World's Major Freshwater Stores 28. It's a Breath of Fresh Air ● **26, 34**		

AE: Environment

The following World Wide Web sites have been carefully researched and selected to support the articles found in this reader. The sites are cross-referenced by number and the Web icon (●) in the topic guide. In addition, it is possible to link directly to these Web sites through our DUSHKIN ONLINE support site at *http://www.dushkin.com/online/*.

The following sites were available at the time of publication. Visit our Web site—we update DUSHKIN ONLINE regularly to reflect any changes.

General Sources

1. Britannica's Internet Guide
http://www.britannica.com
This site presents extensive links to material on world geography and culture, encompassing material on wildlife, human lifestyles, and the environment.

2. EnviroLink
http://envirolink.netforchange.com
One of the world's largest environmental information clearinghouses, EnviroLink is a grassroots nonprofit organization that unites organizations and volunteers around the world and provides up-to-date information and resources.

3. Library of Congress
http://www.loc.gov
Examine this extensive Web site to learn about resource tools, library services/resources, exhibitions, and databases in many different subfields of environmental studies.

4. SocioSite: Sociological Subject Areas
http://www.pscw.uva.nl/sociosite/TOPICS/
This huge sociological site from the University of Amsterdam provides many discussions and references of interest to students of the environment, such as the links to information on ecology and consumerism.

5. U.S. Geological Survey
http://www.usgs.gov
This site and its many links are replete with information and resources in environmental studies, from explanations of El Niño to discussion of concerns about water resources.

The Global Environment: An Emerging World View

6. Earth Science Enterprise
http://www.earth.nasa.gov
This site will direct you to information about NASA's Mission to Planet Earth program and its Science of the Earth System. Surf here to learn about satellites, El Niño, and even "strategic visions" of interest to environmentalists.

7. National Geographic Society
http://www.nationalgeographic.com
This site provides links to National Geographic's huge archive of maps, articles, and other documents. There is a great deal of material related to the atmosphere, the oceans, and other environmental topics.

8. Santa Fe Institute
http://acoma.santafe.edu
This home page of the Santa Fe Institute—a nonprofit, multidisciplinary research and education center—will lead to many interesting links related to its primary goal: to create a new kind of scientific research community, pursuing emerging science.

9. United Nations
http://www.unsystem.org
Visit this official Web site Locator for the United Nations System of Organizations to get a sense of the scope of international environmental inquiry today. Various UN organizations concern themselves with everything from maritime law to habitat protection to agriculture.

10. United Nations Environment Programme
http://www.unep.ch
Consult this home page of UNEP for links to critical topics of concern to environmentalists, including desertification, migratory species, and the impact of trade on the environment. The site will direct you to useful databases and global resource information.

The World's Population: People and Hunger

11. The Hunger Project
http://www.thp.org
Browse through this nonprofit organization's site to explore the ways in which it attempts to achieve its goal: the sustainable end to global hunger through leadership at all levels of society. The Hunger Project contends that the persistence of hunger is at the heart of the major security issues that are threatening our planet.

12. Penn Library Resources
http://www.library.upenn.edu/resources/websitest.html
This vast site is rich in links to information about virtually every subject you can think of in environmental studies. Its extensive population and demography resources address such concerns as migration, family planning, and health and nutrition in various world regions.

13. World Health Organization
http://www.who.int
This home page of the World Health Organization will provide links to a wealth of statistical and analytical information about health and the environment in the developing world.

14. WWW Virtual Library: Demography & Population Studies
http://demography.anu.edu.au/VirtualLibrary/
This is a definitive guide to demography and population studies. A multitude of important links to information about global poverty and hunger can be found here.

Energy: Present and Future Problems

15. Alternative Energy Institute, Inc.
http://www.altenergy.org
On this site created by a nonprofit organization, you can learn about the impacts of the use of conventional fuels on the environment. Also learn about research work on new forms of energy.

16. Communications for a Sustainable Future
http://csf.colorado.edu
This site will lead to information on topics in international environmental sustainability. It pays particular attention to the political economics of protecting the environment.

17. Energy and the Environment: Resources for a Networked World
http://zebu.uoregon.edu/energy.html
This University of Oregon site points to an extensive array of materials having to do with energy sources—both renewable and nonrenewable—as well as other topics of interest to students of the environment.

18. Institute for Global Communication/EcoNet
http://www.igc.org/igc/gateway/
This environmentally friendly site provides links to dozens of governmental, organizational, and commercial sites having to do with energy sources. Resources address energy efficiency, renewable generating sources, global warming, and more.

19. U.S. Department of Energy
http://www.energy.gov
Scrolling through the links provided by this Department of Energy home page will lead to information about fossil fuels and a variety of sustainable/renewable energy sources.

Biosphere: Endangered Species

20. Friends of the Earth
http://www.foe.co.uk/index.html
Friends of the Earth, a nonprofit organization based in the United Kingdom, pursues a number of campaigns to protect the Earth and its living creatures. This site has links to many important environmental sites, covering such broad topics as ozone depletion, soil erosion, and biodiversity.

21. GORP: Great Outdoor Recreation Pages
http://www.gorp.com/gor/resource/Us_National_Park/AK/wild_den.htm
This GORP program is an example of an organization that is now offering wildlife watching. Visit this site to investigate what these specialized tours have to offer.

22. Smithsonian Institution Web Site
http://www.si.edu
Looking through this site, which will provide access to many of the enormous resources of the Smithsonian, offers a sense of the biological diversity that is threatened by humans' unsound environmental policies and practices.

23. Tennessee Green
http://korrnet.org/tngreen/
Visit this site to find a wealth of information related to sustainability and ways that we can "lighten our load on the environment." It provides links to other environmental sites and guidance to articles and books.

24. World Wildlife Federation
http://www.wwf.org
This home page of the WWF leads to an extensive array of links to information about endangered species, wildlife management and preservation, and more. It provides many suggestions for how to take an active part in protecting the biosphere.

Resources: Land, Water, and Air

25. Global Climate Change
http://www.puc.state.oh.us/consumer/gcc/index.html
PUCO (Public Utilities Commission of Ohio) aims for this site to serve as a clearinghouse of information related to global climate change. Its extensive links provide an explanation of the science and chronology of global climate change, acronyms, definitions, and more.

26. National Oceanic and Atmospheric Administration
http://www.noaa.gov
Through this home page of NOAA, part of the U.S. Department of Commerce, find information about coastal issues, fisheries, climate, and more.

27. National Operational Hydrologic Remote Sensing Center
http://www.nohrsc.nws.gov
Flood images are available at this site of the NOHRSC, which works with the U.S. National Weather Service to track weather-related information.

28. Virtual Seminar in Global Political Economy/Global Cities & Social Movements
http://csf.colorado.edu/gpe/gpe95b/resources.html
This site of Internet resources is rich in links to subjects of interest in regional environmental studies, covering topics such as sustainable cities, megacities, and urban planning. Links to many international nongovernmental organizations are included.

29. Websurfers Biweekly Earth Science Review
http://shell.rmi.net/~michaelg/index.html
This is a biweekly compilation of Internet sites devoted to the terrestrial and planetary sciences. It includes a list of hyperlinks to related earth science sites and news items.

Pollution: The Hazards of Growth

30. IISDnet
http://iisd1.iisd.ca
This site of the International Institute for Sustainable Development, a Canadian organization, presents information through links on business and sustainable development, developing ideas, and Hot Topics.

31. Persistant Organic Pollutants
http://irptc.unep.ch/pops/
Visit this site to learn more about persistant organic pollutants (POPs) and the issues and concerns surrounding them.

32. School of Labor and Industrial Relations: Hot Links
http://www.lir.msu.edu/hotlinks/
This Michigan State University SLIR page goes to sites regarding industrial relations throughout the world. It has links to U.S. government statistics, newspapers and libraries, international intergovernmental organizations, and more.

33. Space Research Institute
http://arc.iki.rssi.ru/Welcome.html
For a change of pace, browse through this home page of Russia's Space Research Institute for information on its Environment Monitoring Information Systems, the IKI Satellite Situation Center, and its Data Archive.

34. Worldwatch Institute
http://www.worldwatch.org
The Worldwatch Institute is dedicated to fostering the evolution of an environmentally sustainable society in which human needs are met without threatening the health of the natural environment. This site provides access to *World Watch Magazine* and *State of the World 2000*. Click on Alerts and Press Briefings for discussions of current problems.

We highly recommend that you review our Web site for expanded information and our other product lines. We are continually updating and adding links to our Web site in order to offer you the most usable and useful information that will support and expand the value of your Annual Editions. You can reach us at: *http://www.dushkin.com/annualeditions/*.

Unit 1

Unit Selections

1. **Environmental Surprises: Planning for the Unexpected,** Chris Bright
2. **The Nemesis Effect,** Chris Bright
3. **Harnessing Corporate Power to Heal the Planet,** L. Hunter Lovins and Amory B. Lovins
4. **Crossing the Threshold: Early Signs of an Environmental Awakening,** Lester R. Brown

Key Points to Consider

❖ Why are environmental changes so difficult to predict and how can human planning systems develop mechanisms to adjust to continued environmental change?

❖ How are environmental systems linked together and how do they tend to converge? Illustrate an example of a synergistic environmental problem such as the link between climate change and forest fires.

❖ What is meant by the term "natural capitalism" and how does the application of principles of this form of economics promise a new creation of appropriate environmental strategies? How do the concepts of natural capitalism fit into the present methods for dealing with the world's resources and systems?

❖ What are some social thresholds of environmental problems? How do environmental events like hurricanes increase both public awareness and government and corporate action on environmental issues?

 **Links**   www.dushkin.com/online/

6. **Earth Science Enterprise**
   http://www.earth.nasa.gov
7. **National Geographic Society**
   http://www.nationalgeographic.com
8. **Santa Fe Institute**
   http://acoma.santafe.edu
9. **United Nations**
   http://www.unsystem.org
10. **United Nations Environment Programme**
    http://www.unep.ch

These sites are annotated on pages 4 and 5.

The Global Environment: An Emerging World View

More than three decades after the celebration of the first Earth Day in 1970, public apprehension over the environmental future of the planet has reached levels unprecedented even during the late 1960s and early 1970s "Age of Aquarius." No longer are those concerned about the environment dismissed as "ecofreaks" and "tree-huggers." Many serious scientists have joined the rising clamor for environmental protection, as have the more traditional environmentally conscious public interest groups. There are a number of reasons for this increased environmental awareness. Some of these reasons arise from environmental events; it is, for example, becoming increasingly difficult to deny the effects of global warming. But more arise simply from the increase in global information systems and the maturation of concepts about the global nature of environmental processes. For example, the raising of the Iron Curtain has allowed information and ideas to pass more freely between East and West.

Much of what has been learned through this increased information flow, particularly by Western observers, has been of an environmentally ravaged Eastern Europe and Russia—a chilling forecast of what other industrialized nations will become in the near future unless strict international environmental measures are put in place. For perhaps the first time ever, countries are beginning to recognize that environmental problems have no boundaries and that international cooperation is the only way to solve them.

The subtitle of this first unit, "An Emerging World View," is an optimistic assessment of the future: a future in which less money is spent on defense and more on environmental protection and cleanup. The authors of the Worldwatch Institute's *State of the World* have recently described a new world order in which political influence will be based more upon leadership in environmental and economic issues than upon military might. Perhaps it is far too early to make such optimistic predictions, to name the decade of the 1990s "The Decade of the Environment," or to conclude that the world's nations—developed and underdeveloped—will begin to recognize that Earth's environment is a single unit. Nevertheless, there is growing international realization—aided by the "information superhighway"—that we are all, as environmental activists have been saying for decades, inhabitants of "Spaceship Earth" and will survive or succumb together.

The articles selected for this unit have been chosen to illustrate this increasingly global perspective on environmental problems and the degree to which their solutions must be linked to political, economic, and social problems and solutions. In the lead piece of the unit, "Environmental Surprises: Planning for the Unexpected," Christopher Bright of the Worldwatch Institute describes how environmental futurists are discovering new ways to identify environmental problems and to plan solutions for them around the complex web of politics, economies, and societies. He notes that the process of environmental decline need not be slow and gradual but may come abruptly. This means that planning for change has to be capable of adjusting quickly to new environmental conditions. And since environmental conditions are never stable, Bright concludes, "plan to keep on planning."

The second selection in the unit is also directed toward the interconnectedness of environmental systems and toward new concepts and ways of thinking about the environment and human impact. "The Nemesis Effect," also written by Christopher Bright, describes the manner in which the growing number of overlapping stresses on ecosystems might cause those systems to decline rapidly and unexpectedly. Bright presents "a spreading matrix of trouble" in which he lists 13 of the worst pressures we are inflicting on the planet and on ourselves and shows how these corrosive forces interact. (He suggests that complex systems of policies that are more effective than any of their constituent parts should be developed to deal with systems of environmental problems that are more complex than the problems of any of their discrete parts.)

In "Harnessing Corporate Power to Heal the Planet," L. Hunter Lovins and Amory B. Lovins of the Rocky Mountain Institute, a non-profit environmental policy center, introduce the concept of "natural capitalism," a new form of economics that takes into account the value of natural resources and, more importantly, the ecosystem services such as water and nutrient recycling, atmospheric and ecological stability, and biodiversity. Traditional capitalistic accounting has dealt with both resources and services as "free" and, hence, have paid little attention to their depletion or deterioration. The Lovinses suggest that treating the environment as something of tangible monetary value can increase resource productivity, eliminate the concept of waste, create better environmental services, and promote investment in natural capital—in other words, natural capitalism could result in the reversal of the worldwide deterioration of the ecosystem.

The social prospects of arriving at such an economic solution, given the global nature of the problem, is the focus of the fourth article in this section. In "Crossing the Threshold: Early Signs of an Environmental Awakening," Lester R. Brown, president of the Worldwatch Institute, notes that in spite of the spate of environmental disruptions capturing headlines (global warming, storms, floods, forest fires), the world may be approaching a social threshold that could change our way of looking at the environment. Ecologists speak of thresholds as crucial parameters beyond which ecosystems change dramatically. Brown sees equally dramatic potential changes in social systems on the horizon. Among such changes are breakthroughs in support for alternative energy sources such as wind and solar power, the shifting views on material use and population, and the recognition that resources must be sustainable. What is most important, Brown claims, is that while changes in attitude and perception have, in the past, taken place among members of the general population, now they are taking place in government agencies and corporate boardrooms.

Article 1

Environmental Surprises: Planning for the Unexpected

Environmental futurists are finding new ways to anticipate the effects of complex trends.

By Chris Bright

The process of environmental decline usually seems gradual and predictable. We are comforted by the thought that even if we have not turned the trends around there will be time for our children to rise to the challenge.

But this way of thinking is like sleepwalking. To understand why, you have to look at decline close up. Here is how it has happened in one small country, with big implications.

The Honduran Predicament

In the early 1970s, Honduras was caught up in a drive to build agricultural exports. Landowners in the south increased their production of cattle, sugarcane, and cotton. This more-intensive farming reduced the soil's water absorbency, so more and more rain ran off the fields and less remained to evaporate back into the air. The drier air reduced cloud cover and rainfall. The region grew a lot warmer.

As the land became less productive, people began to leave. Many moved north to work on newly developed plantations or to carve their own small farms out of the area's rain forests. Much of this northern agriculture was devoted to export crops, too, primarily bananas, melons, and pineapples.

But it is difficult to mass-produce big, succulent fruits near rain forests—even badly fragmented rain forests—because there are so many insects and fungi around to eat them. So the plantations came to rely heavily on pesticides. From 1989 to 1991, Honduran pesticide imports increased more than fivefold, to about 8,000 tons.

The steaming, ragged forest was a perfect habitat for malaria mosquitoes. Around the plantations, the insecticide drizzle suppressed them for a time, but they eventually acquired resistance to a whole spectrum of chemicals. As a result, the mosquitoes were basically released from human control. When their populations bounced back, they encountered a landscape stocked with their favorite prey: people. And since these people were from an area where malaria infection had become rare, their immunity to the disease was low. Malaria rapidly reasserted itself: From 1987 to 1993, the number of cases in Honduras jumped from 20,000 to an estimated 90,000.

The situation was brought to light in 1993 by a group of researchers concerned about the public health implications of environmental decline. But their primary interest was not in what had already happened—it was in what might happen next. Some very nasty surprises might be tangled up somewhere in this web of pressures. They argued, for example, that deforestation and changing patterns of disease had made the country vulnerable to climate change.

They were right. In October 1998, Hurricane Mitch slammed into the Gulf Coast of Central America and stalled there for four days. Nightmarish mudslides obliterated entire villages; half the population of Honduras was displaced, and the country lost 95% of its agricultural production.

Mitch was the fourth-strongest hurricane to enter the Caribbean in the twentieth century, but much of the damage was caused by deforestation: If forests had been gripping the soil on those hills, fewer villages

1. Environmental Surprises: Planning for the Unexpected

would have been buried in mudslides. And in the chaos and filth of Mitch's wake there followed tens of thousands of additional cases of malaria, cholera, and dengue fever.

Complexities of Change

It is hard to shake the feeling that "normal change"—even change for the worse—should not happen this way. In the first place, too many trends are spiking. Instead of gradual change, the picture is full of *discontinuities*—very rapid shifts that are much harder to anticipate. There is a rapid warming in the south, then an abrupt expansion in deforestation in the north, as plantations are developed. Then malaria infections jump. Then the mudslides, in addition to killing thousands of people, cause a huge increase in the rate of topsoil loss.

There also seem to be too many overlapping pressures—too many *synergisms*. The mudslides were not the work of Mitch alone; they were caused by Mitch plus the social conditions that encouraged the farming of upland forests. The malaria emerged not just from the mosquitoes, but from the movement of a low-immunity population into a mosquito-infested area, and from heavy pesticide use.

Discontinuities and synergisms frequently catch us by surprise. They tend to subvert our sense of the world because we so often assume that a trend can be understood

The future of a trend—any trend—depends on the behavior of the entire system in which it is embedded.

in isolation. It is tempting, for example, to believe that a smooth line on a graph can be used to see into the future: All you have to do is extend the line. But the future of a trend—any trend—depends on the behavior of the entire system in which it is embedded. When we isolate a phenomenon in order to study it, we may actually be preventing ourselves from knowing the most important things about it.

Such a fragmented form of inquiry is becoming increasingly dangerous—and not just because we might miss problems in small, poor countries like Honduras. After all, there is nothing special about the pressures in the Honduran predicament. Deforestation, climate change, chemical contamination, and many other forms of environmental corrosion are at work on a global scale. Each has engendered its own minor research industry. But even as the publications pile up, we may actually be missing the biggest problem of all: What might the inevitable convergence of these forces do?

"When one problem combines with another problem, the outcome may be not a double problem, but a super-problem," writes Norman Myers, an Oxford-based ecologist who is one of the most active pioneers in the field of environmental surprise. We have hardly begun to identify those potential super-problems, but in the planet's increasingly stressed natural systems, the possibility of rapid, unexpected change is pervasive and growing. Important theaters of surprise include coral reefs, the atmosphere, and an ecosystem I will discuss in detail: tropical rain forests.

Fires in the Forest

Eight thousand years ago, before people began to clear land on a broad scale, forests covered more than 6 billion hectares (14.8 billion acres), or around 40% of the planet's land surface. Today, the earth's natural forests (as opposed to tree plantations) amount to 3.6 billion hectares

Three Types of Environmental Surprise

Type of Surprise	Example
A **discontinuity** is an abrupt shift in a trend or a previously stable state. The abruptness is not necessarily apparent on a human scale; what counts is the time frame of the processes involved.	Overfishing has pushed some fish species into a population crash rather than a gradual decline. As recently as the 1970s, for example, the white abalone occurred along the coast of northern Mexico and southern California at densities of up to 10,000 per hectare. Today its total population is probably no more than a few dozen, and its extinction is imminent.
A **synergism** is a change in which several phenomena combine to produce an effect that is greater than would have been expected from adding up their effects taken separately.	The monstrous 1998 flood of China's Yangtze River did $30 billion in damage, displaced 223 million people, and killed another 3,700. The damage was a synergism caused not just by heavy rains, but by extremely dense settlement of the floodplain and by deforestation—the Yangtze basin has lost 85% of its forest cover.
An **unnoticed trend**, even if it produces no discontinuities or synergisms, may still do a surprising amount of damage before it is discovered.	In the United States, where natural areas are monitored with much greater attention than in most parts of the world, aggressive nonnative weeds may still have to be in the country for 30 years or have spread to thousands of hectares before they are even discovered. In the United States and elsewhere, such weeds displace native plants, upset fire and water cycles, and do billions of dollars in agricultural damage every year.

Sources: TREE (Trends in Ecology and Evolution); U.S. Congress, Office of Technology Assessment; Worldwatch Institute.

(8.9 billion acres) at most. Every year, at least another 14 million hectares are lost. Among the many thousands of species that are believed to go extinct every year, the majority are forest creatures, primarily tropical insects, who have lost their habitat.

Currently, well over 90% of forest loss is occurring in the tropics—on a scale so vast that it might appear to have exceeded its capacity to surprise us. In 1997 and 1998, fires set to clear land in Amazonia claimed more than 5.2 million hectares of Brazilian forest, brush, and savanna—an area nearly 1.5 times the size of Taiwan. In Indonesia, some 2 million hectares of forest were torched during 1997 and 1998.

All this is certainly news, but if you are interested in conservation, it is the kind of dreadful news you have come to expect. And yet our expectations may not be an adequate guide to the sequel, assuming the destruction continues at its current pace. A substantial portion of the damage is hidden—it does not show up in the conventional analysis. But once you take the full extent of the damage into account, you can begin to make out some of the surprises it is likely to trigger.

Consider, for example, the destruction of Amazonia. Over half the world's remaining tropical rain forest lies within the Amazon basin, where more forest is being lost than anywhere else on earth. Deforestation statistics for the area are intended primarily to track the conversion of forest into farms and ranches. Typically, the process begins with the construction of a road, which opens up a new tract of forest to settlement.

In June 1997, for example, some 6 million hectares of forest were officially released for settlement along a major new highway, BR-174, which runs from Manaus, in central Amazonia, over 1,000 kilometers (620 miles) north to Venezuela. Ranchers and subsistence farmers clear cut patches of forest along the road and burnt the slash during the July—November dry season. (The farmers generally have few other options: Brazil has large numbers of poor, land-hungry people, and the plots they cut from the forest lose their fertility rapidly, so there is a constant demand for fresh soil.)

But the damage to the forest generally extends much farther than the areas that are "deforested" in this conventional sense, because of the way fire works in Amazonia, In the past, major fires have not been a frequent enough occurrence to promote any kind of adaptive "fire proofing" in the region's dominant tree species. Some temperate-zone and northern trees, by contrast, are "fire-adapted" in one way or another—they may have especially thick bark, for example, or the ability to resprout after burning. The lack of such adaptations in Amazonian trees means that even a small fire can begin to unravel the forest.

During the burning season, the flames often escape the cuts and sneak into neighboring forest. Even in intact forest, there will be patches of forest floor that are dry enough to allow a small "surface fire" to feed on the dead leaves. Surface fires do not climb trees and become crown fires. They crackle along the forest floor in patches of flame, going out at night when the temperature drops and rekindling the next day. They are fatal to most of the smaller trees they touch. Overall, an initial surface fire may kill 10% of the living forest biomass.

The damage may not seem dramatic, but another tract of forest may already be doomed by an incipient positive feedback loop of fire and drying. After a surface fire, the amount of shade is reduced from about 90% to 60%, and the dead and injured trees rain debris down on the floor. So a year or two later, the next fire in that spot finds more tinder and a warmer, drier floor. Some 40% of forest biomass may die in the second fire. At this point, the forest integrity is seriously damaged; grasses and vines invade and contribute to the accumulation of combustible material. The next dry season may eliminate the forest entirely. Once the original forest is gone, the scrubby second growth or pasture that replaces it will almost certainly burn too frequently to allow the forest to restore itself.

In a recent survey, researchers cross-checked satellite maps with field observations and concluded that conventional deforestation estimates for Brazilian Amazonia were missing some 1 to 1.5 million hectares of severe forest damage done by logging every year. Surface fire damage is harder to quantify, but the same researchers did a fire survey and found that the amount of standing forest that had suffered a surface fire in 1994 and 1995 was 1.5 times the area fully deforested in those years. Overall, they suggested, the area of Amazonian forest attacked by surface fire every year may be roughly equivalent to the area deforested outright. And in some parts of the basin, the extent of this cryptic damage is so great that the conventional measurements may no longer be all that useful. In one region, around Paragominas in eastern Amazonia, the researchers found that, although 62% of the land was classified as forested, only about one-tenth of this consisted of undisturbed forest.

Other Tropical Surprises

As the Amazonian forest dwindles, a surprising second-order effect may emerge as the hydrological cycle changes. Because trees exhale so much water vapor, a forest to some degree creates its own climate. Much of this water vapor condenses below the canopy and drips back into the soil; some of it rises higher before falling back in as rain. Researchers estimate that most of the Amazon's rainfall comes from water vapor exhaled by the forest. Widespread deforestation will therefore tend to make the region substantially drier, and that will accelerate the feedback loop created by the fires.

Other kinds of surprises are lurking in tropical forests as well. As developing countries industrialize, some forest maladies better known in the industrial world are likely to appear in these countries, too. Acid rain, for example, is already reported to be affecting the forests of southern China. In parts of South Asia, Indonesia, South America, and West Africa, this form of pollution is bound to increase as industrialization proceeds and cities enlarge. The soil in these areas tends to be fairly acidic already, which would make it incapable of buffering large doses of additional acid. At least in some of these places, acid-induced decline may therefore be much more abrupt than in the temperate zone.

Other development pressures may be unique to the tropics. Increased hunting pressure, for instance, has reduced animal populations in a good number of tropical forests. Many forest-dwelling peoples have armed themselves with shotguns and rifles, which are far more lethal than traditional weapons. And logging is bringing additional hunters into forest interiors. Hunting often supplements the logging: It feeds the loggers, and the surplus is sold as bush meat in towns and cities farther from the frontier. Such hunting is typically very indiscriminate; almost any creature of any size is potential game. Hundreds—even thousands—of animals may fall prey to a single camp.

In the Republic of Congo, for example, the annual take in a single camp of 648 people was found to be 8,251 animals, amounting to 124 tons of meat. In tropical forests the world over, mammals, birds, and reptiles play critical roles in pollinating trees, dispersing their seeds, and eating other creatures that prey on the seeds. If hunting continues at its present rate, some tropical tree species are liable to disappear along with the animals themselves.

One last forest surprise: Recent research suggests that the Central African bush meat trade may have sparked the AIDS epidemic. It is

1. Environmental Surprises: Planning for the Unexpected

Nature has no reset button. Environmental corrosion is not just killing off individual species—it is setting off system-level changes that are, for all practical purposes, irreversible.

likely that the original host of the HIV virus was a population of chimpanzees in Cameroon and Gabon. Chimps in that area are commonly hunted for their meat; now, apparently, one of their diseases is hunting us.

An Agenda for the Unexpected

Human pressures on the earth's natural systems have reached a point at which they are more and more likely to engender problems that we are less and less likely to anticipate. Dealing with this predicament is obviously going to require more than simply reacting to problems as they appear. We need to forge a new ethic for managing our relationship with nature—one that emphasizes minimal interference in the lives of wild beings and in the broad natural processes that sustain all living things. Such an ethic might begin with three basic principles.

First, nature is a system of unfathomable complexity. Our predominant response to that complexity has been specialization, in both the sciences and public policy. Learning a lot about a little is a form of progress, but it comes at a cost. Experience is seductive: It is easy for specialists to get into the habit of thinking that they understand all the consequences of a plan. But in a complex, highly stressed system, the biggest consequences may not emerge where the experts are in the habit of looking. This inherent unpredictability condemns us to some degree of error, so it is important to err on the side of minimal disruption whenever possible.

Second, nature gives away nothing for free. You cannot get an appreciable quantity of anything out of nature without sacrificing something in the process. Even sustainable resource management is a trade-off—it's simply one we regard as acceptable. In our dealings with nature, as with any other sort of transaction, we need to know the full cost of the goods before deciding whether they are worth the price, or whether there is a better way to pay for them.

Third, nature has no reset button. Environmental corrosion is not just killing off individual species—it is setting off system-level changes that are, for all practical purposes, irreversible. For example, even if all the world's coral reef species were miraculously to survive the impending bout of rapid climate change, that does not mean that our descendants will be able to reconstruct reef communities. The near impossibility of restoring complex systems to some previous state is another strong argument for minimal disruption.

These are basic features of the natural world: We will never understand it completely, it will not do our bidding for free, and we cannot put it back the way it was. A policy ethic sensitive to these facts of life might emphasize the following themes.

• **Monoculture technologies are brittle, so plan for diversity.** Huge, uniform sectors generally exhibit a kind of superficial efficiency because they generate economies of scale. You can see this in fossil-fuel-based power grids, megadam projects, and even in woodpulp plantations. But because they are beholden to vast quantities of invested capital—both financial and

An Alternative View: Privatization Proponents See Positive Trends

If you're tired of hearing only bad news about the environment, then *Earth Report 2000* may be the book for you. Published by the Competitive Enterprise Institute (CEI), a think tank in Washington, D.C., *Earth Report* lives up to its billing as a rebuttal to environmental doomsayers:

"For years, The Worldwatch Institute and the environmental establishment have generated publications trumpeting the inevitable destruction of the world's forests, rivers, wetlands, and wildlife. Each year finds the predictions more and more dire, yet somehow the catastrophe never seems to materialize," write CEI researchers.

In contrast, *Earth Report 2000* presents a drumbeat of optimism, arguing that deforestation, declining world fisheries, global warming, population stress, and the loss of biodiversity are simply not the mega-tragedies described by alarmists, but rather problems that can be solved through human ingenuity, privatization of environmental resources, and competition unfettered by government interference.

In a chapter called "Fishing for Solutions," for example, CEI research associate Michael De Alessi cites the fast-growing aquaculture industry as a positive response to the stagnation of the world's ocean fish catch. Private ownership is key, he notes, because resource entrepreneurs have a stronger incentive to maintain environmental quality than do government bureaucrats. However, in applying the privatization scenario to the waning oyster beds of the Chesapeake Bay, it isn't clear whether De Alessi would include among the property rights of oystermen the power to restrict landowners' rights to use fertilizer and pesticides on farms in the surrounding Maryland and Delaware watershed.

The emphasis on privatization in *Earth Report 2000* tends to reject the environmentalists' concept of the *commons*, in which citizens who do not own ecosystems are recognized as having a stake in environmental quality and resource use. According to CEI researchers, government environmental regulations such as the Clean Air Act have been at best ineffective, at worst restrictive of private enterprise, and too costly. It should be noted, however, that, while *Earth Report 2000* criticizes government subsidies for renewable energy research in solar, wind power, and biomass, it does not mention longstanding federal subsidies that benefit the grazing, timber, and mining industries.

Earth Report 2000 and Worldwatch's *State of the World 2000* represent nothing less than a clash of world views—over science, government environmental policy, economics, and politics. Reading and considering both books may not settle the debate, but it's a lively place to start. —*Dan Johnson*

Source: *Earth Report 2000: Revisiting the True State of the Planet* by Ronald Bailey et al. McGraw-Hill. 2000. 362 pages. Paperback. Available from the Futurist Bookstore for $19.95 ($17.95 for Society members), cat no. B-2329.

---

political—industrial monocultures are extremely difficult to reform when their hidden costs begin to catch up with them. More diverse technologies are liable to be more adaptable because their investors are not all "betting" on exactly the same future. And whether the goal is the production of irrigation water, paper, or electricity, a more-adaptable system is likely to be more durable over the long term.

• **Direct opposition to a natural force is generally counterproductive, so plan to work with nature.** Heavy-handed approaches sometimes exacerbate a problem, as when intensive pesticide use causes a population explosion of resistant pests. And some successes are often worse than outright failures. Dams and levees, for instance, may end up controlling a flood-prone river—and largely killing the riverine ecosystem in the process. Sound management often tends to be more oblique than direct. Restoration of water-absorbing floodplain ecosystems can make for more-effective flood control than dams. Cropping systems that mimic natural floral diversity make it harder for any particular pest to dominate a field.

• **You can never have just one effect, so plan to have several.** Thinking through the likely "ripple effects" of a plan will help locate not just the risks, but additional opportunities. Encouraging organic agriculture in the U.S. Midwest, for example, could help reduce nutrient leakage into the Mississippi River. That, in turn, could ease the stress on reefs and other marine communities in the Gulf of Mexico. Marine conservation may therefore "overlap" with agricultural reform; it might even be possible to extend this overlap to include reform of the North American diet.

Environmental policy is full of such latent positive synergisms. In many countries, for instance, there may be a powerful overlap between the need for meaningful employment and the need to replace the "throwaway" economy with one that emphasizes durable goods.

• **Solutions are almost never permanent, so plan to keep on planning.** In the 1950s, organochlorine pesticides were hailed as a permanent "fix" for insect pest problems; given the pervasive ecological damage that these chemicals are now known to cause, the idea of a permanent chemical solution to anything may seem rather naïve today. But because our relationship with nature is in a constant state of flux, even realistic fixes will need regular revision. The 1986 Montreal Protocol (calling for a phase-out of ozone-depleting chemicals) is not a permanent patch for the ozone layer, in part because climate change will probably exacerbate ozone loss. The Green Revolution is not a permanent answer to world hunger, in part because conventional agriculture is overtaxing aquifers.

The growing strain on the earth's natural systems will probably force an increase in the tempo of policy revision—so it makes sense to take full advantage of the powerful new information and communications technologies. Because of their ability to bring together enormous quantities of data from different areas and disciplines, such technologies could help counter the blinkering effects of specialization.

• **None of us may find the answer alone, but together we probably can.** In social as well as natural systems, there is a potent class of properties that exists only on the system level—properties that cannot be directly attributed to any particular component. In a political system, for example, institutional pluralism can create a public space that no single institution could have created alone. One of the most important policy activities may therefore be to encourage innovation outside policy institutions.

Policy may need to become increasingly a matter of creating not so much solutions per se as the conditions from which solutions can arise. In the face of the unexpected, our best hopes may lie in our collective imagination.

About the Author

Chris Bright is senior editor of *World Watch* magazine and a research associate focusing on biodiversity issues at the Worldwatch Institute, 1776 Massachusetts Avenue, N.W., Washington, D.C. 20036. Telephone 1-202-452-1999; fax 1-202-296-7365; e-mail cbright@worldwatch.org; Web site www.worldwatch.org.

This article draws on his chapter, "Anticipating Environmental Surprise," in *State of the World 2000* by Lester Brown et al. W.W. Norton. 2000. 276 pages. Paperback. Available from the Futurist Bookstore for $19.95 ($17.95 for Society members), cat. no. B-2138.

The Nemesis Effect

Burdened by a growing number of *overlapping* stresses, the world's ecosystems may grow increasingly susceptible to rapid, unexpected decline.

by Chris Bright

In 1972, a dam called the Iron Gates was completed on a stretch of the Danube River between Romania and what is now Serbia. It was built to generate electricity and to prevent the river from visiting some 26,000 square kilometers of its floodplain. It has done those things, but that's not all it has done.

The Danube is the greatest of the five major rivers that run into the Black Sea. For millennia, these rivers have washed tons of dead vegetation into this nearly landlocked ocean. As it sinks into the sea's stagnant depths, the debris is decomposed by bacteria that consume all the dissolved "free" oxygen (O_2), then continue their work by pulling oxygen out of the sulfate ions (SO_4) that are a normal component of seawater. That process releases hydrogen sulfide gas (H_2S), which is one of the world's most poisonous naturally occurring substances. One deep breath of it would probably kill you. The sea's depths contain the largest reservoir of hydrogen sulfide in the world, and the dissolved gas forces virtually every living thing in the water to cling to the surface or die. The Black Sea is alive only along its coasts, and in an oxygenated surface layer that is just 200 meters thick at most—less than a tenth of the sea's maximum depth.

The Danube contributes 70 percent of the Black Sea's fresh water and about 80 percent of its suspended silicate—essentially, tiny pieces of sand. The silicate is consumed by a group of single-celled algae called diatoms, which use it to encase themselves in glassy coats. The diatoms fuel the sea's food web, but any diatoms that don't get eaten eventually die and sink into the dead zone below, along with any unused silicate. Fresh contributions of silicate are therefore necessary for maintaining the diatom population. But when the Iron Gates closed, most of the Danube's silicate began to settle out in the still waters of the vast lake behind the dam. Black Sea silicate concentrations fell by 60 percent.

The drop in silicate concentrations coincided with an increase in nitrogen and phosphorus pollution from fertilizer runoff and from the sewage of the 160 million people who live in the Black Sea drainage. Nitrogen and phosphorus are plant nutrients—which is why they're in fertilizer. In water, this nutrient pollution promotes explosive algal blooms. The Black Sea diatoms began blooming, but the lack of silicate limited their numbers and prevented them from consuming all the nutrient. That check created an opportunity for other types of algae, formerly suppressed by the diatoms. Some of these were dinoflagellate "red tide" organisms, which produce powerful toxins. Soon after the Iron Gates closed, red tides began to appear along the sea's coasts.

In the early 1980s, a jellyfish native to the Atlantic coast of the Americas was accidentally released into the sea from the ballast tank of a ship. The jellyfish population exploded; it ate virtually all the zooplankton, the tiny animals that feed on the algae. Liberated from their predators, the algae grew even thicker, especially the dinoflagellates. In the late 1980s, during the height of the jellyfish infestation, the dinoflagellates seemed to be summoning the death from below. Their blooms consumed all the oxygen in the shallows and the rotten-egg stench of hydrogen sulfide haunted the streets of Odessa. Carpets of dead fish—asphyxiated or poisoned—bobbed along the shores.

The jellyfish nearly ate the zooplankton into oblivion, then its population collapsed too. But it's still in the Black Sea and there's probably no way to remove it. The red tides have increased six-fold since the early 1970s, and it doesn't look as if antipollution efforts are going to put the dinoflagellates back under the control of the diatoms. The fisheries are in a dismal state—overharvested, starved of zooplankton, periodically suffocated and poisoned. The rest of the ecosystem isn't faring much better. The mollusks, sponges, sea urchins, even the marine worms are disappearing. The shallows, where vast beds of seagrass once breathed life into the waters, are regularly fouled in a fetid algal soup laced with a microbe that thrives in such conditions: cholera.

2. Nemesis Effect

COULD IT HAVE BEEN PREDICTED THAT THE DAM ON THE DANUBE would end up triggering this spasm of ecological chaos? The engineers who designed the Iron Gates were obviously attempting to make nature more orderly and productive (in a very narrow sense of those terms). Could they have foreseen this form of disorder, which has no obvious relationship to the dam itself? Here is what they would have had to anticipate: that the dam would cause a downstream change in water chemistry which would combine with an increase in a certain type of pollution to produce an effect that neither change would probably have had on its own—and that effect would then be magnified by something that was going to be pumped out of a ship's ballast tank.

It seems absurd even to entertain the idea that such things could be foreseen. Yet this is precisely the kind of foresight that is now required of anyone who is concerned, professionally or otherwise, with the increasingly dysfunctional relationship between our societies and the environment. The forces of ecological corrosion—pollution, overfishing, the invasion of exotic species like that jellyfish—such forces interact in all sorts of ways. Their effects are determined, not just by the activities that initially produced them, but *by each other and by the way ecosystems respond to them.* They are, in other words, parts of an enormously complex system. And unless we can learn to see them *within the system,* we have no hope of anticipating the damage they may do.

A system is a set of interrelated elements in which some sort of change is occurring, and even very simple systems can behave in unpredictable ways. Three elements are enough to do it, as Isaac Newton demonstrated three centuries ago, when he formulated the "N body problem." Is it possible to define the gravitational interaction between three or more moving objects with complete precision? No one has been able to do it thus far. The unpredictable dynamics of system behavior have inspired an entire mathematical science, variously known as complexity or systems theory. (The most famous type of complexity is "chaos.") Systems theory is useful for exploring several other sciences, including ecology. It's also useful for exploring the ways in which we can be surprised.

Suppose, for example, that you were a marine biologist studying Black Sea plankton in the early 1970s. Had you confined your observations solely to the plankton themselves, you would have had no basis for predicting the explosion of red tides that followed the closing of the Iron Gates. Such "nonlinear" events usually come as a surprise, not because they're unusual—they're actually common—but because of a basic mismatch between our ordinary perceptions and system behavior. Most people, most of the time, just aren't looking upriver: we have a strong intuitive tendency to assume that incremental change can be used to predict further incremental change—that the gradual rise or fall of a line on a graph means more of the same. But that's not true. The future of a trend—any trend—depends on the behavior of the system as a whole.

In 1984, the sociologist Charles Perrow published a book, *Normal Accidents: Living with High-Risk Technologies,* in which he explored the highly complex industrial and social systems upon which we've become increasingly dependent. David Ehrenfeld, an ecologist at Rutgers University in New Jersey, has observed that much of what Perrow said of nuclear reactors, air traffic, and so forth could also apply to ecosystems—or more precisely, to the ways in which we interact with them. Here are some of the criteria that Perrow uses to define complex systems:

- many common mode connections between components . . . not in a production sequence [that is, elements may interact in ways that won't fit into a predictable sequence];
- unfamiliar or unintended feedback loops;
- many control parameters with potential interactions [that is, we have many ways to influence the system but we can't be sure what the overall result of our actions will be];
- indirect or inferential information sources [we can't always see what's happening directly];
- limited understanding of some processes.

There's something ominous in Perrow's rather bland, clinical terminology—it's like a needle pointing the wrong way on an instrument panel. "Limited understanding of some processes!" No ecologist could have put it better. Ehrenfeld wrote a paper on Perrow's relevance to ecology; he was fascinated with Perrow's treatment of nuclear accidents. What is it like to be a nuclear plant operator during a Three Mile Island event? You watch the monitors, you try to second-guess your equipment, you make inferences about the state of the core. Perrow says, "You are actually creating a world that is congruent with your interpretation, even though it may be the wrong world. It may be too late before you find that out."

Into the Theaters of Surprise

"NUCLEAR. MORE THAN YOU EVER IMAGINED." THAT'S THE SLOGAN OF the Nuclear Energy Institute, a nuclear power industry association based in Washington, D.C. To me, at least, the phrase isn't very reassuring, and I would bet that it will sound like a joke to most of the people who read this article. My guess, in other words, is that your imagination already operates well beyond the stage settings of nuclear industry PR. But how much farther are you willing to push it?

Throughout most of our species' existence, the bounds of our collective imagination have not been a survival issue in the way that they are today. Either our societies were rather loosely coupled to their environment, so there was more "give" in the system, or when we got into trouble, it was a local or regional predicament rather than a global one. But today, our rapport with the environment is growing increasingly analogous to the task of managing a nuclear power plant. We live within a set of systems

1 ❖ THE GLOBAL ENVIRONMENT: AN EMERGING WORLD VIEW

that are "tightly coupled," requiring constant attention, not entirely predictable, and capable of various types of meltdown.

Consider, for example, two representative theaters of surprise. See if you find here more than you ever imagined.

1. The Forests of Eastern North America

As far as conservation is concerned, the woodlands of eastern North America might seem about as far as you can get from the highly publicized tropical scenario, with its poorly understood and rapidly disappearing forests, its desperate agrarian poverty and rapacious logging. For this scorched confusion, substitute some of the most thoroughly studied ecosystems in the world, growing over the heads of some of the world's wealthiest, best-educated, and most information-saturated people. These are highly populated woodlands too—138 million people live beneath the trees or within a few hours' drive of them.

Virtually all of the original "old growth" in the eastern United States was cut long ago, but these forests comprise one of the few large regions anywhere in the world that could be thought of as undergoing some sort of ecological renaissance. With the exception of northern New England, the loggers had done their worst to the region a century ago or more, and moved west in search of bigger timber. And over the course of the 19th century, fewer and fewer fields were being tortured by the plow, as the nation's agriculture shifted to the lavish fertility of the midwest. So the eastern second growth has quietly spread and matured, absorbing hundreds of old cutover wood lots and anonymous, abandoned farmsteads. But today these forests are in the throes of a quiet agony—a pathology that is harder to read than tropical deforestation, but which may lead to a form of degradation that is just as profound. The air they are breathing is poisoning them, the water bathes them in acid, the soil is growing toxic, they are gnawed by exotic pests, and the climate to which they are adapted is likely to shift.

A primary cause of this agony involves changes in the "nitrogen cycle." Nitrogen is an essential nutrient of plants and it's the main constituent of the atmosphere: 78 percent of the air is nitrogen gas. But plants can't metabolize this pure, elemental nitrogen directly. The nitrogen must be "fixed" into compounds with hydrogen or oxygen before it can become part of the biological cycle. In nature this process is accomplished by certain types of microbes and by lightning strikes, which fuse atmospheric oxygen and nitrogen into nitrogen oxides.

Humans have radically amplified this process. Farmers boost the nitrogen level of their land through fertilizers and the planting of nitrogen-fixing crops (actually, it's symbiotic microbes that do the fixing). The burning of forests and the draining of wetlands release additional quantities of fixed nitrogen that had been stored in vegetation and organic debris. And fossil-fuel combustion releases still more fixed nitrogen, partly from fuel contaminants, and partly through the production of nitrogen oxides in the same way that lightning works. Natural processes probably incorporate around 140 million tons of nitrogen into the terrestrial nitrogen cycle every year. (The ocean cycle is largely a mystery.) Thus far, human activity has at least doubled that amount.

As in much of the industrialized world, eastern North America is bathed in the nitrogen oxides pumped into the air from car exhaust and coal-burning power plants. In the presence of sunlight, one of these chemicals, nitric oxide (NO), produces ozone (O_3). Ozone is good in the stratosphere, where it filters out harmful ultraviolet (UV) radiation, but it's very bad in the troposphere, the thick blanket of air at the Earth's surface. Ozone is a primary component of smog. Clean air laws, understandably, aim to cut ozone levels to a point at which they are unlikely to harm people (or at least, healthy people). But the problem for the forests is that leaf tissue is far more sensitive to ozone than human lung tissue. Ozone "bleaches" leaves. According to Charles Little, a seasoned chronicler of North American forests, you might as well be spraying them with Clorox. Ozone also reduces flower, pollen, and seed production, thereby hindering reproduction.

In this region, you can just about name the tree, and ozone is probably injuring it somewhere. Ozone combines with UV radiation to burn and scar the needles of white pine, the region's tallest conifer. Ozone exposure correlates strongly with hickory and oak die-off. Ozone is hard on the tulip tree, a major canopy species especially where white oak has declined. It's injuring native magnolias as well. Nor is it just the obviously smoggy urban areas that suffer. In the Great Smoky Mountains National Park of North Carolina, researchers have found ozone damage to some 90 plant species.

In rural West Virginia, ozone is apparently working a weird, nonlinear form of forest decline: continual ozone exposure can reduce photosynthesis to the point at which the tree can't grow enough roots to support itself. Apparently minor but chronic leaf damage eventually provokes catastrophic failure of the roots, then death. This is one of several mechanisms underlying the syndrome known as the "falling forest." Reasonably healthy-looking trees just keel over and die.

Airborne nitrogen oxides also produce nitric acid, which contributes to acid rain. The other major constituent of acid rain is sulfuric acid, which derives from the sulfur dioxide released by coal-burning power plants and metal smelters. (Sulfur is a common contaminant of coal and metal ores.) Smoke stack "scrubbers" and a growing preference for low-sulfur coal and natural gas have helped reduce sulfur dioxide emissions in the United States, Canada, and Western Europe. U.S. emissions, for example, fell from nearly 30 million tons in 1970 to 16 million in 1995. (The global picture isn't so encouraging: world sulfur dioxide emissions rose from about 115 million tons a year in 1970 to around 140 million tons by 1988 and have remained relatively stable since then.)

Even in the United States, the amount of acid aloft is still substantial by ecological standards. On the fog-drenched slopes of Mount Mitchell, north on the Appalachian spine from the Smoky Mountains, the pH of the dew and ice sometimes drops as low as 2.1, which is more acid than lemon juice. The acid treatment, combined with insect attack and drought, has killed up to 80 percent of mature red spruce and Fraser fir on the most exposed slopes.

But the problem is not just the acid in the air today. Decades of acid rain have begun to leach out the soil's stock of calcium and magnesium, both essential plant nutrients. Replenishing those minerals, a process dependent on the weathering of rock, may take centuries. In the meantime, the legacy of coal is likely to be stunted forests, at least where the leaching is well advanced, as in some areas of New England. Recent studies at the Hubbard Brook Experimental Forest in the mountains of New Hampshire, for instance, have identified minerals leaching as the main reason the vegetation there has shown no overall growth for nearly a decade.

This slowing of the trees' metabolism is not just a matter of gradual, overall decline—there are nonlinear effects here too. Acid rain is making the New England winters lethal to red spruce and balsam fir, two of the region's most important conifers. Like most conifers, these species don't lose their leaves—their "needles"—in winter, so they can't just go dormant when it gets cold. They have to maintain a metabolic rate high enough to keep the needles functioning properly. In cold weather, conifers close the stomata in their needles when light dims, in order to protect the needles from freezing. (The stomata are the microscopic pores in leaf tissue, where gas exchange occurs.) The mineral-starved trees can't readily perform this function, so sometimes the cells in the needles freeze solid. That kills needles; when enough of the needles die, the tree dies. At higher elevations in Vermont's Green Mountains, three-quarters of mature red spruce have frozen to death.

The acid rain hasn't just made the soils less nutritious—it has also made them toxic. In calcium-rich soils, the acid is generally neutralized, since calcium is alkaline. But as the calcium level drops, more and more acid accumulates and that tends to release aluminum from its mineral matrix. Aluminum is a common soil constituent; when it's bonded to other minerals it's biologically inert, but free aluminum is toxic to both plants and animals. In some Appalachian streams, you can find stones covered with a silvery-whitish tinge—that's aluminum released by acid rain. This burden of "mobilized" metal is compounded by the traces of cadmium, lead, and mercury that the air brings in along with the acid and ozone.

The metals poisoning may create a kind of synergistic overlap with ozone pollution. In some dying red spruce stands in Vermont, researchers have found elevated levels of phytochelatins, a class of chemicals that plants produce to bind to toxic metals and render them inert. But to make the phytochelatins, the spruces have to draw down their stocks of another substance, glutathione, which is used to counteract ozone. So exposure to one kind of poison leaves the spruces more vulnerable to another.

There's another big overlap here as well: the trees' ability to fight off stresses is also being weakened by nitrogen pollution. Plants don't have the same kind of immune system that animals do. Instead of killer cells and antibodies, they produce an immense arsenal of chemicals. Some of these, like phytochelatins, neutralize toxins; others kill pathogens or make leaves less palatable to pests. Excess nitrogen tends to clog the cellular machinery that produces these chemicals. Farmers don't have to worry about this problem when they apply fertilizer to crops, because crops are intensively managed for pest control and because they're generally harvested at the end of a single growing season. But trees that are exposed to high nitrogen year after year will inevitably absorb more of the material than they can possibly metabolize. So the nitrogen builds up in their tissues, where it tends to alter the recipes for all those defensive chemicals. As the chemicals lose their punch, toxins aren't effectively neutralized; soil pathogens permeate the roots, and the leaves grow more susceptible to insect attack. It has been estimated that nitrogen pollution in the eastern United States is triple the level that forests can tolerate over the long term. Nitrogen pollution can cause a kind of botanical equivalent of AIDS.

This weakening of the forests' immune system is likely to upset the balance between the trees and their pathogens. Another reason for West Virginia's "falling forests," for example, is a fungal infection called *Armillaria* root rot. *Armillaria* is a widespread type of fungus, common in forest soils all over the world. In healthy stands, it usually satisfies itself with the occasional diseased or very old tree. But in a badly stressed stand, it becomes a subterranean monster—a huge, amorphous disease organism, sprouting rootlike tentacles that probe the soil for victims. It picks away at the stand, gradually killing it, tree by tree.

But it's not just the native pests that are taking advantage of the forests' weakened state. The forests are crawling with a host of exotic insects and diseases as well. The American chestnut and the American elm succumbed to exotic pathogens earlier in the century and are now functionally extinct. (They have not disappeared completely but they are no longer functioning components of their native ecosystems.) Today many other species are in trouble. The Canadian hemlock, for example, is being attacked by an Asian insect, the hemlock wooly adelgid; in parts of New England, the adelgid is wiping out entire stands. Nitrogen pollution puts the adelgid on the insect equivalent of steroids: the excess nitrogen makes the leaves much more nutritious and can boost adelgid densities five-fold. Oaks are the principal victims of the gypsy moth, a European insect whose occasional population explosions defoliate thousands of hectares. In the nitrogen-poisoned stands, the moth droppings produce a weak solution of nitric acid on the forest floor, leaching out soil nutrients as the moth gnaws away at the canopy.

1 ❖ THE GLOBAL ENVIRONMENT: AN EMERGING WORLD VIEW

Exotic fungal pathogens are attacking the butter nut, the American beech, and the eastern dogwood. The dogwood has a very broad range, which covers most of the eastern United States, and the fungus that is killing it has spread throughout that range in little more than a decade—a phenomenal rate of spread for a tree pathogen. Acid rain appears to be part of the reason for the dogwood's susceptibility, and the dogwood die-off is liable to reinforce the effects of acid rain on the soil. The dogwood is very efficient at pulling calcium out of the soil and depositing it, through its leaf litter, on the forest floor. That process reduces calcium leaching, so the disappearance of this tree could deal an additional blow to calcium-starved forests.

This is the condition of what is, by world standards, an upper middle-class forest: conifer die-offs of 70 to 80 percent in the southern Appalachians, sugar maple mortality at 35 percent in Vermont; the butternut, eastern dogwood, and red mulberry in widespread decline. The American beech and Canadian hemlock in trouble over large parts of their range. The elm and the chestnut already gone. And besides the pests and pollution, decades of fire suppression have eliminated plant communities dependent on fire for renewing themselves. Other stands are now giving way to asphalt and suburbia. Over all, according to a survey of five eastern states, tree mortality may now stand at three to five times historical levels.

Last year, climate scientists discovered that North American broadleaf forests were probably absorbing far more carbon from the atmosphere than had been previously assumed. The continent's eastern forests, it turns

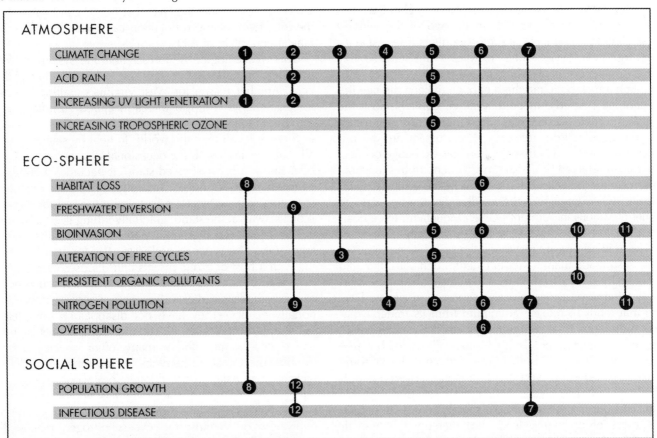

A Spreading Matrix of Trouble

Below are 13 of the worst pressures that we are inflicting on the planet and ourselves. The lines show a few of the ways in which these corrosive forces interact. See the numbered key for each of the combinations indicated. Note that neither the list of pressures nor the set of interactions is inclusive—if your background is in environmental studies, you will almost certainly be able to extend the matrix. We welcome your thoughts.

KEY TO THE MATRIX

1 Climate change + UV: Greenhouse-forced warming of the lower atmosphere may cause a *cooling* of the stratosphere, especially over the Arctic. (Major air currents may shift, and block the warmer surface air from moving North and up.) A cooling stratosphere will exacerbate damage to the ozone layer because the colder it is, the more effective CFCs become at breaking down ozone. The ozone layer over the Arctic could grow progressively thinner as warming proceeds.

(Continued)

out, are an important part of the "missing carbon sink"—the heretofore unexplained hole in the calculations that attempt to define the global carbon budget. But if these forests continue to sicken, their appetite for carbon will eventually falter. That is likely to speed up the processes of climate change. And climatic instability will add yet another stress to a region that is already exhibiting a kind of paradoxical system effect: it is covered with new growth but many of its forests appear to be dying.

2. Coral Reefs

Coral reefs are perhaps the greatest collective enterprise in nature. Reefs are the massed calcareous skeletons of millions of coral—small, sedentary, worm-like animals that live on the reef surface, filtering the water for edible debris. Reefs form in shallow tropical and subtropical waters, and host huge numbers of plants and animals. The reef biome is small in terms of area—less than 1 percent of the earth's surface—but it's the richest type of ecosystem in the oceans and the second richest on earth, after tropical forests. One-quarter of all ocean species thus far identified are reef-dwellers, including at least 65 percent of marine fish species.

Coral is extremely vulnerable to heat stress and the unusually high sea surface temperatures (SSTs) of the past two decades may have damaged this biome just as badly as the unusual fires have damaged the tropical forests. Much of the ocean warming is related to El Niño the weather pattern that begins with shifting currents and air pressure cells in the tropical Pacific region and ends by

---

2 Climate change + acid rain + UV: In eastern Canada, two decades of mild drought and a slight warming trend have reduced streamflow into many of the region's lakes. The lake water has grown clearer, since the weakened streams are washing in less organic debris. The clearer water allows UV radiation to penetrate more deeply—at a time when more UV light is striking the lakes in the first place, because of the deterioration of the ozone layer. (UV light can injure fish and other aquatic organisms just as it injures humans.) Acid rain, which affects northern lakes in both Canada and Eurasia, causes even more organic matter to precipitate out of the water, further opening the lakes to UV light. In some lakes, the overall effect may be to increase the depth of UV penetration from 20–30 centimeters to over 3 metres.

3 Climate change + alteration of fire cycles: The fire ecology of forests all over the world is in a profound state of flux; we have introduced fire into some tropical rainforests that do not naturally burn at all, while in many temperate forests, where fire is essential for maintaining the native plant community, we have suppressed it. Climate change will probably cause further instability in fire cycles, as some regions become drier and others wetter. The results cannot be predicted, but are unlikely to favor original forest composition. If the overall rate of burning increases, that could create a positive feedback loop in the climate cycle, by releasing ever greater quantities of heat-trapping carbon into the atmosphere.

4 Climate change + N pollution: As a factor in the decline of some temperate-zone forests, nitrogen pollution is probably reducing their capacity to absorb carbon from the atmosphere.

5 Climate change + acid rain + UV + trospheric ozone + bioinvasion + alteration of fire cycles + N pollution: This complex of pressures is pushing eastern North American forests into decline. (See text.)

6 Climate change + habitat loss + bioinvasion + N pollution + overfishing: This set of pressures is pushing the world's coral reefs into decline. (See text.)

7 Climate change + N pollution + infectious disease: Cool weather often limits the ranges of mosquitoes and other insects that carry human pathogens. Even relatively slight increases in minimum temperatures can admit a pest into new areas. Warm coastal ocean water, especially when it's nitrogen-polluted, creates habitat for cholera.

8 Habitat loss + population growth: Last year, the floodings of China's Yangtze River did $30 billion in damages, displaced 223 million people, and killed another 3,700. The flooding was not wholly a natural event: with 85 percent of its forest cover gone, the Yangtze basin no longer had the capacity to absorb the heavy rains. (Forests are like immense sponges—they hold huge quantities of water.) And the densely settled floodplain guaranteed that the resulting monster flood would find millions of victims. (See "Record Year for Weather-Related Disasters," 27 November 1998, at www.worldwatch.org/alerts/index.)

9 Freshwater diversion + N pollution: Extensive irrigation can turn an arid region into productive cropland, but chemical fertilization is likely to follow and make the fields a source of nitrous oxide.

10 Bioinvasion + POPs: In the Great Lakes, exotic zebra mussels are ingesting dangerous organochlorine pesticides and other persistent organic chemicals that have settled into the loose, lake-bottom muck. Once in the zebra mussels, the chemicals may move elsewhere in the food web. Over the past decade or so, poisoning with such chemicals is also thought to be a factor in the growing susceptibility of marine mammals to the various epidemics that have emerged here and there throughout the world's oceans.

11 Bioinvasion + N pollution: Nitrogen pollution of grassland tends to favor the spread of aggressive exotic weeds. Nitrogen pollution of forests tends to weaken tree defenses against pests, both exotic and native.

12 Population growth + infectious disease. Over the next half-century, the centers of population growth will be the crowded, dirty cities of the developing world. These places are already breeding grounds for most of humanity's deadliest pathogens: cholera, malaria, AIDS, and tuberculosis among them. As the cities become more crowded, rates of infection are likely to grow and "overlapping infections" are likely to increase mortality rates.

rearranging a good deal of the planet's weather. El Niños appear to be growing more frequent and more intense; many climate scientists suspect that this trend is connected with climate change. It's very difficult to sort out the patterns, but there is probably also a general SST warming trend in the background, behind the El Niños. That too is a likely manifestation of climate change.

When SSTs reach the 28–30° C range, the coral polyp may expel the algae that live within its tissues. This action is known as "bleaching" because it turns the coral white. Coral usually recovers from a brief bout of bleaching, but if the syndrome persists it is generally fatal because the coral depends on the algae to help feed it through photosynthesis. Published records of bleaching date back to 1870, but show nothing comparable to what began in the early 1980s, when unusually warm water caused extensive bleaching throughout the Pacific. Coral bleached over thousands of square kilometers. By the end of the decade, mass bleaching was occurring in every coral reef region in the world. The full spectrum of coral species was affected in these events—a phenomenon that had never been observed before.

In the second half of this decade, SSTs set new records over much of the coral's range and the bleaching has become even more intense. Last year saw the most extensive bleaching to date. Over a vast tract of the Indian Ocean, from the African coast to southern India, 70 percent of the coral appears to have died. Some authorities think that a shift from episodic events to chronic levels of bleaching is now under way.

The bleaching has triggered outbreaks of the crown-of-thorns starfish, a coral predator that is chewing its way through reefs in the Red Sea, off South Africa, the Maldives, Indonesia, Australia, and throughout much of the Pacific. The starfish are normally kept at bay by antler-like "branching corals," which have stinging cells and host various aggressive crustaceans. But as the branching corals bleach and die, the more palatable "massive corals" growing among them become ever more vulnerable to starfish attack. Over the course of a year, a single adult crown-of-thorns can consume 13 square meters of coral.

Overfishing is also promoting these outbreaks, by removing the fish that eat starfish. Overfishing also helps another enemy of the reefs: various types of algae that compete with coral. Floating algae can starve corals for light; macro-algae—"seaweeds"—can colonize the reefs themselves and displace the coral directly. Because reefs are shallow-water communities, they generally occur in coastal zones, where they are likely to be exposed to nitrogen-rich agricultural runoff and sewage. Nitrogen pollution is as toxic to reefs as it is to temperate-zone forests, because nitrogen fertilizes algae. Remove the algae-eating fish under these conditions, and you might as well have poisoned the coral directly. This overlap is the main reason Jamaica's reefs never recovered from Hurricane Allen in 1980; 90 percent of the reefs off the island's northwest coast are now just algae-covered humps of limestone.

In the Caribbean, over-fishing seems to have played a role in yet another complication for the reefs: the population collapse of an algae-eating sea urchin, *Diadema antillarum*. This urchin appears to have been the last line of defense against the algae after the progressive elimination of other algae-eating creatures. The first to go may have been the green sea turtle. Now endangered, the turtle once apparently roamed the Caribbean in immense herds, like bison on the Great Plains. Its Caribbean population may have surpassed 600 million. Christopher Columbus's fleet reportedly had to reef sail for a full day to let a migrating herd pass. By the end of the 18th century, the turtles had nearly all been slaughtered for their meat. In the following two centuries, essentially the same operation was repeated with the algae-eating fish.

The removal of its competitors must have given the urchin a great deal of room, and for most of this century it was one of the reefs' most common denizens. But its abundance seems to have set it up for the epidemic that struck during the El Niño of the early 1980s. In roughly a year, a mysterious pathogen virtually eliminated *D. antillarum* from the Caribbean; some 98 percent of the species disappeared over an area of more than 2.5 million square kilometers. Contemporary history offers no precedent for a die-off of that magnitude in a marine animal. The urchin is reportedly back in evidence, at least in some areas of its former range, but until its relationship with the pathogen is better understood, it won't be possible to define its long-term appetite for algae.

With the algae, the pollution, and the warming waters, the Caribbean is becoming an increasingly hostile environment for the organism that has shaped so much of its biological character. And now the coral itself is sickening; the Caribbean has become a caldron of epidemic coral diseases. The first such epidemic, called black-band disease, was detected in 1973 in Belizean waters. Black band is caused by a three-layer complex of "blue-green algae" (actually, cyanobacteria), each layer consisting of a different species. The bottom layer secretes highly toxic sulfides which kill the coral. The complex creeps very slowly over a head of coral in a narrow band, leaving behind only the bare white skeleton.

Black band has since been joined by a whole menagerie of other diseases: white-band, yellow-band, red-band, patchy necrosis, white pox, white plague type I and II, rapid-wasting syndrome, dark spot. The modes of action are as various as the names. White pox, for example, is caused by an unknown pathogen that almost dissolves the living coral tissue. Infected polyps disintegrate into mucous-like strands that trail off into the water, and bare, dead splotches appear on the reefs, giving them a kind of underwater version of the mange. Rapid wasting syndrome probably starts with aggressive biting by spotlight parrotfish; the wounds are then infected by some sort of fungus that spreads out from the wound site. On the reefs off Florida, the number of diseases has increased from five or six to 13 during the past decade. In 1996, nine of the 44 coral species occurring on these reefs were diseased; a year later the number of infected species had climbed to 28. Nor are the Caribbean reefs the only ones under attack; coral epidemics are turning up here and there throughout the Pacific and Indian Oceans, in the Persian Gulf and in the Red Sea.

For most of these diseases, a pathogen has yet to be identified; it's not even clear whether each of those names really refers to a distinct syndrome. But it's not likely that the diseases are "new" in the sense of being caused by pathogens that have recently evolved. It's much more likely that the coral's vulnerability to them is new. Take, for example, the disease that's killing sea-fan coral around the Caribbean. In this case, the pathogen is known: it's *Aspergillus sydowii,* a member of a very common genus of terrestrial fungi. The last time you threw something out of your refrigerator because it was moldy—there's a good chance you were looking at an *Aspergillus* species. In a very bizarre form of invasion, *A. sydowii* breached the land-sea barrier, and found a second home in the ocean. But it evidently took the plunge decades ago and has only been killing sea-fans for some 15 years or so. Why? Part of the answer is probably the higher SSTs: *A. sydowii* likes warmer water. Other coral diseases appear to do especially well in nutrient-laden waters.

Disease lies at one end of the spectrum of threat. Pathogens create a kind of microscopic pressure, but there are macroscopic pressures too: the ecosystems allied in one way or another with the reef biome are also deteriorating. The stretch of shallow, protected water between a reef and the coast often nurtures beds of seagrass. These beds filter out sediment and effluent that would injure the reefs, and the seagrass provides crucial cover for young fish. Seagrass is the major nursery for many fish species that spend their adult lives out on the reefs. Perhaps 70 percent of all commercially important fish spend at least part of their lives in the seagrass. But the tropical seagrass beds are silting up under tons of sediment from development, logging, mining, and the construction of shrimp farms. They are suffocating under algal blooms in nitrogen-polluted waters; they are being poisoned by herbicide runoff. According to one estimate, half of all seagrass beds within about 50 kilometers of a city have disappeared.

If you follow the seagrass-choking sediment back the way it came, you're increasingly likely to find a shoreline denuded of mangroves. In the warmer regions of the world, mangroves knit the land and sea together. These stilt-rooted trees trap sediment that would otherwise leak out to sea and they stabilize coastlines against incoming storms. Like the seagrass beds and the reefs, the mangrove ecosystem is incredibly productive—in the mangroves' case, with both terrestrial and aquatic organisms. (Mangrove roots are important fish nurseries too.)

The mangroves' importance as a sediment filter is perhaps greatest in the center of reef diversity, the Indonesian archipelago and adjoining areas. About 450 coral species are known to grow in the Australasian region; the Caribbean, by comparison, contains just 67 species. Australasia is correspondingly rich in fish too: a quarter of the world's fish species inhabit these waters. It is estimated that half of all the sediments received by oceanic waters are washed from the Indonesian archipelago alone. Nearby areas of Southeast Asia are also major contributors of sediment. But throughout the region, logging and shrimp farming are obliterating the mangroves that once filtered this tremendous burden of silt. Southeast Asia has lost half its mangrove stands over the past half century. A third of the mangrove cover is gone from Indonesian coasts, three-quarters from the Philippines.

About 10 percent of the world's coral reefs may already have been degraded beyond recovery. If we can't find a way to ease the reefs' afflictions, nearly three-quarters of the ocean's richest biome may have disappeared 50 years from now. Such a prospect gives new meaning to the term "natural disaster," but it's also a social disaster in the making. Reef fish make up perhaps 10 percent of the global fish catch; one estimate puts their contribution to the catch of developing countries at 20 to 25 percent.

And there's much more at stake here than just fisheries. The death of the coral would also jeopardize the reef *structures*—leaving them unable to repair storm damage. If the reefs give way, wave erosion of the coasts behind them will increase. The coasts are already facing some unavoidable degree of damage from climate change, as sea-levels rise.

(Warming water expands; that physical effect will combine with runoff from melting glaciers to push sea levels up.) Rising seas, like the crumbling reefs, will allow storm surges to reach farther inland. About one-sixth of the world's coasts are shielded by reefs, and some of these coasts, like the ones in South and Southeast Asia, support some of the densest human populations in the world. The disintegration of the reefs would leave a large portion of humanity hungrier, poorer, and far more vulnerable to the vagaries of a changing climate.

CORAL REEFS AND TEMPERATE-ZONE FORESTS—IN BOTH OF THESE theaters of surprise, the familiar could rapidly become something else. But you can begin to see similar system effects just about anywhere, and emerging from just about any form of environmental pressure:

- Nitrogen pollution has tripled the occurrence of low-oxygen dead zones in coastal ocean waters over the past 30 years. As in the Black Sea, excess nitrogen appears generally to be promoting the emergence of red tide organisms. (Over the past decade, the num-

ber of algae species known to be toxic has increased from around 20 to at least 85.)
- Organochlorine pollutants seem to be creating immunodeficiencies in marine mammals, triggering a growing number of viral epidemics. (Exposure to the red-tide toxins may also depress the immune systems of some marine mammals and sea turtles.)
- The hunting of birds and primates in tropical forests may become another form of deforestation, because these creatures are so important in pollinating tree flowers and dispersing seeds.
- Powerful storms, which may grow more common as the climate changes, tend to magnify invasions of exotic plants by dispersing their seeds over huge areas.
- And a whole spectrum of threats appears to underlie the global decline in amphibians: habitat loss, pollution, disease, exotic predators, and higher levels of UV exposure resulting from the disintegration of the ozone layer. (See the table, "A Spreading Matrix of Trouble," for some additional system effects.)

Given the pressures to which the global environment is now subject, the potential for surprise is, for all practical purposes, unlimited. We have stepped into a world in which our assumptions and prejudices are more and more likely to betray us. We are confronting a demon in a hall of mirrors. At this point, a purely reactive approach to our tormentor will lead inevitably to exhaustion and failure.

Towards a Complexity Ethic

OUR PREDICAMENT, ESSENTIALLY, IS THIS: ENVIRONMENTAL PRESSURES ARE converging in ways that are likely to create a growing number of unanticipated crises. Each of these crises will demand some sort of fix, and each fix will demand money, time, and political capital. Yet no matter how many fixes we make, we've no realistic expectation of reducing the potential for additional crises—if "fixing" is all we do. The key to controlling that demon is to do a better job of managing systems in their entirety. And whether the system in question is the global trading network, a national economy, or a single natural area, many of the same operating principles will apply. Here, in my view, are four of the most important ones.

Monoculture technologies are brittle.

Huge, uniform sectors generally exhibit an obvious kind of efficiency because they generate economies of scale. You can see this in fossil fuel-based power grids, car-dominated transit systems, even in the enormous woodpulp plantations that are an increasingly important part of the developing world's forestry sector. But this efficiency is usually superficial because it doesn't account for all sorts of "external" social and environmental costs. Thus, for instance, that apparently cheap fossil-fuel electricity is purchased with the literally incalculable risks of climatic dislocation, with acid rain and ozone pollution, with mine runoff, and in the countries that rely most heavily on coal—China, for instance, and South Africa—with a heavy burden of respiratory disease.

Yet even when the need for change is obvious and alternative technologies are available, industrial monocultures can be extremely difficult to reform. In energy markets, solar and wind power are already competitive with fossil fuel for many applications, even by a very conventional cost comparison. And when you bring in all those external costs, there's really no comparison at all. But with trillions of dollars already invested in coal and oil, the global energy market is responding to renewables in a very slow and grudging way.

More diverse technologies—in energy and in any other field—will encourage more diverse investment strategies. That will tend to make the system as a whole more adaptable because investors will not all be "betting" on exactly the same future. And a more adaptable system is likely to be more durable over the long term.

Direct opposition to a natural force usually invites failure—or a form of success that is just as bad.

In the "Iron Gates" brand of development, it is sometimes difficult to distinguish success from failure. Less obvious, perhaps, is the fact that even conservation activities can run afoul of natural forces. Take, for example, the categorical approach to forest fire suppression. A no-burn policy may increase a forest's fuel load to the point at which a lightning strike produces a huge crown fire. That's outright failure: a catastrophic "artificial" fire may consume stands that survived centuries of the natural fire cycle. On the other hand, if the moisture regime favors rapid decomposition of dead wood, the policy could eliminate fire entirely. Without burning, the fire-tolerant tree species would probably also begin to disappear, as they are replaced by species better adapted to the absence of fire. That's "success." Either way, you lose the original forest.

Sound policy often tends to be more "oblique" than direct. A vaccine, for instance, turns the power of the pathogen against itself; that's why, when there's a choice, immunization is usually a better tactic for fighting disease than quarantine. Restoration of floodplain ecosystems can be a more effective form of flood control than dams and levees, because wetlands and forests function as immense sponges. (The catastrophic flooding last year in China's Yangtze river basin was largely the result of deforestation.)

An oblique approach might also help reduce demand for especially energy- or materials-intensive goods: if large numbers of people can be convinced to "transfer" their demand from the goods themselves to the services that the goods provide, then it might be possible to encourage consumption patterns that do less environmental damage. For example, joint ownership of cars, especially in cities, could satisfy needs for occasional private transportation, with a little coordination.

Since you can never have just one effect, always plan to have several.

Thinking through the likely systemic effects of a plan will help locate the risks, as well as indirect opportunities. Every day, for example, I ride the car pool lanes into Washington D.C., and my conversations with other commuters have led me to suspect that this environmentally correct ribbon of asphalt could actually *increase* pollution and sprawl, by contributing to a positive feedback loop. Here's how I think it may work: as the car pool lanes extended outward from the city, commute times dropped; that would tend to promote the development of bedroom communities in ever more remote areas. Eventually, the new developments will cause traffic congestion to rebound, and that will create political pressure for another bout of highway widening. A more "system sensitive" policy might have permitted the highway projects only when a county had some realistic plan to limit sprawl. (According to one recent estimate, metropolitan Washington is losing open space faster than any other area in the United States outside of California's central valley.) Car pool lanes might then have become a means of conserving farmland, instead of a possible factor in its demise.

For environmental activists, "system sensitivity" could help locate huge political constituencies. Look, for instance, at the potential politics of nitrogen pollution. Since a great deal of the nitrogen that is threatening coral reefs is likely to be agricultural runoff, and since much of that runoff is likely to be the result of highly mechanized "factory farming," it follows that anyone who cares about reefs should also care about sustainable agriculture. Obviously, the reverse is true as well: if you're trying to encourage organic farming in the Mississippi basin, you're conserving Caribbean reefs. The same kind of political reciprocity could be built around renewable energy and forest conservation.

I don't know the answer and neither do you, but together we can probably find one.

A system can have qualities that exist only *on the system level*—qualities that cannot be attributed directly to any of the components within. No matter how hard you look, for example, at the individual characteristics of oxygen, nitrogen, hydrogen, carbon, and magnesium, you will never find grounds for inferring the amazing activities of chlorophyll—the molecule that powers photosynthesis. There are system properties in political life as well: institutional pluralism can create a public space that no single institution could have created alone. That's one objective of the "balance of powers" aimed at in constitutional government.

It should also be possible to build a "policy system" that is smarter and more effective than any of its component groups of policy makers. Consider, for example, the recent history of the U.S. Forest Service. For decades, environmental activists have accused the service of managing the country's forests almost exclusively for timber production, with virtually no regard for their inherent natural value. Distrust of the service has fueled a widespread, grassroots forest conservation movement, which has grown increasingly sophisticated in its political and legal activities, and now even undertakes its own scientific studies on behalf of the forests. This movement, in turn, has attracted the interest and sympathy of a growing number of officials within the service. Many environmentalists (including this author) would argue that things are nowhere near what they should be inside the service, but it's possible that what we are witnessing here is the creation of a new space for conservation—a space that even a much more ecologically enlightened Forest Service couldn't have created on its own.

It remains to be seen whether this forum will prove powerful enough to the save the forests that inspired it. But in the efforts of the people who are building it, I think I can see, however dimly, a future in which the world's dominant cultures re-experience the shock of living among forests, prairies, and oceans—instead of among "natural resources." After all, the forests and prairies are where we came from and they're where we are going. We are the children of a vast natural complexity that we will never fathom.

Chris Bright is a research associate at the Worldwatch Institute, senior editor of WORLD WATCH, and author of *Life Out of Bounds: Bioinvasion in a Borderless World* (New York: W.W. Norton & Co., 1998).

A FEW KEY SOURCES

Harvard Ayers, Jenny Hager, and Charles E. Little, eds., *An Appalachian Tragedy: Air Pollution and Tree Death in the Eastern Forests of North America* (San Francisco: Sierra Club Books, 1998).

Osha Gray Davidson, *The Enchanted Braid: Coming to Terms with Nature on the Coral Reef* (New York: John Wiley, 1998).

Paul Epstein et al., *Marine Ecosystems: Emerging Diseases as Indicators of Change,* Health Ecological and Economic Dimensions (HEED) of the Global Change Program (Boston: Center for Health and Global Environment, Havard Medical School, December 1998).

Robert Jervis, *System Effects: Complexity in Political and Social Life* (Princeton, NJ: Princeton University Press, 1997).

Charles Perrow, *Normal Accidents: Living with High-Risk Technologies* (New York: Basic Books, 1984).

HEALTH OF THE EARTH

Harnessing Corporate Power to Heal the Planet

Pioneering companies in sectors ranging from wire to plastic films and planned residential communities have already demonstrated that today's environmental challenges hold many profit-enhancing opportunities.

L. Hunter Lovins and Amory B. Lovins

The late twentieth century witnessed two great intellectual shifts: the fall of communism, with the apparent triumph of market economics; and the appearance, in a rapidly growing number of businesses, of the end of the war against the earth, and the emergence of a new form of economics we call natural capitalism.

This term implies that capitalism as practiced is an aberration, not because it is capitalist but because it is defying its own logic. It does not value, but rather liquidates, the most important form of capital: *natural* capital, in other words the natural resources and, more importantly, the ecosystem services upon which all life depends.

Deficient logic of this sort can't be corrected simply by placing a monetary value on natural capital. Many key ecosystem services have no known substitutes at any price. For example, the $200 million Biosphere II project, despite a great deal of impressive science, was unable to provide breathable air for eight people. Biosphere I, our planet, performs this task daily at no charge for six billion of us.

Ecosystem services give us tens of trillions of dollars' worth of benefits each year, or more than the global economy. But none of this is reflected on anyone's balance sheets.

The best technologies cannot substitute for water and nutrient cycling, atmospheric and ecological stability, pollination and biodiversity, topsoil and biological productivity, and the process of assimilating and detoxifying society's wastes. With the human race increasing by 8,700 people every hour, more people are chasing after fewer resources. The limits to economic growth are coming to be set by scarcities of natural capital.

Sometimes the value of ecosystem services becomes apparent only when they are lost. In China's Yangtze basin in 1998, for example, deforested watersheds fostered flooding that killed 3,700 people, dislocated 223 million, and inundated 60 million acres of cropland. That $30 billion disaster forced a logging moratorium and a $12 billion crash program of reforestation.

This is not to say we're running out of such commodities as copper and oil. Even with recent fluctuations, prices for almost all commodities are near record lows and will fall for some time, in part because of improvements in extraction technologies. But in many instances these technologies impose environmental costs that further degrade the ability of living systems to sustain a growing human population.

Before the Industrial Revolution, from which capitalism emerged, it was inconceivable that people could work more productively. Nonetheless, textile mills introduced in the late 1700s enabled one Lancashire spinner to do the work previously done by 200 weavers, and the mills were only one of many technologies that increased the productivity of workers.

Profit-maximizing capitalists economized on their scarcest production factor: skilled people. They substituted seemingly abundant resources and the ability of the planet to absorb their pollution to enable people to do more work.

Given today's patterns of scarcity and abundance, that same business logic dictates using more people and more brains to wring 4, 10, or even 100 times as much benefit from each unit of energy, water, materials, or anything else borrowed from the

3. Harnessing Corporate Power to Heal the Planet

planet. Success at this will be the basis of competitiveness in the decades to come and will be the hallmark of the next industrial revolution.

The first of four principles of natural capitalism is to increase resource efficiency radically. This not only increases profits, but also solves most of the environmental dilemmas facing the world today. It greatly slows resource depletion at one end of the economic process and discharge of pollution (resources out of place) at the other end. It creates profits by reducing the costs of both resources and pollution. And it also buys time, forestalling the threatened collapse of natural systems.

That time should then be used to implement the other three principles of natural capitalism. These are: (2) eliminate the concept of waste by redesigning the economy based on models that close the loops of materials flows; (3) shift the focus of the economy from processing materials and making things to creating service and flow; and (4) reverse the destruction of the planet now under way by instituting programs of restoration that invest in natural capital.

By applying the four principles of natural capitalism, businesses can behave as if ecosystem services were properly valued and begin to reverse the loss of such services even as they increase profits.

1. Increase resource productivity

It is relatively easy to profit by using resources more efficiently because they are used incredibly wastefully now. The stuff that drives the industrial metabolism of the United States currently amounts to, for each American, more than 20 times your body weight every day, or more than a million pounds per year.

Globally, this is a flow of half a trillion tons per year. But only about 1 percent of all the materials mobilized in the economy is ever embodied in a product that endures six months after sale. Cutting such waste represents a vast business opportunity.

Nowhere are the opportunities for savings easier to see than in energy. The United States has already cut its annual energy bills by $150 billion relative to what they would have been if savings had not been implemented since the first oil shock in 1973. However, we still waste $300 billion worth of energy each year in an economy whose total energy bill is more than $500 billion ($516 billion in 1995). Just the energy thrown away by U.S. power stations as waste heat, for example, equals the total energy used by Japan.

Fortunately, we already have ample examples of companies that have shown how to improve energy efficiency (i.e., reduce waste) and increase profits.

Southwire Corporation—an energy-intensive maker of cable, rod, and wire—halved its energy per pound of product in six years. The savings roughly equaled the company's profits during that period, and company officials estimated that the energy-efficiency effort probably secured 4,000 jobs at 10 plants in six states that were jeopardized by competitive market forces. The company then went on to save even more energy, achieving two-year paybacks despite all the earlier energy-efficiency improvements.

Dow's Louisiana division (now Dow Louisiana Operations) implemented over 900 worker-suggested energy-saving projects during the period 1981 to 1993, with average annual returns on investment of over 200 percent. Both returns and savings *rose* in later years, even after the accumulated annual savings from the projects had passed $100 million, because the engineers were learning faster than they were using up the cheapest opportunities.

State-of-the-shelf technologies can make old buildings three- to fourfold more energy-efficient and new buildings nearer 10-fold—and cheaper to build. Examples include large and small buildings in climates ranging from well below freezing to

COURTESY OF SOUTHWIRE CORPORATION

■ Southwire Corporation, which nearly went bankrupt due to severe competition in the wire industry, survived by cutting its energy use per a pound of wire in half as it also enhanced the efficiency of all aspects of operations. Today the company is thriving.

Just the energy thrown away by U.S. power stations as waste heat equals the total energy used by Japan.

sweltering, both types of which can be kept comfortable with no heating or cooling systems. Industries can achieve profitable savings in motor systems, process designs, and materials productivity. Rocky Mountain Institute's Hypercar design synthesis for automobiles and other road vehicles will produce huge energy and materials savings—and spell the end of the car, oil, steel, aluminum, coal, and electricity industries as we know them [see "The Car of the Future?" THE WORLD & I, August 1996, p. 148].

How can such savings be captured? An international company recently redesigned a standard industrial pumping loop slated for installation in its new Shanghai factory. The original, supposedly optimized, design needed 95 horsepower for pumping. Dutch engineer Jan Schilham made two embarrassingly simple design changes that cut that 95 hp to only 7 hp—a 92 percent reduction. The redesigned system cost less to build and worked better in all respects.

First, Schilham chose big pipes and small pumps rather than small pipes and big pumps. The friction in a pipe falls inversely as nearly the fifth power of its diameter. In considering how big to make the pipes, normal engineering practice balances the capital cost of the pipe against the ongoing energy costs of pumping fluid through the pipe.

But this textbook optimization ignores the capital cost of the pumping equipment—the pump, motor, variable-speed electronic control, and electrical supply—that must be big enough to fight the pipe friction. Ignoring the potential equipment saving, and optimizing one component (the pipe) in isolation, "pessimizes" the system. Optimizing the *whole system* instead, and counting savings in total capital cost as well as in energy cost, makes it clear that, within a critical range, as pipe size increases the capital cost falls more rapidly for equipment than it rises for the much fatter pipe. The whole system therefore costs less but works better.

Schilham's second innovation was to lay out the pipes first, then the equipment. The normal sequence is the opposite: install the equipment in traditional positions (far apart, at the wrong height, facing the wrong way, with other stuff in between), then tell the pipe fitter to hook it all up. The resulting long, crooked pipes have about three to six times as much friction as short, straight pipes. Using short, straight pipes to minimize friction cuts both capital and operating costs. In this case, it also saved 70 kilowatts of heat loss, because straight pipes are easier to insulate.

This matters because pumping is a major user of electricity worldwide. Optimizing a whole pumping system, at the level of a whole building or a whole factory, can typically yield energy savings of 3- to 10-fold and cost less to operate. But more importantly, the thought process of whole systems thinking applies to almost every technical system that uses resources.

Consider real estate development. Typical tract home developments drain storm water in expensive underground sewers. Village Homes, an early solar housing development in Davis, California, instead installed natural drainage swales. This saved $800 per house and provided more green space. The company then used the saved money to pay for edible landscaping that provided shade, nutrition, beauty, community focus, and crop revenues that paid the homeowners' assessments and paid for a community center. The people-centered site planning (narrower, tree-lined streets, with the housing fronting on the greenways) saved more land and more money. It also cooled off the microclimate, yielding better comfort at lower cost, and it created safe and child-friendly neighborhoods that cut crime 90 percent compared with neighboring subdivisions. Real estate brokers once described the project as weird. It is now the most desirable real estate in town, with market values $11 per square foot over average.

The same approach can diminish the risk of climate change. DuPont is proposing to reduce its greenhouse gas emissions 65 percent from 1990 levels by 2010. In addition, by 2010 DuPont aims to derive a tenth of its energy and a quarter of its raw materials from renewables. It is making these changes in the name of increasing shareholder value. In a similar vein, ST Microelectronics, a manufacturer of microchips, has set a goal of zero new emissions while it implements a 20-fold production increase.

Many executives are realizing that protecting the climate is not costly but profitable, because saving fuel costs less than buying fuel. Using energy in a way that saves money is therefore an important way to strengthen the bottom line and the whole economy, while also resolving the climate problem.

This is why the European Union has already adopted at least a fourfold ("Factor Four") gain in resource productivity as the new basis for sustainable development policy and practice. Some countries, like the Netherlands and Austria, have declared this a national goal. Environment ministers from the OECD (Organization for Economic Cooperation and Development), the government of Sweden, and distinguished industrial and academic leaders in Europe, Japan, and elsewhere have gone even further, adopting "Factor Ten" improvements as their goal. The World Business Council for Sustainable Development and the UN Environment Programme have called for "Factor Twenty." There is growing evidence that such ambi-

3. Harnessing Corporate Power to Heal the Planet

tious goals are feasible and achievable in the marketplace. They may, in fact, offer even greater profits.

2. Eliminate the concept of waste

Resource efficiency is natural capitalism's cornerstone, but only its beginning.

Natural capitalism would eliminate the entire concept of waste by adopting biological patterns, processes, and often materials. This implies eliminating any industrial output that represents a disposal cost rather than a salable product.

Architect Bill McDonough tells the story of being asked by the Steelcase subsidiary DesignTex to design a "green" textile for upholstering office chairs. The fabric it was to replace used such toxic chemicals to treat and dye the cloth that the Swiss government had declared its edge trimmings a hazardous waste. McDonough's team eventually found a chemical firm that would let them explore its textile chemistry in detail.

They screened more than 8,000 chemicals, rejecting any that were toxic, built up in food chains, or caused cancer, mutations, birth defects, or endocrine disruption. The 38 that passed could make all colors. The cloth would look better, feel better in the hand, and last longer, because the natural fibers wouldn't be damaged by harsh chemicals. Fewer and cheaper feedstocks, as well as no health and safety concerns, meant that production cost less. The new fabric was beautiful and won design awards.

The Swiss environmental inspectors who tested the new plant thought their equipment was malfunctioning when the effluent water proved cleaner than the Swiss drinking water input: the cloth itself was acting as a filter. More important, the redesign of the process "took the filters out of the pipes and put them where they belong, in the designers' heads."

Professor Hanns Fischer noticed that the University of Zurich's basic chemistry lab course was turning pure, simple reagents into mainly hazardous wastes, incurring costs at both ends. The students were also learning once-through, linear thinking. So in some of the lessons, the students turned the toxic wastes back into pure, simple reagents. Students volunteered vacation time to repurify the wastes, because it was so much fun. Demand for wastes soon outstripped supply. Waste production declined 99 percent, costs fell by about $20,000 a year, and the students learned the closed-loop thinking that must ultimately save the chemical industry.

This is an archetype for the emerging world where environmental regulation will be an anachronism. In that biological world, the design lessons of nature will improve business—as well as health, housing, mobility, community, and national security. Such a world emerges from the cybernetics of not inflicting on others any emission to which you wouldn't expose yourself: How clean a car would you buy if its exhaust pipe, instead of being aimed at pedestrians, fed directly into the passenger compartment? How clean would a city or factory make the water it discharges if its intake pipes were downstream of its outlets? We all live downwind, downstream.

3. Create service and flow

A further key element of natural capitalism is to shift the structure of the economy from focusing on the production and sale of things to focusing on providing the customer a continuous flow of service and value.

This change in the business model provides incentives for a continuous improvement in the elimination of waste, because it structures the relationships so that *the provider and customer both make money by finding more efficient solutions that benefit both.* That contrasts sharply with the conventional sale or leasing of physical goods in which the vendor wants to provide more things more often—increasing waste—and at a higher price, while the customer has the opposite interests.

For example, Schindler leases vertical transportation services instead of selling elevators, Electrolux/Sweden leases the performance of professional floor-cleaning and commercial food-service equipment rather than the equipment itself, and Dow and Safety-Kleen lease dissolving services rather than selling solvents. Both customer and provider profit from minimizing the flow of energy and materials.

Carrier, the world's largest manufacturer of air conditioners, is experimenting with leases of comfort instead of sales of air conditioners. Making the equipment more efficient or more durable will give Carrier greater profits and its customers better comfort at lower cost. So too, however, will making the building itself more efficient, so that less cooling yields the same comfort. Carrier is therefore starting to team up with other firms that can improve lighting, glazings, and other building systems. Providing a more systemic solution, creating a relationship that continually aligns interests, is obviously better for customers, shareholders and the earth than selling air conditioners.

A striking example is emerging at Atlanta carpet maker Interface. Most broadloom carpet is replaced every decade because it develops worn spots. An office is shut down, furniture removed, and carpet torn up and sent to landfill. (The millions of tons deposited each year will last up to 20,000 years.) New carpet is laid down, the office restored, operations resumed, and workers get sick from the carpet-glue fumes.

Instead, Interface prefers to lease floor-covering services. People want to walk on and look at carpet, not own it. They can obtain these services at much lower cost if Interface owns the carpet and remains responsible for keeping it clean and

Whole-system solutions create more life, more value, and ultimately more profits.

fresh. For a monthly fee, Interface will visit regularly and replace the 10–20 percent of the carpet tiles that show 80–90 percent of the wear. This reduces the mass flow of carpet to landfill by about 80 percent and provides better service at lower cost. It also increases net employment, eliminates the disruption (worn tiles are seldom under furniture), and turns a capital expenditure into an operating lease.

Interface's latest technical innovation goes further. Other manufacturers say that they recycle carpet. Actually they downcycle it—reusing it in lower-grade products. In contrast, Interface's new Solenium product provides floor covering that is almost completely remanufacturable into identical carpet. This will cut their net flow of materials and energy it takes to make them by 97 percent.

It will also provide better service, because the new floor covering, which may be leased or sold, is nontoxic, virtually stainproof, easy to clean with water, four times as durable, one-third less materials-intensive, renewably produced, and otherwise superior in every respect.

Interface's first four years on this systematic quest to turn avoided waste into profit returned doubled revenues, tripled operating profits, and nearly doubled employment. Its latest $250 million revenue came with no increase in energy or materials inputs, from mining internal waste.

Or consider the Films division of DuPont. Once failing, it now leads its 59-firm market because it makes its films thinner, stronger, and better matched to customers' needs. This enables it to produce higher-value products using fewer materials. It also recycles used film, closing the materials loops, getting it back from customers with a process now coming to be known as "reverse logistics," a new topic of study in business schools. Jack Krol, past chairman of DuPont, has remarked that he sees no end to DuPont's ability to profit in this way.

4. Invest in natural capital

The fourth principle of natural capitalism is to invest to reverse the worldwide destruction of the ecosystem.

If natural capital is the most important, valuable, and indispensable form of capital, a true capitalist will restore it where degraded and sustain it where healthy—the better to create wealth and sustain life. Once toxicity and waste are designed out of industry, then forestry, farming, and fishing must be redesigned to be restorative to natural ecosystems. This will be especially important as the primary inputs to industry come to be grown, not mined, and living nanotechnologies replace vast industries.

This will place a premium on understanding biological models and on using nature as model and mentor rather than as a nuisance to be evaded. The incentive will derive not just from the goodwill of corporations, but from the scent of real profits and the promise of long-term corporate survival. No doubt some managers (the commercial about the company that is not interested in e-business comes to mind) will lack the willingness to tackle natural capitalism, and their companies will likely fossilize. Meanwhile, more visionary, adventurous managers will lead the wave of companies that embrace the new competitive grounds set by natural capitalism.

Catching up with centuries of deferred but unbooked planetary maintenance might sound expensive. But whole-system solutions create more life, more value, and ultimately more profits. Production is automatically carried out; people need only create hospitable conditions and do no harm. In this exciting sphere of innovation lie such opportunities as these:

- Dr. Allan Savory, cofounder of the Albuquerque-based Center for Holistic Management, has redesigned ranching to mimic the migration of large herds of native grazers that coevolved with grasslands. This can greatly improve the carrying capacity of even degraded rangelands, which turn out to have been not overgrazed but undergrazed, out of ignorance of how brittle ecosystems evolved.
- The California Rice Industry Association partnered with environmental groups to switch from burning rice straw to flooding the rice fields after harvest. They now flood 30 percent of California's rice acreage, from which they can harvest a more profitable mix of wildfowl, high-silica straw, groundwater recharge, and other benefits, with rice as a by-product.
- Dr. John Todd of Ocean Arks International and Living Technologies, based in Burlington, Vermont, builds biological "Living Machines" that turn sewage into clean water—plus valuable flowers, a tourist venue, and other by-products—with no toxicity, no odor, and reduced capital costs. Such "bioneers" are using living organisms to "bioremediate" toxic pollutants into forms that are harmless or salable or both.

These practices adopt the design experience of nearly four billion years of evolutionary testing in which products that failed were recalled by the Manufacturer. Though many details of such nature-mimicking practices are still evolving, the broad contours of the lessons they teach are already clear [see "The Living Building," THE WORLD & I, October 1999, p. 160].

Some of the most exciting developments are modeled on nature's low-temperature, low-pressure assembly techniques, whose products rival anything man-made. Janine Benyus' book *Biomimicry* points out

Natural capitalism implemented in a company creates an extraordinary outpouring of energy, initiative, and enthusiasm at all levels.

that spiders make silk as strong as Kevlar—but much tougher—from digested crickets and flies, without needing boiling sulfuric acid and high-pressure extruders.

The abalone makes an inner shell twice as tough as ceramics, and diatoms make seawater into glass; neither need furnaces. Trees turn air, sunlight, and soil into cellulose, a sugar stiffer and stronger than nylon. We may never be as skillful as spiders, abalone, diatoms, or trees, but such benign natural chemistry may be a better model than industrialism's primitive approach of "heat, beat, and treat."

Beyond profits: What's in it for us?

Natural capitalism implemented in a company creates an extraordinary outpouring of energy, initiative, and enthusiasm at all levels. It removes the contradiction between what people do at work and what they want for their families when they go home. This makes natural-capitalist firms some of the most exciting places in the world to work.

Civilization in the twenty-first century is imperiled by the dissolution of civil societies into lawlessness and despair; weakened life-support systems; and the dwindling public purse needed to address these problems and reduce human suffering.

These three threats share a common cause—waste.

The leaders in waste reduction will be in the corporate sector. But there remains a vital role for governments and for civil society. It is important to remember markets' purposes and limitations. Markets make a splendid servant but a bad master and a worse religion. Markets produce value, but only communities and families produce values. A society that substitutes markets for politics, ethics, or faith is dangerously adrift. Commerce can create a durable system of production and consumption by properly applying sound market principles. Yet not all value is monetized; not every priceless thing is priced. Nor is accumulating money the same thing as creating wealth or improving people. Many of the best things in life are not the business of business. And as the Russians are finding under "gangster capitalism," unless democratic institutions establish and maintain a level playing field, only the most ruthless can conduct business.

One of government's most powerful tools is tax policy. Such taxes as FICA and other penalties on employment that grew out of the first industrial revolution encourage companies to use more resources and fewer people. Gradual and fair tax shifting and desubsidization can provide more of what we want—jobs and income—and less of what we don't want: environmental and social damage.

But government's power is limited. Today over half the world's 100 largest economic entities are not countries but companies. Corporations may be the only institution with the resources, agility, organization, and motivation to tackle the toughest problems.

Firms that pursue the four principles of natural capitalism—profiting from advanced resource productivity, closing materials loops and eliminating waste, providing their customers with efficient solutions, and reinvesting in natural capital—will gain a commanding competitive advantage. They'll be behaving as if natural and human capital were properly valued. And as Ed Woollard, former chairman of DuPont, once remarked, companies that don't take these principles seriously won't be a problem, because they won't be around.

Perhaps the only problem with capitalism—a system of wealth creation built on the productive flow and expansion of all forms of capital—is that it is only now beginning to be tried.

L. Hunter Lovins and Amory B. Lovins are co-CEOs of the Rocky Mountain Institute (RMI), a nonprofit resource policy center they cofounded in 1982 in Snowmass, Colorado. The Lovinses are coauthors along with Paul Hawken of the book Natural Capitalism: Creating the Next Industrial Revolution *(Little, Brown, 1999), available from RMI at orders@rmi.org.*

Crossing the Threshold

Early Signs of an Environmental Awakening

by Lester R. Brown

At a time when the Earth's average temperature is going off the top of the chart, when storms, floods and tropical forest fires are more damaging than ever before, and when the list of endangered species grows longer by the day, it is difficult to be optimistic about the future. Yet even as these stories of environmental disruption capture the headlines, I see signs that the world may be approaching the threshold of a sweeping change in the way we respond to environmental threats—a social threshold that, once crossed, could change our outlook as profoundly as the one that in 1989 and 1990 led to a political restructuring in Eastern Europe.

If this new threshold is crossed, changes are likely to come at a pace and in ways that we can only begin to anticipate. The overall effect could be the most profound economic transformation since the Industrial Revolution itself. If so, it will affect every facet of human existence, not only reversing the environmental declines with which we now struggle, but also bringing us a better life

Thresholds are encountered in both the natural world and in human society. One of the most familiar natural thresholds, for example, is the freezing point of water. As water temperature falls, the water remains liquid until it reaches the threshold point of 0 degrees Celsius (32 degrees Fahrenheit). Only a modest additional drop produces dramatic change, transforming a liquid into a solid.

The threshold concept is widely used in ecology, in reference to the "sustainable yield threshold" of natural systems such as fisheries or forests. If the harvest from a fishery exceeds that threshold for an extended period, stocks will decline and the fishery may abruptly collapse. When the demands on a forest exceed its sustainable yield and the tree cover begins to shrink, the result can be a cascade of hundreds of changes in the ecosystem. For example, with fewer trees and less leaf litter on the forest floor, the land's water-absorptive capacity diminishes and runoff increases—and that, in turn, may lead to unnaturally destructive flooding lower in the watershed.

In the social world, the thresholds to sudden change are no less real, though they are much more difficult to identify and anticipate. The political revolution in Eastern Europe was so sudden that with no apparent warning the era of the centrally planned economy was over, and those who had formidably defended it for half a century realized it was too late to reverse what had happened. Even the U.S. Central Intelligence Agency failed to foresee the change. And after it happened, the agency had trouble explaining it. But at some point, a critical mass had been reached, where enough people were convinced of the need to change to tip the balance and bring a cascading shift in public perceptions.

In recent months, I have become increasingly curious about such sudden shifts of perception for one compelling reason. If I look at the global environmental trends that we have been tracking since we first launched the Worldwatch Institute 25 years ago, and if I simply extrapolate these trends a few years into the next century, the outlook is alarming to say the least. It is now clear to me that if we are to turn things around in time, we need some kind of *breakthrough*. This is not to discount the many gradual improvements that we have made on the environmental front, such as increased fuel efficiency in cars or better pollution controls on factories. Those are important. But we are not moving fast enough to reverse the trends that are undermining the global economy. What we need now is a rapid shift in consciousness, a dawning awareness in people everywhere that we have to shift quickly to a sustainable economy if we want to avoid damaging our natural support systems beyond repair. The question is whether there is any evidence that we are approaching such a breakthrough.

While shifts of this kind can be shockingly sudden, the underlying causes are not. The conditions for profound social change seem to require a long gestation period. In Eastern Europe, it was fully four decades from the resistance to socialism when it was first imposed until its demise. Roughly 35 years passed between the issuance of

4. Crossing the Threshold

New climate data drew intense media interest, as more than 100 reporters gathered for a WORLD WATCH press conference at the release of the January/February issue—and more than 2,000 newspapers carried our followup study on rising storm damages.

the first U.S. Surgeon General's report on smoking and health—and the hundreds of research reports it spawned—and the historic November 1998 $206 billion settlement between the tobacco industry and 46 state governments. (The other four states had already settled for $45 billion.) Thirty-seven years have passed since biologist Rachel Carson published *Silent Spring*, issuing the wake-up call that gave rise to the modern environmental movement.

Not all environmentalists will agree with me, but I believe that there are now some clear signs that the world does seem to be approaching a kind of paradigm shift in environmental consciousness. Across a spectrum of activities, places, and institutions, the atmosphere has changed markedly in just the last few years. Among giant corporations that could once be counted on to mount a monolithic opposition to serious environmental reform, a growing number of high profile CEOs have begun to sound more like spokespersons for Greenpeace than for the bastions of global capitalism of which they are a part. More and more governments are taking revolutionary steps aimed at shoring up the Earth's long-term environmental health. Individuals the world over have established thriving new markets for products that are distinguished by their compatibility with a sustainable economy. What in the world is going on?

Thomas Kuhn, in his classic work *The Structure of Scientific Revolutions*, observes that as scientific understanding in a field advances, reaching a point where existing theory no longer explains reality, theory has to change. Perhaps history's best known example of this process is the shift from the Ptolemaic view of the world, in which people believed the sun revolved around the Earth, to the Copernican view which argued that the Earth revolved about the sun. Once the Copernican model existed, a lot of things suddenly made sense to those who studied the heavens, leading to an era of steady advances in astronomy.

We are now facing such a situation with the global economy. Although economists have long ignored the Earth's natural systems, evidence that the economy is slowly self-destructing by destroying its natural support systems can be seen on every hand. The Earth's forests are shrinking, fisheries are collapsing, water tables are falling, soils are eroding, coral reefs are dying, atmospheric CO_2 concentrations are increasing, temperatures are rising, floods are becoming more destructive, and the rate of extinction of plant and animal species may be the greatest since the dinosaurs disappeared 65 million years ago.

These ecological trends are driving analysts to a paradigm shift in their view of how the economy will have to work in the future. For years, these trends were marginalized by policymakers and the media as "special interest" topics, but as the trends have come to impinge more and more directly on people's lives, that has begun to change. The findings of these analysts are primary topics now not only for environmentalists, but for governments, corporations, and the media.

Learning From China

If changes in physical conditions are often the driving forces in perceptual shifts, one of the most powerful forces driving the current shift in our understanding of the ecological/economic relationship is the flow of startling information coming from China. Not only the world's most populous country, China since 1980 has been the world's fastest growing economy, raising incomes nearly fourfold. As such, China is in effect telescoping history, showing us what happens when large numbers of people become more affluent.

As incomes have climbed, so has consumption. If the Chinese should reach the point where they eat as much beef as Americans, the production of just that added beef will take an estimated 340 million tons of grain per year, an amount equal to the entire U.S. grain harvest. Similarly, if the Chinese were to consume oil at the American rate, the country would need 80 million barrels of oil a day—more than the entire world's current production of 67 million barrels a day.

What China is dramatizing—to its own scientists and government and to an increasingly worried international community—is that the Western industrial development model will not work for China. And if the fossil-fuel-based, automobile-centered, throwaway economy will not work for it, then it will not work for India, with its billion people, nor for the other two billion in the developing world. And, in an increasingly integrated global economy, it will not work in the long run for the industrial economies either.

1 ❖ THE GLOBAL ENVIRONMENT: AN EMERGING WORLD VIEW

Just how powerfully events in China are beginning to sway perceptions was brought home to me at our press lunch for *State of the World 1998* when I was talking with some reporters sitting on the front row before the briefing began. A veteran reporter, rather skeptical as many seasoned reporters are, said that he had never been convinced by our argument that we need to restructure the global economy—but that the section in *State of the World* on rising affluence in China and the associated rising claims on global resources had now convinced him that we have little choice.

Fortunately, we now have a fairly clear picture of how to do that restructuring. When Worldwatch began to pioneer the concept of environmentally sustainable economic development 25 years ago, we were already aware that instead of being based on fossil fuels, the new model would be based on solar energy. Instead of having a sprawling automobile-centered urban transportation system, it would be based on more carefully designed cities, with shorter travel distances and greater reliance on rail, bicycles, and walking. Instead of a throwaway economy, it would be a reuse/recycle economy. And its population would have to be stable.

When we described our model in the early days, it sounded like pie in the sky—as the reporter's skepticism reminded me. Now, with the subsequent advances in solar and wind technologies, gains in recycling, mounting evidence of automobile-exacerbated global warming, and the growing recognition that oil production will decline in the not-too-distant future, it suddenly becomes much more credible, a compelling alternative. Just as early astronomers were limited in how far they could go in understanding the heavens with the Ptolemaic model, so, too, we are limited in how long we can sustain economic progress with the existing economic model. As a result, in each of the four major areas of that model—renewable energy, efficient urban transport, materials recycling, and population stability—I believe public vision is shifting rapidly.

Shifting Views of Energy

A decade ago, there were plenty of avid afficionados of renewable energy, but the subject was of only marginal interest to the global public. That has changed markedly, as escalating climate change has thrust questions about the climate-disrupting effects of burning fossil fuels into the center of public debate. In 1998, not only did the Earth's average temperature literally go off the top of the chart we have been using to track global temperature for many years, but storm-related weather damage that year climbed to a new high of $89 billion. This not only exceeded the previous record set in 1996 by an astonishing 48 percent, but it exceeded the weather-related damage for the entire decade of the 1980s.

When Worldwatch issued a brief report in late 1998 noting the record level of weather-related damage during the year, it was picked up by some 2,000 newspapers worldwide—an indication that energy issues were beginning to hit home, literally. Closely related to the increase in storms and floods was a dramatic rise in the number of people driven from their homes, for days or even months, as a result of more destructive storms and floods. Almost incomprehensibly, 300 million people—a number that exceeds the entire population of North America—were forced out of their homes in 1998.

If the news were only that fossil fuels are implicated in escalating damages, I'm not sure I'd see signs of a paradigm change. But along with the threats of rising damages, there were the data we released in 1998 indicating that the solutions to these threats have been coming on strong. Not only are fossil-fuel-exacerbated damages escalating, but technological alternatives—wind and solar power—are booming. While oil and coal still dominate the world energy economy, the new challengers are expanding at the kind of pace that makes venture capitalists reach for their phones. From 1990 to 1997, coal and oil use increased just over 1 percent per year, while solar cell sales, in contrast, were expanding at roughly 15 percent per year. In 1997 they jumped over 40 percent.

An estimated 500,000 homes, most of them in remote third world villages not linked to an electrical grid, now get their electricity from solar cells. The use of photovoltaic cells to supply electricity has recently gotten a big boost from the new solar roofing tiles developed in Japan. These "solar shingles," which enable the roof of a building to become its own power plant, promise to revolutionize electricity generation worldwide, making it easier to forget fossil fuels.

The growth in wind power has been even more impressive, a striking 26 percent per year since 1990. If you are an energy investor and are interested in growth', it is in wind, not oil. The U.S. Department of Energy's Wind Resource Inventory indicates that three states—North Dakota, South Dakota, and Texas—have enough harnessable wind energy to satisfy national electricity needs. And China could double its current electricity generation with wind alone.

Shifting Views of Urban Transport

In Bangkok, the average motorist last year sat in his car going nowhere for the equivalent of 44 working days. And in London, the average speed of a car today is little better than that of a horse-drawn carriage a century ago. Clearly, the automobiles that once provided much-needed mobility for rural societies cannot do the same for a society that will soon be largely urban. As a result, more and more national and city governments are beginning to confront the inherent conflict between the automobile and the city—a sign that we may be approaching a threshold of revolutionary change in how we view the very nature of urban life.

4. Crossing the Threshold

While the automobile industry still promotes the vision of a world with a car in every garage, some national and many city governments are emphasizing alternatives to the automobile, ones that center on better public rail transport and the bicycle. This movement in Europe is led by the Netherlands and Denmark, where bicycles account for 30 percent and 20 percent respectively of daily trips in cities. In Germany, policies encouraging bicycle use have raised the share of urban trips by bike nationwide from 8 percent in 1972 to 12 percent in 1995.

In Beijing, where air pollution is a health issue and where traffic conditions worsen by the month, the official enthusiasm for the car—dominant model of a few years ago seems to have cooled. A group of eminent scientists in China have directly challenged the government's plans to develop a Western-style, automobile-centered transportation system. They observe that China does not have enough land both to feed its people and to build the roads, highways, and parking lots needed for the automobile. They also argue that the automobile will increase traffic congestion, worsen urban air pollution—already the worst in the world—and force a growing dependence on imported oil.

The Chinese scientists argue that the country should develop "a public transportation network that is convenient, complete, and radiating in all directions." The effort to convince Party leaders to reverse their policy is being led by one of China's most venerated scientists, physicist He Zuoxiu, who worked on the country's first atomic bomb. He says that China "just simply cannot sustain the development of a car economy."

In the United States, scores of cities are beginning to develop more bicycle-friendly transportation systems. More than 300 U.S. cities now have part of their police force on bicycles. Not long ago I found myself standing on a street corner in downtown Washington, D.C., next to a police officer on a bicycle. As we waited for the light to change, I asked him why there were now so many officers on bicycles. He indicated that it was largely a matter of efficiency, since an officer on a bike can respond to some 50 percent more calls in a day than one in a squad car. The fiscal benefits are obvious. He also indicated that the bicycle police make many more arrests, because they are both more mobile and less conspicuous.

Bicycle transport, like solar or wind power, may still seem to many to be a marginal indicator. But I see the

NGO power is growing fast. In Korea, Lester Brown met with Yul Choi of the 50,000-member Korean Federation for Environmental Movement—one of hundreds of such groups taking root around the world.

same kind of signs of quiet, revolutionary change in the bicycle as in the modern wind turbine: the unthinkable consequences of continuing the existing system, combined with recent sales trends. Bicycle use is growing much faster than automobile use, not only because it is more affordable but because it has a range of environmental and social advantages: it uses far less land (a key consideration in a world where the cropland area has shrunk to barely one-half acre per person); it does not contribute to pollution; it helps reduce traffic congestion; it does not contribute to CO_2 emissions; and, for an increasingly desk-bound work-force, it offers much needed exercise. Indeed, during the past three decades, in which annual car sales worldwide increased from 23 million to 37 million, the number of bicycles sold jumped from 25 million to 106 million.

If cars were used in a future world of 10 billion people at the rate they are currently used in the United States (one car for every two people), that would mean a global fleet of 5 billion cars—10 times the existing, already dangerously burdensome, number. That prospect is inconceivable. Although the automobile industry is not abandoning its global dream of a car in every garage, it is *this* dream that now has a distinctly pie-in-the-sky feel.

Shifting Views of Materials Use

There are few areas in which individuals have participated so actively as in the effort to convert the throwaway economy into a reuse/recycle economy. At the individual level, efforts are concentrated on recycling paper, glass, and aluminum. But there are also important shifts coming in basic industries. For example, in the United States, not always a global leader in recycling, 56 percent of the steel produced now comes from scrap. Steel mills built in recent years are no longer located in western Pennsylvania, where coal and iron ore are in close proximity, but are scattered about the country—in North Carolina, Nebraska, or California—feeding on local supplies of scrap. These new electric arc steel furnaces produce steel with much less energy and far less pollution than that produced in the old steel mills from virgin iron ore.

A similar shift has taken place in the recycling of paper. At one time, paper mills were built almost exclusively in heavily forested areas, such as the northwestern United States, western Canada, or Maine, but now they are often built near cities, feeding on the local supply of scrap paper. The shift in *where* these industries are may prefigure a shift in our understanding of *what* they are.

This new economic model can be seen in the densely populated U.S. state of New Jersey where there are now 13 paper mills running only on waste paper. There are also eight steel mini-mills, using electric arc furnaces to manufacture steel largely from scrap. These two industries, with a combined annual output in excess of $1 billion, have developed in a state that has little forest cover and no iron mines. They operate almost entirely on material already in the system, providing a glimpse of what the reuse/recycle economy of the future looks like.

Shifting Views of Population

No economic system is sustainable with continual population growth, or with continual population declines either. Fortunately, some 32 countries containing 14 percent of the world's people have achieved population stability. All but one (Japan) are in Europe. In another group of some 40 countries, which includes the United States and China, fertility has dropped below two children per

Family planning services—such as the simple expedient of making condoms readily available—are gaining ground in much of the world despite concerted campaigns to suppress them.

woman, which means that these countries are also headed for population stability over the next few decades—assuming, of course, that those fertility trends don't reverse.

Unfortunately, many developing countries are facing huge population increases. Pakistan, Nigeria, and Ethiopia are projected to at least double their populations over the next half-century. India, with a population expected to reach 1 billion this August, is projected to add another 500 million people by 2050. If these countries do not stabilize their populations soon enough by reducing fertility, they will inevitably face a rise in mortality, simply because they will not be able to cope with new threats such as HIV or water and food shortages.

What is new here is that as more people are crowded onto the planet, far more are becoming alarmed about the potentially disastrous consequences of that crowding. In India, for example, the *Hindustan Times,* one of India's leading newspapers, recently commented on the fast-deteriorating water situation, where water tables are falling almost everywhere and wells are going dry by the thousands: "If our population continues to grow as it is now ... it is certain that a major part of the country would be in the grip of a severe water famine in 10 to 15 years." The article goes on to reflect an emerging sense of desperation: "Only a bitter dose of compulsory family planning can save the coming generation from the fast-approaching Malthusian catastrophe." Among other things, this comment appears to implicitly recognize the emerging conflict between the reproductive rights of the current generation and the survival rights of the next generation.

Corporate Converts

Corporations have been endorsing environmental goals for some three decades, but their efforts have been too often centered in the public relations office, not in corporate planning. Now this is beginning to change, as the better informed, more prescient CEOs recognize that the shift from the old industrial model to the new environmentally sustainable model of economic progress represents the greatest investment opportunity in history. In May 1997, for example, British Petroleum CEO John Browne broke ranks with the other oil companies on the climate issue when he said, "The time to consider the policy dimensions of climate change is not when the link between greenhouse gases and climate change is conclusively proven, but when the possibility cannot be discounted and is taken seriously by the society of which we are a part. We in BP have reached that point."

Browne then went on to announce a $1 billion investment by BP in the development of wind and solar energy. In effect he was saying, "we are no longer an oil company; we are now an energy company." Within a matter of weeks Royal Dutch Shell announced that it was committing $500 million to development of renewable energy sources. And in early 1998, Shell announced that it was leaving the Global Climate Coalition, an industry-supported group in Washington, D.C. that manages a disinformation campaign designed to create public confusion about climate change.

These commitments to renewable energy by BP and Shell are small compared with the continuing investment of vast sums in oil exploration and development, but they are investments in energy sources that cannot be depleted, while those made in oil fields can supply energy only for a relatively short time. In addition, knowing that world oil production likely will peak and begin to decline within the next 5 to 20 years, oil companies are beginning to look at the alternatives. This knowledge, combined with mounting concern about global warming, helps explain why the more forward-looking oil companies are now investing in wind and solar cells, the cornerstones of the new energy economy.

4. Crossing the Threshold

Ken Lay, the head of Enron, a large Texas-based national gas supplier with annual sales of $20 billion that is fast becoming a worldwide energy firm, sees his company, and more broadly the natural gas industry, playing a central role in the conversion from a fossil-fuel-based energy economy to a solar/hydrogen energy economy. As the cost of wind power falls, for example, cheap electricity from wind at wind-rich sites can be used to electrolyze water, producing hydrogen, a convenient means of both storing and transporting wind energy or other renewable energy resources. The pipeline network and storage facilities used for natural gas can also be used for hydrogen. George H.B. Verberg, the managing director of Gasunie in the Netherlands, has publicly outlined a similar role for his organization with its well developed natural gas infrastructure.

In the effort to convert our throwaway economy into a reuse/recycle economy, too, I see signs that new initiatives are coming not just from eco-activists but from industry. In Atlanta, Ray Anderson, the head of Interface, a leading world carpet manufacturer with sales in 106 countries, is starting to shift his firm from the sale of carpets to the sale of carpeting services. With the latter approach, Interface contracts to provide carpeting service to a firm for its offices for say a 10-year period. This service involves installing the carpet, cleaning, repairing and otherwise maintaining the quality of carpeting desired by the client. The advantage of this system is that when the carpet wears out, Interface simply takes it back to one of its plants and recycles it in its entirety into new carpeting. The Interface approach requires no virgin raw material to make carpets, and it leaves nothing for the landfill.

Perhaps one of the most surprising—and significant—signs of impending change came last year from the once notorious MacMillan Bloedel, a giant forest products firm operating in Canada's western-most province of British Columbia. "MacBlo," as it is called, startled the world—and other logging firms—when it announced that it was giving up the standard forest industry practice of clear-cutting. Under the leadership of a new chief executive, Tom Stevens, the company affirmed that clear-cutting will be replaced by selective cutting, leaving trees to check runoff and soil erosion, to provide wildlife habitat, and to help regenerate the forest. In doing so, it acknowledged the growing reach of the environmental movement. MacMillan Bloedel was not only being pressured by local groups, but it also had been the primary target of a Greenpeace campaign to ban clear-cutting everywhere.

Governments Catching On

At the national level, too, there are signs of major changes. Six countries in Europe—Denmark, Finland, the Netherlands, Sweden, Spain, and the United Kingdom—began restructuring their taxes during the 1990s in a process known as tax shifting—reducing income taxes while

1 ❖ THE GLOBAL ENVIRONMENT: AN EMERGING WORLD VIEW

CIA investigation following up Lester Brown's WORLD WATCH analysis signified that intelligence agencies are taking environmental threats more seriously now.

offsetting these cuts with higher taxes on environmentally destructive activities such as fossil fuel burning, the generation of garbage, the use of pesticides, and the production of toxic wastes. Although the reduction in income taxes does not yet exceed 3 percent in any of these countries, the basic concept is widely accepted. Public opinion polls on both sides of the Atlantic show 70 percent of the public supporting tax shifting.

In mid 1998, the new government taking over in Germany, a coalition of Social Democrats and Greens, announced a massive restructuring of the tax system, one that would simultaneously reduce taxes on wages and raise taxes on CO_2 emissions. This shift, the largest yet contemplated by any government, was taken unilaterally, not bogging down in the politics of the global climate treaty, or contingent on steps taken elsewhere. The framers of the new tax structure argued that this tax restructuring would help strengthen the German economy by creating additional jobs and at the same time reducing air pollution, oil imports, and the rise in atmospheric CO_2—the principal threat to climate stability. With Germany taking this bold initiative unilaterally, other countries may follow.

Over the past generation, the world has relied heavily on regulation to achieve environmental goals, but in most instances using tax policy to restructure the economy is far more likely to be successful because it permits the market to operate, thus taking advantage of its inherent efficiency in linking producers and consumers. Restructuring taxes to achieve environmental goals also minimizes the need for regulation.

In effect, the governments moving toward tax shifting have decided that the emphasis on taxing wages and income from investments discourages both work and saving,

activities that should be encouraged, not discouraged. They believe we should be discouraging environmentally destructive activities by taxing them instead. Since tax shifting does not necessarily change the overall level of taxation, and thus does not materially alter a country's competitive position in the world market, it can be undertaken unilaterally.

Environmental leadership does not always come from large countries. At the December 1997 Kyoto conference on climate, President José Maria Figueres of Costa Rica announced that by the year 2010, his country planned to get all of its electricity from renewable sources. In Copenhagen, the Danish government has banned the construction of coal-fired power plants.

In the U.S. government, no longer a leader on the environmental front, there are signs of a breakthrough in at least some quarters. The Forest Service announced in early 1998 that after several decades of building roads in the national forests to help logging companies remove timber, it was imposing an 18-month moratorium on road building. Restricting this huge public subsidy, which had built some 380,000 miles of roads to facilitate clear-cutting on public lands, signals a fundamental shift in the management of national forests. The new chief of the Forest Service, Michael Dombeck, responding to a major shift in public opinion and no longer intimidated by the "wise-use" movement of the early Clinton years, said the service was focusing on the use of national forests for recreation, for wildlife protection, to supply clean water, and as a means of promoting tourism as well as supplying timber. The shift in opinion seems to reflect a growing public recognition of the environmental consequences of clear-cutting, including more destructive flooding, soil erosion, silting of rivers, and in the Northwest, the destruction of salmon fisheries.

In mid-August 1998, after several weeks of near-record flooding in the Yangtze river basin, Beijing acknowledged for the first time that the flooding was not merely an act of nature, but that it had been greatly exacerbated by the deforestation of the upper reaches of the watershed. Premier Zhu Rongji personally issued orders to not only halt the tree-cutting in the upper reaches of the Yangtze basin and elsewhere in China, but also to convert some state timbering firms into tree-planting firms. The official view in Beijing now is that trees are worth three times as much standing as they are cut, simply because of the water storage and flood retention capacity of forests.

Meanwhile, back in Washington, even the U.S. intelligence community is beginning to realize that environmental trends can adversely affect the global economy on a scale that could lead to political instability. The National Intelligence Council, the organizational umbrella over the CIA, DIA, and other U.S. intelligence agencies, was provoked by the article, "Who Will Feed China?" that I published in WORLD WATCH in 1994. It was concerned that projected losses of cropland and irrigation water in China could lead to soaring grain imports, rising world grain prices and, ultimately, to widespread political instability in third world cities. In response, the Council assembled a team of prominent U.S. scientists to undertake an exhaustive interdisciplinary analysis of China's long-term food prospect.

This analysis, completed in late 1997, showed horrendous water deficits emerging in the water basins of the northern half of China, deficits that could decimate the grain harvest in some regions even as the demand for grain continues to climb. It concluded that China will likely need to import 175 million tons of grain by 2025, an amount that approaches current world grain exports of 200 million tons. When the U.S. intelligence community, which was for half a century fixated on the Communist threat, now raises an alarm about an environmental threat in a Communist country—that is indeed a sign that we are approaching a new threshold.

NGOs as Catalysts

Among the signs that new perceptions are overtaking old institutions is the robust proliferation of nongovernmental organizations (NGOs). The formation of environmental NGOs is a response of civil society to the immobility of existing institutions and specifically to their lack of a timely response to spreading environmental destruction. The new economic model outlined earlier originated not in the halls of academe or in the councils of government but within the research groups among the environmental NGOs. There are hundreds of international and national environmental groups and literally thousands of local single-issue groups.

At the international level, groups like Greenpeace, the International Union for Conservation of Nature, and the Worldwide Fund for Nature have become as influential in shaping environmental policies as national governments. The budgets of some of the individual environmental groups, such as the 1.2 million-member U.S. World Wildlife Fund ($82 million) or Greenpeace International ($60 million), begin to approach the $105 million budget of the United Nations Environment Programme, the U.N. agency responsible for environmental matters. In fact, much of the impetus toward a global consciousness of environmental threats—and much of the hard work of establishing the new mechanisms needed to build an environmentally sustainable economy—have come from NGOs. The research that underpinned the UN-sponsored Earth Summit in Rio de Janeiro in 1992, notably, came largely from organizations like the Wuppertal Institute in Germany and the U.S.-based World Resources Institute and Worldwatch Institute.

Almost every industrialized country now has a number of national environmental groups, many with memberships measured in the hundreds of thousands. Some developing countries, too, now have strong environmental groups. In Korea, for example, the Korean Federation for Environmental Movement, a group with a membership that recently passed 50,000 and a full-time staff of 60, has become a force to be reckoned with by the government.

At the grassroots, thousands of local single-issue groups work on objectives ranging from preventing construction of a nuclear power plant in Japan's Niigata prefecture to protecting the Amazonian rainforest from burning by cattle ranchers so that the forest products can continue to be harvested by local people. The little-heralded work of small groups like this on every continent is quietly helping to move us within reach of a major shift in public awareness.

Approaching the Threshold

One reason more people are aware of the environmental underpinnings of their lives now is that many more have been directly affected by environmental disruptions. And even when events don't impinge directly, media coverage is more likely to expose the damage now than a decade ago. Among the events that are mobilizing public concern, and therefore support for restructuring the economy, are fishery collapses, water shortages, rainforests burning uncontrollably, sudden die-offs of birds, dolphins, and fish, record heat waves, and storms of unprecedented destructiveness.

Weather-related damages are now so extensive that insurance companies can no longer use linear models from the past to calculate risks in the future. When the cost of insuring property rises sharply in the future, as now seems inevitable, millions of people may take notice—including many who have not before.

1 ❖ THE GLOBAL ENVIRONMENT: AN EMERGING WORLD VIEW

Are we indeed moving toward a social threshold which, once crossed, will lead to a dizzying rate of environmentally shaped economic change, on a scale that we may not now even imagine? No one knows for sure, but some of the preconditions are clearly here. An effective response to any threat depends on a recognition of that threat, which is broad enough to support the response. There is now a growing worldwide recognition outside the environmental community that the economy we now have cannot take us where we want to go. Three decades ago, it was only environmental activists who were speaking out on the need for change, but the ranks of activists have now broadened to include CEOs of major corporations, government ministers, prominent scientists, and even intelligence agencies.

Getting from here to there quickly is the challenge. But at least we have a clear sense of what has to be done. The key to restructuring the global economy, as noted earlier, is restructuring the tax system. Seven European countries, led by Germany, are advancing on this front.

New institutional initiatives, too, are helping set the stage for the economic restructuring. For example, ecological labeling of consumer products is being implemented as a means of raising awareness—and shifting purchasing priorities—in several industries. Consumers who want to protect forests from irresponsible logging practices now have the option of buying only products that come from those forests that are being managed in a certifiably responsible way. In the United States, even electric power can now be purchased from "green" sources in some areas, if the consumer so chooses. Public awareness of the differences among energy sources is raised significantly, as each power purchaser is confronted with the available options.

Another institutional means for expressing public preferences is government procurement policy. If national or local governments decide to buy only paper that has a high recycled content, for example, they provide market support for economic restructuring. And governments, like individual users, can become "green" consumers by opting for climate-benign sources of electricity.

Trying times require bold responses, and we are beginning to see some, such as the decision by Ted Turner, the founder of Turner Broadcasting and Cable News Network (CNN), now part of the Time Warner complex, to contribute $1 billion to the United Nations to be made available at $100 million per year over the next ten years. Not only is Turner committing a large part of his personal fortune to dealing with some of the world's most pressing population, environmental, and humanitarian problems, but he is also urging other billionaires, of whom there are now more than 600 in the world, not to wait until their deaths to put money in foundations that might work on these issues. He argues, quite rightly, that time is of the essence, that right now we are losing the war to save the future.

In a world where the economy has expanded from $6 trillion in output in 1950 to $39 trillion in 1998, new collisions between the expanding economy as now structured and its environmental support systems are occurring somewhere almost daily. Time is running out. The Aral Sea has died. Its fisheries are gone. The deterioration of Indonesia's rainforests may have reached the point of no return. We may not be able to save the glaciers in Glacier National Park.

The key to quickly gaining acceptance of the new economic model is to accelerate the flow of information about how the old model is now destroying its natural support systems. Some governments are now doing this. For example, beginning in late summer of 1997, the Clinton White House began holding press briefings, regularly reporting new climate findings. On June 8, 1998, Vice President Al Gore held a press conference announcing that for the world 1997 "was the warmest year on record and we've set new temperature records every month since January." He went on to say, "This is a reminder once again that global warming is real and that unless we act, we can expect more extreme weather in the year ahead."

Even China is taking steps toward more open dissemination of information. In early 1998, Beijing became the 39th Chinese city to start issuing weekly air quality reports since the beginning of 1997. These reports, providing data on such indicators as the levels of nitrous oxides from car exhaust and particulate matter from coal burning, reveal that Chinese urban dwellers breathe some of the world's most polluted air. Air pollution is estimated to cause 178,000 pre-mature deaths per year, more than four times the number of automobile fatalities in the United States. "Who Will Feed China?," initially banned in China, is now being promoted on Central Television. This new openness by the government is expected to enhance public support for taking the steps needed to control air pollution, whether it be restricting automobile traffic, closing the most polluting factories, or shifting to clean sources of energy. Information on how the inefficient use of water could lead to food shortages can boost support for water pricing.

Media coverage of environmental trends and events is also increasing, indicating a rising appreciation of their importance. One could cite thousands of examples, but let me mention just two. First is the media coverage given to the 1997/98 El Niño, the periodic rise in the surface temperature of water in the eastern Pacific that affects climate patterns world-wide. This is not a new phenomenon. It has occurred periodically for as far back as climate records exist. But the difference is in the coverage. In 1982/83 there was an El Niño of similar intensity, but it did not become a household word. In 1997/98, it did largely because a more enlightened community of television meteorologists who report daily weather events understood better how El Niño was affecting local climate. Public recognition of the importance of "El Niño" was perhaps most amusingly demonstrated for me last winter, when a large automobile dealer in my area advertised that

4. Crossing the Threshold

it was having an "El Niño" sale. It was going to be a big one!

At a more specific level, in the fall of 1997, *Time* magazine produced a special issue of its international edition under the headline "Our Precious Planet:Why Saving the Environment Will be the Next Century's Biggest Challenge." As the title implies, the issue recognized—in a way few major news organizations have in the past—the extraordinary dimensions of the challenge facing humanity as we try to sustain economic progress in the next century.

More and more people in both the corporate and political worlds are now beginning to share a common vision of what an environmentally sustainable economy will look like. If the evidence of a global awakening were limited to one particular indicator, such as growing membership in environmental groups, it might be dubious. But with the evidence of growing momentum now coming from a range of key indicators simultaneously, the prospect that we are approaching the threshold of a major transformation becomes more convincing. The question is, if it does come, whether it will come soon enough to prevent the destruction of natural support systems on a scale that will undermine the economy.

As we prepare to enter the new century, no challenge looms greater than that of transforming the economy into one that is environmentally sustainable. This Environmental Revolution is comparable in scale to the Agricultural Revolution and the Industrial Revolution. The big difference is in the time available. The Agricultural Revolution was spread over thousands of years. The Industrial Revolution has been underway for two centuries.The Environmental Revolution, if it succeeds, will be compressed into a few decades. We study the archeological sites of civilizations that moved onto economic paths that were environmentally destructive and could not make the needed course corrections either because they did not understand what was happening or could not summon the needed political will. We *do* know what is happening. The question for us is whether our global society can cross the threshold that will enable us to restructure the global economy before environmental deterioration leads to economic decline.

Lester Brown is president of Worldwatch Institute.

Unit 2

Unit Selections

5. **The Population Surprise,** Max Singer
6. **Population and Consumption: What We Know, What We Need to Know,** Robert W. Kates
7. **Food for All in the 21st Century,** Gordon Conway
8. **Escaping Hunger, Escaping Excess,** Gary Gardner and Brian Halweil

Key Points to Consider

❖ What is the role of fertility rates in predicting future population growth? Why do fertility rates rise and fall and where have fertility rates dropped below replacement levels?

❖ Describe the IPAT formula for calculating environmental impact and tell why it may no longer be appropriate to use when attempting to describe the role of population growth, affluence, and technology in environmental deterioration.

❖ Why are people in the more developed countries of the world generally ignorant of the true dimensions of the world's food problem? How can increased awareness of food scarcity and misallocation lead to solutions for both food production and environmental protection?

❖ Explain the differences between "malnutrition" in a country like the United States and a nation in sub-Saharan Africa. To what extent is the world's food problem more one of distribution and availability of food than of total supply?

 **Links**   www.dushkin.com/online/

11. **The Hunger Project**
    http://www.thp.org
12. **Penn Library Resources**
    http://www.library.upenn.edu/resources/websitest.html
13. **World Health Organization**
    http://www.who.int
14. **WWW Virtual Library: Demography & Population Studies**
    http://demography.anu.edu.au/VirtualLibrary/

These sites are annotated on pages 4 and 5.

The World's Population: People and Hunger

One of the greatest setbacks on the road to the development of more stable and sensible population policies came about as a result of inaccurate population growth projections made in the late 1960s and early 1970s. The world was in for a population explosion, the experts told us back then. But shortly after the publication of the heralded works *The Population Bomb* (Paul Ehrlich, 1975) and *Limits to Growth* (D. H. Meadows et al., 1974), the growth rate of the world's population began to decline slightly. There was no cause and effect relationship at work here. The decline in growth was simply demographic transition at work, a process in which declining population growth tends to accompany increasing levels of economic development. Unfortunately, since the alarming predictions did not come to pass, the world began to relax a little. However, two facts still remain—population growth in biological systems must be limited by available resources, and the availability of Earth's resources is finite.

Consider the following: In developing countries, high and growing rural population densities have forced the use of increasingly marginal farmland once considered to be too steep, too dry, too wet, too sterile, or too far from market for efficient agricultural use. Farming this land damages soil and watershed systems, creates deforestation problems, and adds relatively little to total food production. In the more developed world, farmers also have been driven—usually by market forces—to farm more marginal lands and to rely more on environmentally harmful farming methods utilizing high levels of agricultural chemicals (such as pesticides and artificial fertilizers). These chemicals create hazards for all life and rob the soil of its natural ability to renew itself. The increased demand for food production has also created an increase in the use of precious groundwater reserves for irrigation purposes, depleting those reserves beyond their natural capacity to recharge and creating the potential for once-fertile farmland and grazing land to be transformed into desert. The continued demand for higher production levels also contributes to a soil erosion problem that has reached alarming proportions in all agricultural areas of the world, whether high or low on the scale of economic development. The need to increase the food supply and its consequent effects on the agricultural environment are not the only results of continued population growth. For industrialists, the larger market creates an almost irresistible temptation to accelerate production, requiring the use of more marginal resources and resulting in the destruction of more fragile ecological systems, particularly in the tropics. For consumers, the increased demand for products means increased competition for scarce resources, driving up the cost of those resources until only the wealthiest can afford what our grandfathers would have viewed as an adequate standard of living.

The articles selected for this second unit all relate, in one way or another, to the theory and reality of population growth (and its relationship to food supply).

In the first selection, "The Population Surprise," Max Singer notes that the pattern of population dynamics in the more developed nations of the world, what we call "the demographic transition," indicates that when population growth begins to slow down—as it has worldwide over the last few decades—the decline in growth does not stop at the replacement rate but often continues beneath that point: in other words, populations begin to decline. Rather than doubling the world's population by the middle of the twenty-first century, as some believe, Singer expects the global number of people to begin to decline.

The unit's second article moves from a discussion of population dynamics to one of how to analyze the relationship between population growth and environmental degradation. In "Population and Consumption: What We Know, What We Need to Know," Robert W. Kates notes that growth in population, affluence, and technology are all contributors to deteriorating environments. While we understand the mechanisms of population growth, however, we do not understand those influencing the growth of consumption—other than that they relate to affluence and technology. Population problems will not be solved until we know more about the innate tendency of human societies to consume resources. The third article in this unit also addresses the issue of consumption, but in a nontheoretical context different from that of Kates. Gordon Conway, an agricultural ecologist who serves as president of the Rockefeller Foundation in New York City, relates consumption to the issues of poverty, noting that while most of the world consumes food at adequate levels, the allocation of food resources is far from equal—even within the world's wealthier countries. In "Food for All in the 21st Century," Conway argues that it is possible to balance consumption with supply, but only through a revolution in agriculture and natural resource production will that be both equitable and sustainable.

Finally, in the section's concluding piece, Gary Gardner and Brian Halweil of the Worldwatch Institute continue to discuss the issue of inequitable food distribution and consumption. In "Escaping Hunger, Escaping Excess," Garner and Halweil note that it is not just the scarcity of food in poor countries that should be viewed with concern but the conspicuous overconsumption in the richer countries. Malnourishment can just as easily be brought on by too much consumption as by too little, and both forms of malnutrition are as costly for national economic activities as they are for personal health. All the authors of the selections in this unit make it clear that the global environment is being stressed by population growth and that more people means more pressure and more poverty. But while it should be evident that we can no longer afford to permit the unplanned and unchecked growth of the planet's dominant species, it should also be apparent that doomsday predictions of population and food imbalances are less commonplace now than they were just a few years ago.

The Population Surprise

The old assumptions about world population trends need to be rethought. One thing is clear: in the next century the world is in for some rapid downsizing

by Max Singer

FIFTY years from now the world's population will be declining, with no end in sight. Unless people's values change greatly, several centuries from now there could be fewer people living in the entire world than live in the United States today. The big surprise of the past twenty years is that in not one country did fertility stop falling when it reached the replacement rate—2.1 children per woman. In Italy, for example, the rate has fallen to 1.2. In Western Europe as a whole and in Japan it is down to 1.5. The evidence now indicates that within fifty years or so world population will peak at about eight billion before starting a fairly rapid decline.

Because in the past two centuries world population has increased from one billion to nearly six billion, many people still fear that it will keep "exploding" until there are too many people for the earth to support. But that is like fearing that your baby will grow to 1,000 pounds because its weight doubles three times in its first seven years. World population was growing by two percent a year in the 1960s; the rate is now down to one percent a year, and if the patterns of the past century don't change radically, it will head into negative numbers. This view is coming to be widely accepted among population experts, even as the public continues to focus on the threat of uncontrolled population growth.

As long ago as September of 1974 *Scientific American* published a special issue on population that described what demographers had begun calling the "demographic transition" from traditional high rates of birth and death to the low ones of modern society. The experts believed that birth and death rates would be more or less equal in the future, as they had been in the past, keeping total population stable after a level of 10–12 billion people was reached during the transition.

Developments over the past twenty years show that the experts were right in thinking that population won't keep going up forever. They were wrong in thinking that after it stops going up, it will stay level. The experts' assumption that population would stabilize because birth rates would stop falling once they matched the new low death rates has not been borne out by experience. Evidence from more than fifty countries demonstrates what should be unsurprising: in a modern society the death rate doesn't determine the birth rate. If in the long run birth rates worldwide do not conveniently match death rates, then population must either rise or fall, depending on whether birth or death rates are higher. Which can we expect?

The rapid increase in population during the past two centuries has been the result of lower death rates, which have produced an increase in worldwide life expectancy from about thirty to about sixty-two. (Since the maximum—if we do not change fundamental human physiology—is about eighty-five, the world has already gone three fifths as far as it can in increasing life expectancy.) For a while the result was a young population with more mothers in each generation, and fewer deaths than births. But even during this population explosion the average number of children born to each woman—the fertility rate—has been falling in modernizing societies. The prediction that world population will soon begin to decline is based on almost universal human behavior. In the United States fertility has been falling for 200 years (except for the blip of the Baby Boom), but partly because of immigration it has stayed only slightly below replacement level for twenty-five years.

Obviously, if for many generations the birth rate averages fewer than 2.1 children per woman, population must eventually stop growing. Recently the United Nations Population Division estimated that 44 percent of the world's people live in countries where the fertility rate has already fallen below the replacement rate, and fertility is falling fast almost everywhere else. In Sweden and Italy fertility has been below replacement level for so long that the population has become old enough to have more deaths than births. Declines in fertility will eventually increase the average age in the world, and will cause a decline in world population forty to fifty years from now.

Because in a modern society the death rate and the fertility rate are largely independent of each other, world population need not be stable. World population can be stable only if fertility rates around the world average out to 2.1 children per woman. But why should they average 2.1, rather than 2.4, or 1.8,

5. Population Surprise

or some other number? If there is nothing to keep each country exactly at 2.1, then there is nothing to ensure that the overall average will be exactly 2.1.

The point is that the number of children born depends on families' choices about how many children they want to raise. And when a family is deciding whether to have another child, it is usually thinking about things other than the national or the world population. Who would know or care if world population were to drop from, say, 5.85 billion to 5.81 billion? Population change is too slow and remote for people to feel in their lives—even if the total population were to double or halve in only a century (as a mere 0.7 percent increase or decrease each year would do). Whether world population is increasing or decreasing doesn't necessarily affect the decisions that determine whether it will increase or decrease in the future. As the systems people would say, there is no feedback loop.

WHAT does affect fertility is modernity. In almost every country where people have moved from traditional ways of life to modern ones, they are choosing to have too few children to replace themselves. This is true in Western and in Eastern countries, in Catholic and in secular societies. And it is true in the richest parts of the richest countries. The only exceptions seem to be some small religious communities. We can't be sure what will happen in Muslim countries, because few of them have become modern yet, but so far it looks as if their fertility rates will respond to modernity as others' have.

Nobody can say whether world population will ever dwindle to very low numbers; that depends on what values people hold in the future. After the approaching peak, as long as people continue to prefer saving effort and money by having fewer children, population will continue to decline. (This does not imply that the decision to have fewer children is selfish; it may, for example, be motivated by a desire to do more for each child.)

Some people may have values significantly different from those of the rest of the world, and therefore different fertility rates. If such people live in a particular country or population group, their values can produce marked changes in the size of that country or group, even as world population changes only slowly. For example, the U.S. population, because of immigration and a fertility rate that is only slightly below replacement level, is likely to grow from 4.5 percent of the world today to 10 percent of a smaller world over the next two or three centuries. Much bigger changes in share are possible for smaller groups if they can maintain their difference from the average for a long period of time. (To illustrate: Korea's population could grow from one percent of the world to 10 percent in a single lifetime if it were to increase by two percent a year while the rest of the world population declined by one percent a year.)

World population won't stop declining until human values change. But human values may well change—values, not biological imperatives, are the unfathomable variable in population predictions. It is quite possible that in a century or two or three, when just about the whole world is at least as modern as Western Europe is today, people will start to value children more highly than they do now in modern societies. If they do, and fertility rates start to climb, fertility is no more likely to stop climbing at an average rate of 2.1 children per woman than it was to stop falling at 2.1 on the way down.

In only the past twenty years or so world fertility has dropped by 1.5 births per woman. Such a degree of change, were it to occur again, would be enough to turn a long-term increase in world population of one percent a year into a long-term decrease of one percent a year. Presumably fertility could someday increase just as quickly as it has declined in recent decades, although such a rapid change will be less likely once the world has completed the transition to modernity. If fertility rises only to 2.8, just 33 percent over the replacement rate, world population will eventually grow by one percent a year again—doubling in seventy years and multiplying by twenty in only three centuries.

The decline in fertility that began in some countries, including the United States, in the past century is taking a long time to reduce world population because when it started, fertility was very much higher than replacement level. In addition, because a preference for fewer children is associated with modern societies, in which high living standards make time valuable and children financially unproductive and expensive to care for and educate, the trend toward lower fertility couldn't spread throughout the world until economic development had spread. But once the whole world has become modern, with fertility everywhere in the neighborhood of replacement level, new social values might spread worldwide in a few decades. Fashions in families might keep changing, so that world fertility bounced above and below replacement rate. If each bounce took only a few decades or generations, world population would stay within a reasonably narrow range—although probably with a long-term trend in one direction or the other.

The values that influence decisions about having children seem, however, to change slowly and to be very widespread. If the average fertility rate were to take a long time to move from well below to well above replacement rate and back again, trends in world population could go a long way before they reversed themselves. The result would be big swings in world population—perhaps down to one or two billion and then up to 20 or 40 billion.

Whether population swings are short and narrow or long and wide, the average level of world population after several cycles will probably have either an upward or a downward trend overall. Just as averaging across the globe need not result in exactly 2.1 children per woman, averaging across the centuries need not result in zero growth rather than a slowly increasing or slowly decreasing world population. But the long-term trend is less important than the effects of the peaks and troughs. The troughs could be so low that human beings become scarcer than they were in ancient times. The peaks might cause harm from some kinds of shortages.

One implication is that not even very large losses from disease or war can affect the world population in the long run nearly as much as changes in human values do. What we have learned from the dramatic changes of the past few centuries is that regardless of the size of the world population at any time, people's personal decisions about how many children they want can make the world population go anywhere—to zero or to 100 billion or more.

Max Singer was a founder of the Hudson Institute. He is a co-author, with Aaron Wildavsky, of *The Real World Order* (1996).

Population and Consumption

What We Know, What We Need to Know

by Robert W. Kates

Thirty years ago, as Earth Day dawned, three wise men recognized three proximate causes of environmental degradation yet spent half a decade or more arguing their relative importance. In this classic environmentalist feud between Barry Commoner on one side and Paul Ehrlich and John Holdren on the other, all three recognized that growth in population, affluence, and technology were jointly responsible for environmental problems, but they strongly differed about their relative importance. Commoner asserted that technology and the economic system that produced it were primarily responsible.[1] Ehrlich and Holdren asserted the importance of all three drivers: population, affluence, and technology. But given Ehrlich's writings on population,[2] the differences were often, albeit incorrectly, described as an argument over whether population or technology was responsible for the environmental crisis.

Now, 30 years later, a general consensus among scientists posits that growth in population, affluence, and technology are jointly responsible for environmental problems. This has become enshrined in a useful, albeit overly simplified, identity known as IPAT, first published by Ehrlich and Holdren in *Environment* in 1972[3] in response to the more limited version by Commoner that had appeared earlier in *Environment* and in his famous book *The Closing Circle*.[4] In this identity, various forms of environmental or resource impacts (I) equals population (P) times affluence (A) (usually income per capita) times the impacts per unit of income as determined by technology (T) and the institutions that use it. Academic debate has now shifted from the greater or lesser importance of each of these driving forces of environmental degradation or resource depletion to debate about their interaction and the ultimate forces that drive them.

However, in the wider global realm, the debate about who or what is responsible for environmental degradation lives on. Today, many Earth Days later, international debates over such major concerns as biodiversity, climate change, or sustainable development address the population and the affluence terms of Holdrens' and Ehrlich's identity, specifically focusing on the character of consumption that affluence permits. The concern with technology is more complicated because it is now widely recognized that while technology can be a problem, it can be a solution as well. The development and use of more environmentally benign and friendly technologies in industrialized countries have slowed the growth of many of the most pernicious forms of pollution that originally drew Commoner's attention and still dominate Earth Day concerns.

A recent report from the National Research Council captures one view of the current public debate, and it begins as follows:

For over two decades, the same frustrating exchange has been repeated countless times in international policy circles. A government official or scientist from a wealthy country would make the following argument: The world is threatened with environmental disaster because of the depletion of natural resources (or climate change or the loss of biodiversity), and it cannot continue for long to support its rapidly growing population. To preserve the environment for future generations, we need to move quickly to control global population growth, and we must concentrate the effort on the world's poorer countries, where the vast majority of population growth is occurring.

Government officials and scientists from low-income countries would typically respond:

If the world is facing environmental disaster, it is not the fault of the poor, who use few resources. The fault must lie with the world's wealthy countries, where people consume the great bulk of the world's natural resources and energy and cause the great bulk of its environmental degradation. We need to curtail overconsumption in the rich countries which use far more than their fair share, both to preserve the environment and to allow the poorest people on earth to achieve an acceptable standard of living.[5]

It would be helpful, as in all such classic disputes, to begin by laying out what is known about the relative responsibilities of both population and consumption for the environmental crisis, and what might need to be known to address them. However, there is a profound asymmetry that must fuel the frustration of the developing countries' politicians and scientists: namely, how much people know about population and how little they know about consumption. Thus, this article begins by examining these differences in knowledge and action and concludes with the alternative actions needed to go from more to enough in both population and consumption.[6]

Population

What population is and how it grows is well understood even if all the forces driving it are not. Population begins with people and their key events of birth, death, and location. At the margins, there is some debate over when life begins and ends or whether residence is temporary or permanent, but little debate in between. Thus, change in the world's population or any place is the simple arithmetic of adding births, subtracting deaths, adding immigrants, and subtracting outmigrants. While whole subfields of demography are devoted to the arcane details of these additions and subtractions, the error in estimates of population for almost all places is probably within 20 percent and for countries with modern statistical services, under 3 percent—better estimates than for any other living things and for most other environmental concerns.

Current world population is more than six billion people, growing at a rate of 1.3 percent per year. The peak annual growth rate in all history—about 2.1 percent—occurred in the early 1960s, and the peak population increase of around 87 million per year occurred in the late 1980s. About 80 percent or 4.8 billion people live in the less developed areas of the world, with 1.2 billion living in industrialized countries. Population is now projected by the United Nations (UN) to be 8.9 billion in 2050, according to its medium fertility assumption, the one usually considered most likely, or as high as 10.6 billion or as low as 7.3 billion.[7]

A general description of how birth rates and death rates are changing over time is a process called the demographic transition.[8] It was first studied in the context of Europe, where in the space of two centuries, societies went from a condition of high births and high deaths to the current situation of low births and low deaths. In such a transition, deaths decline more rapidly than births, and in that gap, population grows rapidly but eventually stabilizes as the birth decline matches or even exceeds the death decline. Although the general description of the transition is widely accepted, much is debated about its cause and details.

The world is now in the midst of a global transition that, unlike the European transition, is much more rapid. Both births and deaths have dropped faster than experts expected and history foreshadowed. It took 100 years for deaths to drop in Europe compared to the drop in 30 years in the developing world. Three is the current global average births per woman of reproductive age. This number is more than halfway between the average of five children born to each woman at the post World War II peak of population growth and the average of 2.1 births required to achieve eventual zero population growth.[9] The death transition is more advanced, with life ex-

> **It is possible to break down the projected growth of the next century and to identify policies that would reduce projected populations even further.**

pectancy currently at 64 years. This represents three-quarters of the transition between a life expectancy of 40 years to one of 75 years. The current rates of decline in births outpace the estimates of the demographers, the UN having reduced its latest medium expectation of global population in 2050 to 8.9 billion, a reduction of almost 10 percent from its projection in 1994.

Demographers debate the causes of this rapid birth decline. But even with such differences, it is possible to break down the projected growth of the next century and to identify policies that would reduce projected populations even further. John Bongaarts of the Population Council has decomposed the projected developing country growth into three parts and, with his colleague Judith Bruce, has envisioned policies that would encourage further and more rapid decline.[10] The first part is unwanted fertility, making available the methods and materials for contraception to the 120 million married women (and the many more unmarried women) in developing countries who in survey research say they either want fewer children or want to space them better. A basic strategy for doing so links voluntary family planning with other reproductive and child health services.

Yet in many parts of the world, the desired number of children is too high for a stabilized population. Bongaarts would reduce this desire for large families by changing the costs and benefits of childrearing so that more parents would recognize the value of smaller families while simultaneously increasing their investment in children. A basic strategy for doing so accelerates three trends that have been shown to lead to lower desired family size: the survival of children, their education, and improvement in the economic, social, and legal status for girls and women.

However, even if fertility could immediately be brought down to the replacement level of two surviving children per woman, population growth would continue for many years in most developing countries because so many more young people of reproductive age exist. So Bongaarts would slow this momentum of population growth by increasing the age of childbearing, primarily by improving secondary education opportunity for girls and by addressing such neglected issues as adolescent sexuality and reproductive behavior.

How much further could population be reduced? Bongaarts provides the outer limits. The population of the developing world (using older projections) was expected to reach 10.2 billion by 2100. In theory, Bongaarts found that meeting the unmet need for contraception could reduce this total by about 2 billion. Bringing down desired family size to replacement fertility would reduce the population a billion more, with the remaining growth—from 4.5 billion today to 7.3 billion in 2100—due to population momentum. In practice, however, a recent U.S. National Academy of Sciences report concluded that a 10 percent reduction is both realistic and attainable and could lead to a lessening in projected population numbers by 2050 of upwards of a billion fewer people.[11]

Consumption

In contrast to population, where people and their births and deaths are relatively well-defined biological events, there is no consensus as to what consumption includes. Paul Stern of the National Research Council has described the different ways physics, economics, ecology, and sociology view consumption.[12] For physicists, matter and energy cannot be consumed, so consumption is conceived as transformations of matter and energy with increased entropy. For economists, consumption is spending on consumer goods and services and thus distinguished from their production and distribution. For ecologists, consumption is obtaining energy and nutrients by eating something else, mostly green plants or other consumers of green plants. And for some sociologists, consumption is a status symbol—keeping up with the Joneses—when individuals and households use their incomes to increase their social status through certain kinds of purchases. These differences are summarized in the box below.

In 1977, the councils of the Royal Society of London and the U.S. National Academy of Sciences issued a joint statement on consumption, having previously done so on population. They chose a variant of the physicist's definition:

> *Consumption is the human transformation of materials and energy. Consumption is of concern to the extent that it makes the transformed materials or energy less available for future use, or negatively impacts biophysical systems in such a way as to threaten human health, welfare, or other things people value.*[13]

On the one hand, this society/academy view is more holistic and fundamental than the other definitions; on the other hand, it is more focused, turning attention to the environmentally damaging. This article uses it as a working definition with one modification, *the addition of information to energy and matter,* thus completing the triad of the biophysical and ecological basics that support life.

In contrast to population, only limited data and concepts on the transformation of energy, materials, and information exist.[14] There is relatively good global knowledge of energy transformations due in part to the common units of conversion between different technologies. Between 1950 and today, global energy production and use increased more than fourfold.[15] For material transformations, there are no aggregate data in common units on a global basis, only for some specific classes of materials including materials for energy production, construction, industrial minerals and metals, agricultural crops, and water.[16] Calculations of material use by volume, mass, or value lead to different trends.

Trend data for per capita use of physical structure materials (construction and industrial minerals, metals, and forestry products) in the United States are relatively complete. They show an inverted S shaped (logistic) growth pattern: modest doubling between 1900 and the depression of the 1930s (from two to four metric tons), followed by a steep quintupling with economic recovery until the early 1970s (from two to eleven tons), followed by a leveling off since then with fluctuations related to economic downturns (see Figure 1).[17] An aggregate analysis of all current material production and consumption in the United States averages more than 60 kilos per person per day (excluding water). Most of this material flow is split between energy and related products (38 percent) and minerals for construction (37 percent), with the remainder as industrial minerals (5 percent), metals (2 percent), products of fields (12 percent), and forest (5 percent).[18]

A massive effort is under way to catalog biological (genetic) information and to sequence the genomes of microbes, worms, plants, mice, and people. In contrast to the molecular detail, the number and diversity of organisms is unknown, but a conservative estimate places the number of species on the order

> **In contrast to population, where people and their births and deaths are relatively well-defined biological events, there is no consensus as to what consumption includes.**

What Is Consumption?

Physicist: "What happens when you transform matter/energy"

Ecologist: "What big fish do to little fish"

Economist: "What consumers do with their money"

Sociologist: "What you do to keep up with the Joneses"

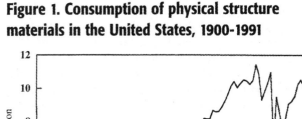

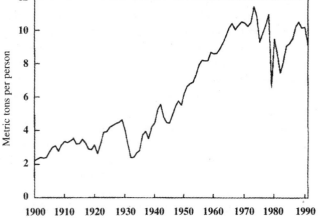

Figure 1. Consumption of physical structure materials in the United States, 1900–1991

SOURCE: I. Wernick, "Consuming Materials: The American Way," *Technological Forecasting and Social Change*, 53 (1996): 114.

of 10 million, of which only one-tenth have been described.[19] Although there is much interest and many anecdotes, neither concepts nor data are available on most cultural information. For example, the number of languages in the world continues to decline while the number of messages expands exponentially.

Trends and projections in agriculture, energy, and economy can serve as surrogates for more detailed data on energy and material transformation.[20] From 1950 to the early 1990s, world population more than doubled (2.2 times), food as measured by grain production almost tripled (2.7 times), energy more than quadrupled (4.4 times), and the economy quintupled (5.1 times). This 43-year record is similar to a current 55-year projection (1995–2050) that assumes the continuation of current trends or, as some would note, "business as usual." In this 55-year projection, growth in half again of population (1.6 times) finds almost a doubling of agriculture (1.8 times), more than twice as much energy used (2.4 times), and a quadrupling of the economy (4.3 times).[21]

Thus, both history and future scenarios predict growth rates of consumption well beyond population. An attractive similarity exists between a demographic transition that moves over time from high births and high deaths to low births and low deaths with an energy, materials, and information transition. In this transition, societies will use increasing amounts of energy and materials as consumption increases, but over time the energy and materials input per unit of consumption decrease and information substitutes for more material and energy inputs.

Some encouraging signs surface for such a transition in both energy and materials, and these have been variously labeled as decarbonization and dematerialization.[22] For more than a century, the amount of carbon per unit of energy produced has been decreasing. Over a shorter period, the amount of energy used to produce a unit of production has also steadily declined.

There is also evidence for dematerialization, using fewer materials for a unit of production, but only for industrialized countries and for some specific materials. Overall, improvements in technology and substitution of information for energy and materials will continue to increase energy efficiency (including decarbonization) and dematerialization per unit of product or service. Thus, over time, less energy and materials will be needed to make specific things. At the same time, the demand for products and services continues to increase, and the overall consumption of energy and most materials more than offsets these efficiency and productivity gains.

What to Do about Consumption

While quantitative analysis of consumption is just beginning, three questions suggest a direction for reducing environmentally damaging and resource-depleting consumption. The first asks: *When is more too much for the life-support systems of the natural world and the social infrastructure of human society?* Not all the projected growth in consumption may be resource-depleting—"less available for future use"—or environmentally damaging in a way that "negatively impacts biophysical systems to threaten human health, welfare, or other things people value."[23] Yet almost any human-induced transformations turn out to be either or both resource-depleting or damaging to some valued environmental component. For example, a few years ago, a series of eight energy controversies in Maine were related to coal, nuclear, natural gas, hydroelectric, biomass, and wind generating sources, as well as to various energy policies. In all the controversies, competing sides, often more than two, emphasized environmental benefits to support their choice and attributed environmental damage to the other alternatives.

Despite this complexity, it is possible to rank energy sources by the varied and multiple risks they pose and, for those concerned, to choose which risks they wish to minimize and which they are more willing to accept. There is now almost 30 years of experience with the theory and methods of risk assessment and 10 years of experience with the identification and setting of environmental priorities. While there is still no readily accepted methodology for separating resource-depleting or environmentally damaging consumption from general consumption

> **While quantitative analysis of consumption is just beginning, three questions suggest a direction for reducing environmentally damaging and resource-depleting consumption.**

or for identifying harmful transformations from those that are benign, one can separate consumption into more or less damaging and depleting classes and *shift* consumption to the less harmful class. It is possible to *substitute* less damaging and depleting energy and materials for more damaging ones. There is growing experience with encouraging substitution and its difficulties: renewables for nonrenewables, toxics with fewer toxics, ozone-depleting chemicals for more benign substitutes, natural gas for coal, and so forth.

The second question, *Can we do more with less?*, addresses the supply side of consumption. Beyond substitution, shrinking the energy and material transformations required per unit of consumption is probably the most effective current means for reducing environmentally damaging consumption. In the 1997 book, *Stuff: The Secret Lives of Everyday Things,* John Ryan and Alan Durning of Northwest Environment Watch trace the complex origins, materials, production, and transport of such everyday things as coffee, newspapers, cars, and computers and highlight the complexity of reengineering such products and reorganizing their production and distribution.[24]

Yet there is growing experience with the three Rs of consumption shrinkage: reduce, recycle, reuse. These have now been strengthened by a growing science, technology, and practice of industrial ecology that seeks to learn from nature's ecology to reuse everything. These efforts will only increase the existing favorable trends in the efficiency of energy and material usage. Such a potential led the Intergovernmental Panel on Climate Change to conclude that it was possible, using current best practice technology, to reduce energy use by 30 percent in the short run and 50–60 percent in the long run.[25] Perhaps most important in the long run, but possibly least studied, is the potential for and value of substituting information for energy and materials. Energy and materials per unit of consumption are going down, in part because more and more consumption consists of information.

The third question addresses the demand side of consumption—*When is more enough?*[26] Is it possible to reduce consumption by more satisfaction with what people already have, by *satiation,* no more needing more because there is enough, and by *sublimation,* having more satisfaction with less to achieve some greater good? This is the least explored area of consumption and the most difficult. There are, of course, many signs of *satiation* for some goods. For example, people in the industrialized world no longer buy additional refrigerators (except in newly formed households) but only replace them. Moreover, the quality of refrigerators has so improved that a 20-year or more life span is commonplace. The financial pages include frequent stories of the plight of this industry or corporation whose markets are saturated and whose products no longer show the annual growth equated with profits and progress. Such enterprises are frequently viewed as failures of marketing or entrepreneurship rather than successes in meeting human needs sufficiently and efficiently. Is it possible to reverse such views, to create a standard of satiation, a satisfaction in a need well met?

Can people have more satisfaction with what they already have by using it more intensely and having the time to do so? Economist Juliet Schor tells of some overworked Americans who would willingly exchange time for money, time to spend with family and using what they already have, but who are constrained by an uncooperative employment structure.[27] Proposed U.S. legislation would permit the trading of overtime for such compensatory time off, a step in this direction. *Sublimation,* according to the dictionary, is the diversion of energy from an immediate goal to a higher social, moral, or aesthetic purpose. Can people be more satisfied with less satisfaction derived from the diversion of immediate consumption for the satisfaction of a smaller ecological footprint?[28] An emergent research field grapples with how to encourage consumer behavior that will lead to change in environmentally damaging consumption.[29]

A small but growing "simplicity" movement tries to fashion new images of "living the good life."[30] Such movements may never much reduce the burdens of consumption, but they facilitate by example and experiment other less-demanding alternatives. Peter Menzel's remarkable photo essay of the material goods of some 30 households from around the world is powerful testimony to the great variety and inequality of possessions amidst the existence of alternative life styles.[31] Can a standard of "more is enough" be linked to an ethic of "enough for all"? One of the great discoveries of childhood is that eating lunch does not feed the starving children of some far-off place. But increasingly, in sharing the global commons, people flirt with mechanisms that hint at such—a rationing system for the remaining chlorofluorocarbons, trading systems for reducing emissions, rewards for preserving species, or allowances for using available resources.

A recent compilation of essays, *Consuming Desires: Consumption, Culture, and the Pursuit of Happiness,*[32] explores many of these essential issues. These elegant essays by 14 well-known writers and academics ask the fundamental question of why more never seems to be enough and why satiation and sublimation are so difficult in a culture of consumption. Indeed, how is the culture of consumption different for mainstream America, women, inner-city children, South Asian immigrants, or newly industrializing countries?

Why We Know and Don't Know

In an imagined dialog between rich and poor countries, with each side listening carefully to the other, they might ask themselves just what they actually know about population and consumption. Struck with the asymmetry described above, they

> **Can people have more satisfaction with what they already have by using it more intensely and having the time to do so?**

might then ask: "Why do we know so much more about population than consumption?"

The answer would be that population is simpler, easier to study, and a consensus exists about terms, trends, even policies. Consumption is harder, with no consensus as to what it is, and with few studies except in the fields of marketing and advertising. But the consensus that exists about population comes from substantial research and study, much of it funded by governments and groups in rich countries, whose asymmetric concern readily identifies the troubling fertility behavior of others and only reluctantly considers their own consumption behavior. So while consumption is harder, it is surely studied less (see Table 1).

The asymmetry of concern is not very flattering to people in developing countries. Anglo-Saxon tradition has a long history of dominant thought holding the poor responsible for their condition—they have too many children—and an even longer tradition of urban civilization feeling besieged by the barbarians at their gates. But whatever the origins of the asymmetry, its persistence does no one a service. Indeed, the stylized debate of population versus consumption reflects neither popular understanding nor scientific insight. Yet lurking somewhere beneath the surface concerns lies a deeper fear.

Consumption is more threatening, and despite the North–South rhetoric, it is threatening to all. In both rich and poor countries alike, making and selling things to each other, including unnecessary things, is the essence of the economic system. No longer challenged by socialism, global capitalism seems inherently based on growth—growth of both consumers and their consumption. To study consumption in this light is to risk concluding that a transition to sustainability might require profound changes in the making and selling of things and in the opportunities that this provides. To draw such conclusions, in the absence of convincing alternative visions, is fearful and to be avoided.

What We Need to Know and Do

In conclusion, returning to the 30-year-old IPAT identity—a variant of which might be called the Population/Consumption (PC) version—and restating that identity in terms of population and consumption, it would be: $I = P*C/P*I/C$, where I equals environmental degradation and/or resource depletion; P equals the number of people or households; and C equals the transformation of energy, materials, and information (see Figure 2 below).

With such an identity as a template, and with the goal of reducing environmentally degrading and resource-depleting influences, there are at least seven major directions for research and policy. To reduce the level of impacts per unit of consumption, it is necessary to separate out more damaging consumption and *shift* to less harmful forms, *shrink* the amounts

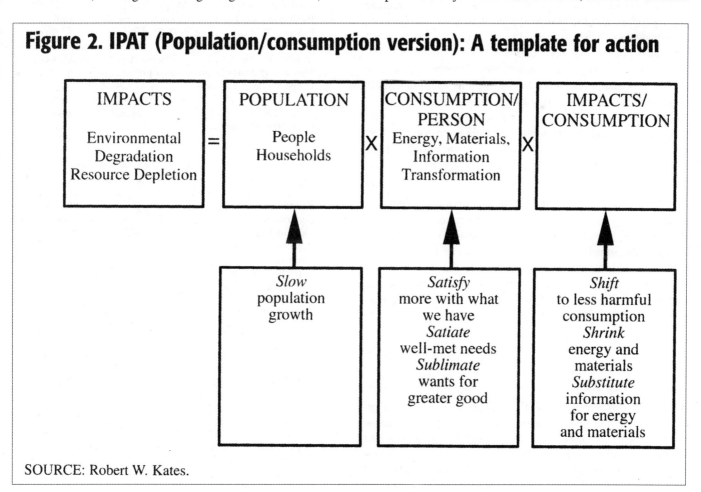

Figure 2. IPAT (Population/consumption version): A template for action

SOURCE: Robert W. Kates.

Table 1. A comparison of population and consumption

Population	Consumption
Simpler, easier to study	More complex
Well-funded research	Unfunded, except marketing
Consensus terms, trends	Uncertain terms, trends
Consensus policies	Threating policies

SOURCE: Robert W. Kates.

of environmentally damaging energy and materials per unit of consumption, and *substitute* information for energy and materials. To reduce consumption per person or household, it is necessary to *satisfy* more with what is already had, *satiate* well-met consumption needs, and *sublimate* wants for a greater good. Finally, it is possible to *slow* population growth and then to *stabilize* population numbers as indicated above.

However, as with all versions of the IPAT identity, population and consumption in the PC version are only proximate driving forces, and the ultimate forces that drive consumption, the consuming desires, are poorly understood, as are many of the major interventions needed to reduce these proximate driving forces. People know most about slowing population growth, more about shrinking and substituting environmentally damaging consumption, much about shifting to less damaging consumption, and least about satisfaction, satiation, and sublimation. Thus the determinants of consumption and its alternative patterns have been identified as a key understudied topic for an emerging sustainability science by the recent U.S. National Academy of Science study.[33]

But people and society do not need to know more in order to act. They can readily begin to separate out the most serious problems of consumption, shrink its energy and material throughputs, substitute information for energy and materials, create a standard for satiation, sublimate the possession of things for that of the global commons, as well as slow and stabilize population. To go from more to enough is more than enough to do for 30 more Earth Days.

Robert W. Kates is an independent scholar in Trenton, Maine; a geographer; university professor emeritus at Brown University; and an executive editor of *Environment*. The research for "Population and Consumption: What We Know, What We Need to Know" was undertaken as a contribution to the recent National Academies/National Research Council report, *Our Common Journey: A Transition Toward Sustainability*. The author retains the copyright to this article. Kates can be reached at RR1, Box 169B, Trenton, ME 04605.

NOTES

1. B. Commoner, M. Corr, and P. Stamler, "The Causes of Pollution," *Environment,* April 1971, 2–19.
2. P. Ehrlich, *The Population Bomb* (New York: Ballantine, 1966).
3. P. Ehrlich and J. Holdren, "Review of The Closing Circle," *Environment,* April 1972, 24–39.
4. B. Commoner, *The Closing Circle* (New York: Knopf, 1971).
5. P. Stern, T. Dietz, V. Ruttan, R. H. Socolow, and J. L. Sweeney, eds., *Environmentally Significant Consumption: Research Direction* (Washington, D.C.: National Academy Press, 1997), 1.
6. This article draws in part upon a presentation for the 1997 De Lange-Woodlands Conference, an expanded version of which will appear as: R. W. Kates, "Population and Consumption: From More to Enough," in *In Sustainable Development: The Challenge of Transition,* J. Schmandt and C. H. Wards, eds. (Cambridge, U.K.: Cambridge University Press, forthcoming), 79–99.
7. United Nations, Population Division, *World Population Prospects: The 1998 Revision* (New York: United Nations, 1999).
8. K. Davis, "Population and Resources: Fact and Interpretation," K. Davis and M. S. Bernstam, eds., in *Resources, Environment and Population: Present Knowledge, Future Options,* supplement to *Population and Development Review,* 1990: 1–21.
9. Population Reference Bureau, *1997 World Population Data Sheet of the Population Reference Bureau* (Washington, D.C.: Population Reference Bureau, 1997).
10. J. Bongaarts, "Population Policy Options in the Developing World," *Science,* 263: (1994), 771–776; and J. Bongaarts and J. Bruce, "What Can Be Done to Address Population Growth?" (unpublished background paper for The Rockefeller Foundation, 1997).
11. National Research Council, Board on Sustainable Development, *Our Common Journey: A Transition Toward Sustainability* (Washington, D.C.: National Academy Press, 1999).
12. See Stern, et al., note 5 above.
13. Royal Society of London and the U.S. National Academy of Sciences, "Towards Sustainable Consumption," reprinted in *Population and Development Review,* 1977, 23 (3): 683–686.
14. For the available data and concepts, I have drawn heavily from J. H. Ausubel and H. D. Langford, eds., *Technological Trajectories and the Human Environment.* (Washington, D.C.: National Academy Press, 1997).

6. Population and Consumption

15. L. R. Brown, H. Kane, and D. M. Roodman, *Vital Signs 1994: The Trends That Are Shaping Our Future* (New York: W. W. Norton and Co., 1994).
16. World Resources Institute, United Nations Environment Programme, United Nations Development Programme, World Bank, *World Resources, 1996–97* (New York: Oxford University Press, 1996); and A. Gruebler, *Technology and Global Change* (Cambridge, Mass.: Cambridge University Press, 1998).
17. I. Wernick, "Consuming Materials: The American Way," *Technological Forecasting and Social Change,* 53 (1996): 111–122.
18. I. Wernick and J. H. Ausubel, "National Materials Flow and the Environment," *Annual Review of Energy and Environment,* 20 (1995): 463–492.
19. S. Pimm, G. Russell, J. Gittelman, and T. Brooks, "The Future of Biodiversity," *Science,* 269 (1995): 347–350.
20. Historic data from L. R. Brown, H. Kane, and D. M. Roodman, note 15 above.
21. One of several projections from P. Raskin, G. Gallopin, P. Gutman, A. Hammond, and R. Swart, *Bending the Curve: Toward Global Sustainability,* a report of the Global Scenario Group, Polestar Series, report no. 8 (Boston: Stockholm Environmental Institute, 1995).
22. N. Nakicénovíc, "Freeing Energy from Carbon," in *Technological Trajectories and the Human Environment,* eds., J. H. Ausubel and H. D. Langford. (Washington, D.C.: National Academy Press, 1997); I. Wernick, R. Herman, S. Govind, and J. H. Ausubel, "Materialization and Dematerialization: Measures and Trends," in J. H. Ausubel and H. D. Langford, eds., *Technological Trajectories and the Human Environment* (Washington, D.C.: National Academy Press, 1997), 135–156; and see A. Gruebler, note 16 above.
23. Royal Society of London and the U.S. National Academy of Science, note 13 above.
24. J. Ryan and A. Durning, *Stuff: The Secret Lives of Everyday Things* (Seattle, Wash.: Northwest Environment Watch, 1997).
25. R. T. Watson, M. C. Zinyowera, and R. H. Moss, eds., *Climate Change 1995: Impacts, Adaptations, and Mitigation of Climate Change—Scientific-Technical Analyses* (Cambridge, U.K.: Cambridge University Press, 1996).
26. A sampling of similar queries includes: A. Durning, *How Much Is Enough?* (New York: W. W. Norton and Co., 1992); Center for a New American Dream, *Enough!: A Quarterly Report on Consumption, Quality of Life and the Environment* (Burlington, Vt.: The Center for a New American Dream, 1997); and N. Myers, "Consumption in Relation to Population, Environment, and Development," *The Environmentalist,* 17 (1997): 33–44.
27. J. Schor, *The Overworked American* (New York: Basic Books, 1991).
28. A. Durning, *How Much Is Enough?: The Consumer Society and the Future of the Earth* (New York: W. W. Norton and Co., 1992); Center for a New American Dream, note 26 above; and M. Wackernagel and W. Ress, *Our Ecological Footprint: Reducing Human Impact on the Earth* (Philadelphia. Pa.: New Society Publishers, 1996).
29. W. Jager, M. van Asselt, J. Rotmans, C. Vlek, and P. Costerman Boodt, *Consumer Behavior: A Modeling Perspective in the Contest of Integrated Asssessment of Global Change,* RIVM report no. 461502017 (Bilthoven, the Netherlands: National Institute for Public Health and the Environment, 1997); and P. Vellinga, S. de Bryn, R. Heintz, and P. Molder, eds., *Industrial Transformation: An Inventory of Research.* IHDP-IT no. 8 (Amsterdam, the Netherlands: Institute for Environmental Studies, 1997).
30. H. Nearing and S. Nearing. *The Good Life: Helen and Scott Nearing's Sixty Years of Self-Sufficient Living* (New York: Schocken, 1990); and D. Elgin, *Voluntary Simplicity: Toward a Way of Life That Is Outwardly Simple Inwardly Rich* (New York: William Morrow, 1993).
31. P. Menzel, *Material World: A Global Family Portrait* (San Francisco: Sierra Club Books, 1994).
32. R. Rosenblatt, ed., *Consuming Desires: Consumption, Culture, and the Pursuit of Happiness* (Washington, D.C.: Island Press, 1999).
33. National Research Council, Board on Sustainable Development, *Our Common Journey: A Transition Toward Sustainability* (Washington, D.C.: National Academy Press, 1999).

Food for All in the 21st Century

by Gordon Conway

For most of the industrialized countries, there does not seem to be a food problem. They produce a surfeit of food, and health problems have more to do with being overweight than with hunger. In the rest of the world there are periodic famines, but few in the industrialized countries realize that millions of people lack enough food most days of their lives.[1]

The Green Revolution was one of the great success stories of the second half of the 20th century. Food production in developing countries kept pace with population growth. Yet today about 800 million people, or some 15 percent of the world's population, get less than 2000 calories per day and live a life of permanent of intermittent hunger and are chronically undernourished.[2] Many of the hungry are women and children. More than 180 million children under five years of age are underweight, that is, they are more than two standard deviations below the standard weight for their age. This represents one-third of the under-fives in the developing countries. Young children crucially need food because they are growing fast and, once weaned, are liable to succumb to infections. Seventeen million children under five die each year, and malnourishment contributes to at least one-third of these children's deaths.

Lack of protein, vitamins, minerals, and other micronutrients in the diet is also widespread. About 100 million children suffer from vitamin A deficiency.[3] As has long been known, lack of this vitamin can cause eye damage. Half a million children become partially or totally blind each year, and many subsequently die. As recent research has shown, lack of vitamin A has an even more serious and pervasive effect, apparently reducing the ability of children's immune systems to cope with infection. Iron deficiency is also common in developing countries, affecting one billion people. More than 400 million women of childbearing age (15–49 years old) are afflicted by anemia caused by iron deficiency. As a result, they tend to produce stillborn or underweight children and are more likely to die in childbirth. Anemia has been identified as a contributing factor in more than 20 percent of all postpartum maternal deaths in Asia and Africa.

Paradoxically, hunger is common despite 20 years of rapidly declining world food prices. Although in many developing countries there is enough food to meet demand, large numbers of people still go hungry. Food prices are low, yet they remain high relative to the earning capacity of the poor. Market demand is satisfied, but there are many who are unable to purchase the food they need and, hence, to them the market is irrelevant.

Not surprisingly, hunger is closely related to poverty. To the casual observer, poverty seems to be worse in the cities but, in reality, the urban poor fare better. To quote one statistic, the incidence of malnutrition is five times higher in the sierra of Peru than in the capital, Lima. About 130 million of the poorest 20 percent of developing country populations live in urban settlements, most of them in slums and squatter settlements. Yet 650 million of the poorest live in rural areas. In sub-Saharan Africa and Asia, most of the poor are rural poor.[4] Some live in rural areas with high agricultural potential and high population densities—the Gangetic plain of India and the island of Java. But the majority, about 370 million, live where the agricultural potential is relatively low and natural resources are poor, such as the Andean highlands and the Sahel.

The first question to ask is: Why should we be concerned? Probably everyone who reads Environment is getting an adequate diet. Does it matter that others are not so fortunate? Does it matter to the industrialized countries that many people in the developing countries are malnourished? Part of the answer to these questions

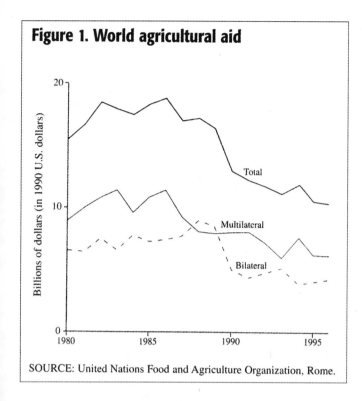

Figure 1. World agricultural aid

SOURCE: United Nations Food and Agriculture Organization, Rome.

2020, there will be about an extra 1.5 billion mouths to feed. If the proportion of the population of the developing countries deprived of an adequate diet remains the same, the number undernourished 20 years from now could be well over one billion.

What is the prognosis for feeding the world's population in the 21st century? Producing forecasts of world food production is complicated. Econometric models are reasonably optimistic.[5] They show that over the next 20 years, the world population growth rate will be matched by a similar growth in food production, and food prices will continue to decline. However, developing countries as a whole will not be able to meet their market demand. In the International Food Policy Research Institute (IFPRI) model, the total shortfall by 2020 is some 190 million tons, which will have to be imported from the developed countries (see Figure 2 and the box, "The Risks of Biotechnology").

Inevitably, models of this kind raise more questions than they answer. Most important, the food needs of the poor and hungry are omitted. As in the real world, they are simply priced out of the market, and their needs are "hidden." By 2020, the total numbers of malnourished children will have declined slightly to 155 million, but in sub-Saharan Africa they will have increased by nearly 50 percent. Probably, close to three-quarters of a billion people will be chronically undernourished.

These models also make optimistic predictions about crop yields and production. But there is evidence, albeit largely anecdotal, of increasing production problems in those places where yield growth has been most marked. For example, in the Punjab, although wheat yields are still increasing, this achievement is now being seriously threatened.[6] Of greatest concern is the increasing scarcity of water. In some of the most intensively cultivated districts, the groundwater table has fallen to a depth of 9–15 meters and continues to fall at about a half a meter a year. This and other, albeit largely anecdotal, evidence from Luzon, Java, and Sonora suggest there are serious and growing threats to the sustainability of the yields of the Green Revolution lands.[7]

There is also widespread evidence of declines in the rates of yield growth (see Figure 3).[8] A combination of causes is responsible.[9] In parts of Asia, declining prices for cereals are causing farmers to invest more in higher value cash crops. But more important, there has been little or no increase in yield ceilings of rice and maize in recent years. A third factor is the cumulative effect of environmental degradation, which is partly caused by agriculture itself. Virtually all long-term experiments with cereal crops in the developing countries exhibit marked downward trends in yields.

is political. The end of the Cold War has not brought about an increase in global stability. While conflict between East and West has declined, there is a fast-growing divide between peoples, countries, and regions who "belong" in global power terms and those who are excluded. Yet this potentially explosive inequity receives relatively little attention in the industrialized countries. The volume of agricultural aid going to developing countries is stagnating in real terms (see Figure 1). People need to recognize that unless developing countries are helped to realize sufficient food, employment, and shelter for their growing populations or helped to gain the means to purchase the food internationally, the political stability of the world will be still further undermined. In today's world, poverty and hunger, however remote, affect everyone.

At the same time, the growing interconnectedness of the world—the process commonly referred to as globalization—holds the promise of alleviating, if not eliminating, poverty and hunger. While globalization threatens to concentrate power and increase division, it also has the economic and technological potential to transform the lives of rich and poor alike. Much depends on where priorities lie and, in particular, whether there is sufficient access by the poor to the economic opportunities created by the products of the new technologies.

Prospects for the Year 2020

If nothing new is done, the numbers of those who are poor and hungry will grow. Most populations in the developing world are still increasing rapidly. By the year

Agriculture and the Environment

The litany of environmental loss is familiar.[10] Soils are eroding and losing their fertility, precious water supplies

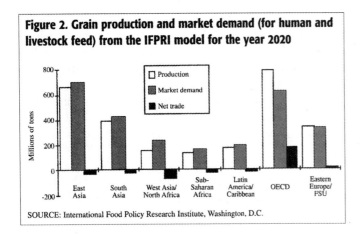

Figure 2. Grain production and market demand (for human and livestock feed) from the IFPRI model for the year 2020

SOURCE: International Food Policy Research Institute, Washington, D.C.

are being squandered, rangeland overgrazed, forests destroyed, and fisheries overexploited. The heavy use of pesticides has caused severe problems. There is growing human morbidity and mortality while, at the same time, pest populations are becoming resistant and escaping from natural control. In the intensively farmed lands of both the developed and developing countries, heavy fertilizer applications are producing nitrate levels in drinking water that approach or exceed permitted levels, increasing the likelihood of government restrictions on the use of fertilizer.

Other agricultural pollutants have the potential for damage on a much larger scale. While industry is often to blame, agriculture is becoming a major contributor to regional and global pollution, producing significant levels of methane, carbon dioxide, and nitrous oxide (see Figure 4).[11] Natural processes generate these gases, but the intensification of agriculture in both the developed and developing countries has increased the rates of emission. Individually or in combination, these gases are contributing to acid deposition, the depletion of stratospheric ozone, the buildup of ozone in the lower atmosphere, and global warming. The effects on the natural environment and human well-being are well known, but in each case there are significant adverse effects on agriculture. In relation to global pollution, agriculture is both culprit and victim.

The Doubly Green Revolution

In theory, the industrialized countries could feed the world. However, this would require several hundred

---

The Risks of Biotechnology

In practice it is difficult to draw a distinct line between traditional plant breeding techniques (through which we have been redesigning nature for thousands of years) and biotechnology. But the capacity of genetic engineering to move genes across genera and families as well as between animals and plants may give rise to unanticipated interactions within the genome with unknown effects. It is a new technology with which we have had limited experience. While we gain experience we need to move cautiously.

The most serious environmental risk is the likelihood of transgenes escaping from cultivated crops into wild relatives (or contaminating organic varieties off nearby farms). This is a justified concern. Genes from existing commercial crops can and do pass to organic crops, and vice versa, and genes from both transfer to wild relatives. Even self-pollinated crops, such as rice, will cross with wild rices. The question is whether the genes remain in the wild relatives and whether they result in adverse ecological effects, such as the production of superweeds. Only extensive, well-designed, and monitored field tests will provide answers.

A potential solution to this problem is to incorporate the gene in the plastid's genome. In most crops, plastids are maternally inherited only and not transmitted via pollen.

Another potential hazard arises from plants containing genes from viral pathogens that confer resistance to these same pathogens. Expressing the viral genes in plants somehow disrupts the virus infection process. But exchange of these genes with other viral pathogens may be possible, creating entirely new virus strains with unknown properties.

A third significant risk—the potential to evolve resistances to the toxins produced by *Bacillus thuringiensis* (Bt) genes—is well known, as are some of the counterstrategies. One answer is to employ refuges of non-Bt crop plants. Another uses two or more toxin genes each with a different molecular target. Experience indicates the need to anticipate the eventual breakdown of control.

Introduction of Bt into a wide range of crops implies a much higher selection pressure than from spraying the insecticide on a single crop. Insect populations need to be carefully monitored for resistance and alternative strategies continuously developed.

The most publicized health risk is that genetically modified (GM) crops carrying antibiotic genes used as selectable markers may generate antibiotic resistance in livestock or humans. The likelihood is fairly small, but alternative selection technologies are now available and should be used. There is also concern that transgenes may increase allergies, through the introduction of new proteins to foodstuffs.

Other fears have less scientific basis. There is no reason to suppose that the process of gene transfer itself confers a health risk. Neither is there any *a priori* reason why ingesting pieces of transgenic DNA is likely to be hazardous, any more than the large quantities of DNA from numerous sources ingested every day in normal diets.

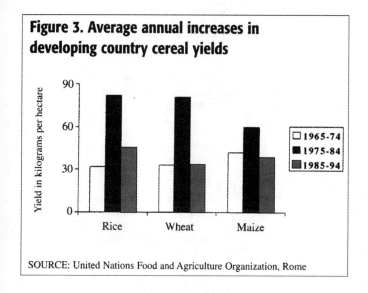

Figure 3. Average annual increases in developing country cereal yields

SOURCE: United Nations Food and Agriculture Organization, Rome

million tons of food aid, many times what they now supply. It would place heavy burdens on both the donors and the recipients. The environmental costs for the developed countries would be high, and for the developing countries the availability of free or subsidized aid in such large quantities would depress local prices and add to existing disincentives for local food production. More importantly, this scenario implies that a large proportion of the population in the developing world would fail to participate in global economic growth.

The alternative scenario is for developing countries to undertake an accelerated, broad-based growth, not only in food production but also in agricultural and natural resource development. This would be part of a larger development process aimed at meeting most of their own food needs, including the needs of the poor.

Implicitly, this scenario recognizes that food security is not a matter solely of producing sufficient food. For the rural poor, food security depends as much on employment and incomes as it does on food production, and agricultural and natural resource development is crucial in both respects. Food security, so defined, is also a key determinant of family size. The greater the degree of security and the higher the level of education, the more women will take advantage of new opportunities and plan ahead for themselves and their families. Appropriate agricultural and natural resource development can also significantly contribute to greater environmental protection and conservation. Finally, vigorous agricultural and economic growth can stimulate world trade and provide significant benefits for all countries, both developed and developing.

These arguments, taken together, point to the need for a second Green Revolution, a revolution that does not simply reflect the successes of the first. The technologies

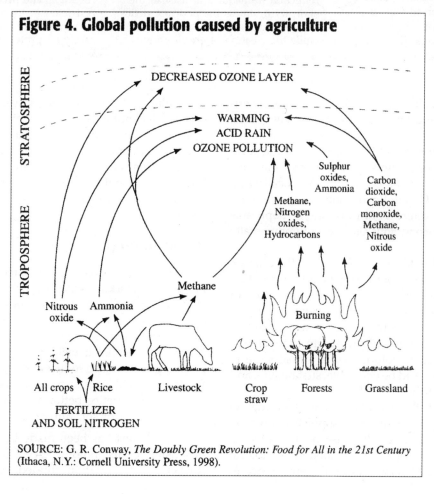

Figure 4. Global pollution caused by agriculture

SOURCE: G. R. Conway, *The Doubly Green Revolution: Food for All in the 21st Century* (Ithaca, N.Y.: Cornell University Press, 1998).

Policies for a Doubly Green Revolution

There is no single recipe for successful agricultural development, though there is a broad consensus on many of the essential ingredients. These include:

- economic policies that do not discriminate against agriculture, forestry, or fisheries;
- liberalized markets for farm inputs and outputs with major private sector involvement;
- efficient rural financial institutions, including adequate access by all types of farmers to credit, inputs, and marketing services;
- in some cases, land reform or redistribution;
- adequate rural infrastructure, including irrigation, transport, and marketing;
- investments in rural education, clean water, health, nutrition programs, and family planning;
- specific attention to satisfying the needs of women and of ethnic and other minority groups and securing their legal rights; and
- effective development and dissemination of appropriate agricultural technologies in partnership with farmers.

Although listed separately above, they are intimately interconnected. While economic liberalization within developing countries and reform of international trading policies are necessary prerequisites for significant agricultural growth, they are not sufficient by themselves. Accelerated growth in agricultural output cannot be maintained without adequate investments in rural infrastructure and in agricultural research and extension. Indeed, without such investment, the results of liberalization policies may well fall short of expectations and set governments against market-oriented approaches.

They will not contribute significantly to poverty alleviation and the reduction in inequity, at least in the short term, unless the poor are deliberately targeted. Essential steps include the creation of employment for the land poor and landless, increased production on small and medium-sized, and large farms, provision of nearby input and output markets, and, recognizing where the rural poor are mostly located, attention to regions of lower agroclimatic and resource potential, not just the best.

This means that agricultural innovation, at least in the developing countries, cannot be simply left to market forces. Inevitably, private research focuses on the major high-value crops, on labor-saving technologies, and on the needs of capital-intensive farming. By contrast, research to feed the poor is less attractive. It frequently involves long lead times, for example, in developing new plant types of minor staples. It is risky, particularly when focused on heterogeneous environments that are subject to high climatic or other variability. Moreover, the beneficiaries have little capacity to pay for the research. The products cannot be restricted to those who can pay, and intellectual property rights can rarely be protected.

---

of the first Green Revolution were developed on experiment stations that were favored with fertile soils, well-controlled water sources, and other factors suitable for high production. There was little perception of the complexity and diversity of farmers' physical environments, let alone the diversity of the economic and social environments. The new Green Revolution must not only benefit the poor more directly, but must also be applicable under highly diverse conditions and be environmentally sustainable.

In effect, the need is for a Doubly Green Revolution, a revolution that is even more productive than the first Green Revolution and even more "green" in terms of conserving natural resources and the environment (see the box, "Policies for a Doubly Green Revolution").[12] During the next three decades, it must aim to repeat the successes of the Green Revolution on a global scale in many diverse localities and be equitable, sustainable, and environmentally friendly.

The complexity of these challenges is daunting, in many respects of a greater order of sophistication than those encountered before. Yet, because of the potential of two key, recent developments in modern biological science, it seems possible. The first is the emergence of molecular and cellular biology, a discipline, with its associated technologies, that is having far-reaching consequences on the ability to understand and manipulate living organisms. (For more on this, see L. Levidow's article, "Regulating Bt Maize in the United States and Europe: A Scientific-Cultural Comparison," in the December issue of Environment.)

Biotechnology

Hitherto, the success of the Green Revolution has depended on working with blueprints of "creating" desirable new plant and animal types through painstaking conventional plant breeding. Biotechnology, and especially genetic engineering, offers a faster route. Moreover, it will be essential if yield ceilings are to be raised, excessive pesticide use reduced, the nutrient value of basic foods increased, and farmers on less favored lands provided with varieties better able to tolerate drought, salinity, and lack of soil nutrients.

A good start has been made in improving rice varieties using biotechnology. Over the past 15 years, the Rockefeller Foundation funded some $100 million of plant biotechnology research and trained more than 400 scientists from Asia, Africa, and Latin America. At several locations in Asia, there is now a critical mass of talent applying the new tools of biotechnology to rice improvement. To date, most of the new varieties are the result of tissue culture and marker-aided selection techniques. For example, a rice variety resulting from tissue culture, called La Fen Rockefeller, is providing farmers in the Shanghai region with 5–15 percent increases in yield. Scientists at the West Africa Rice Development Association have also used tissue culture to cross the high-yielding Asian rices with the traditional African rices. The result is a new plant type that looks like African rice during its early stages of growth (it grows in dry conditions and is able to shade out weeds) but becomes more like Asian rice as it reaches maturity, resulting in higher yields with fewer inputs.

Marker-aided selection is being used in rice to pyramid two or more genes, on the one hand, for resistance to the same pathogen, thereby increasing resistance to pathogens and, on the other hand, to make rice plants more drought tolerant. For some time to come, this is likely to be the most productive use of biotechnology for cereals. However, progress is being made in the production of transgenic cereals for developing countries. As in the industrialized countries, the focus has been largely on traits for disease and pest resistance, but Mexican scientists have added genes to rice and maize that confer tolerance to aluminum toxicity. Indian scientists have added two genes to rice that together appear to help the plant tolerate prolonged submergence.

The realization that the genes of the major cereals are allelic versions that evolved from a common set of genes in a common ancestor means there is enormous potential for sharing genetic information across the cereals and for moving alleles from one cereal to another to modify traits. For example, the dwarfing genes in wheat, maize, and other plants were recently shown to be alleles (any of the alternative forms of a gene that may occur at a given locus).[13] In the long run, there is the possibility of increasing yield ceilings, such as through more efficient photosynthesis or the improved regulation of stomata.[14]

To date, the most exciting development has been the introduction of genes that produce beta-carotene in the rice grain.[15] Beta-carotene is present in the leaves of the rice plant, but conventional plant breeding has been unable to put it into the grain. The transgenic rice grain has a light golden-yellow color and contains sufficient beta-carotene to meet human vitamin A requirements from rice alone. This "golden" rice offers an opportunity to complement vitamin A supplementation programs, particularly in rural areas that are difficult to reach. The same scientists have also added genes to rice that increase its bioavailable iron content over threefold.

Over the next decade, much greater progress in multiple gene introductions that focus on output traits or on difficult-to-achieve input characteristics is likely.[16] The

Figure 5. Effects of applying manure and inorganic fertilizers on sorghum in Burkina Faso over 30 years

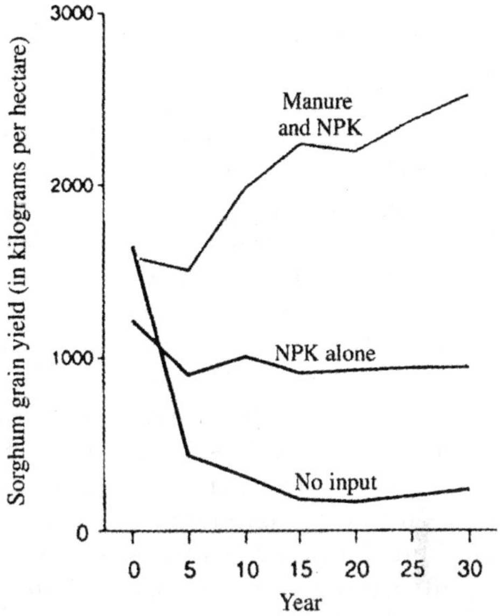

NOTE: NPK = nitrogen, phosphorous, and potassium

SOURCE: International Livestock Research Institute, Nairobi

potentials for genetic engineering are almost endless. But alongside the benefits are risks, some real, some imagined. An impassioned debate in Europe is raising genuine concerns about ethics, the environment, and the potential impact on human health.[17] Developed countries are clearly better equipped to assess such hazards. They can call on a wide range of expertise, and most have now set up regulatory bodies and are insisting on closely monitored trials to identify the likely risks before releasing genetically engineered crops and livestock into the environment. So far, few developing countries have such regulation in place. Perhaps the hazards are often overstated, but if the evident benefits are to be realized for the developing countries, it is the responsibility of all involved to ensure that the hazard assessments are as rigorous as in developed countries.

More important than the potential hazards is the question of who benefits from biotechnology. So far, the focus of biotechnology companies has been on developed country markets where potential sales are large, patents are well protected, and the risks are lower. But they are now turning their attention to the developing countries and embarking on an aggressive policy of identifying and patenting potentially useful genes. Part of the answer to this challenge lies in public-private partnerships whereby genomic information and technologies are donated to public plant breeders and agreements are struck that ensure that new varieties are freely available to poor farmers in developing countries.

The Application of Ecology

The second development is the emergence of modern ecology, an equally powerful discipline that is rapidly increasing understanding of the structure and dynamics of agricultural and natural ecosystems and providing clues to their productive and sustainable management.

The widely successful application of integrated pest management (IPM) to control rice pests in Southeast Asia is proof of what can be achieved. IPM looks at each crop and pest situation as a whole and then devises a program that integrates the various control methods in the light of all factors present. As practiced today, it combines modern technology (including the application of synthetic, yet selective, pesticides and the engineering of pest resistance) with natural methods of control, including agronomic practices and the use of natural predators and parasites. The outcome is sustainable, efficient pest control that is often cheaper than conventional pesticides.

A recent, highly successful example involves the brown planthopper and other rice pests in Indonesia. Under the program, farmers are trained to recognize and regularly monitor the pests and their natural enemies. They then use simple, yet effective, rules to determine the minimum necessary use of pesticides. The outcome is a reduction in the average number of sprayings from more than four to less than one per season, while yields have grown from 6 to nearly 7.5 tons per hectare. Since 1986, rice production has increased 15 percent while pesticide use has declined 60 percent, saving $120 million a year in subsidies. The total economic benefit up to 1990 was estimated to be more than $1 billion.[18] The farmers' health has improved and a not insignificant benefit has been the return of fish to the rice fields.

The next challenge is to extend the principles of integration established in IPM to other subsystems of agriculture, nutrient conservation, and the management of soil, water, and other natural resources such as rangeland. Of great potential value is the development of highly integrated crop-livestock systems, where soil structure and nutrients benefit both from livestock manure and the nitrogen-fixing capacity of forage crops. As African scientists have shown, use of inorganic fertilizers alone can lead to stagnation of crop yields but, if combined with manure, can sustain steady yield increases (see Figure 5). Careful ecological management of crop-livestock systems can create virtuous circles:

> Cowpea thus feeds people and animals directly while also yielding more milk and meat, better soils through nitrogen fixation, and high quality manure, which, used as fertilizer, further improves soil fertility and increases yields.[19]

Forages identified by the International Livestock Research Institute (ILRI) for intercropping have led to 30–100 percent wheat yield increases and up to 300 percent increases in fodder protein while fixing 55–155 kg of nitrogen per hectare.[20]

Participation of Farmers

However, a successful Doubly Green Revolution will not come from the application of biology alone. The first Green Revolution started with the biological challenge inherent in producing new high-yielding food crops and then looked at how the benefits could reach the poor. But this new revolution has to reverse the chain of logic, starting with the socioeconomic demands of poor households and then seeking to identify the appropriate research priorities.

Biologists will have to listen as well as instruct. There will be no easy solutions and few, if any, miracles in the new revolution. Greater food production will come from targeting local agroecosystems and from making the most of indigenous resources, knowledge, and analysis. More than ever before, it will be important to forge genuine partnerships between biologists and farmers.

It will not be enough simply to test new varieties on farmers' fields at the end of the breeding process. Experiments in many parts of the developing world are showing effective ways of involving farmers right at the beginning, in the design of new varieties and in the breeding process itself.

In Rwanda, a five-year experiment involved farmers very early in the breeding process.[21] Beans (*Phaseolus vulgaris* L.) are a key component of the Rwandan diet, and there is an extraordinary range of local varieties—more than 550 have been identified. Farmers (mostly women) are adept at developing local mixtures that breeders have difficulty in bettering. In the experiment, farmers assessed 80 breeding lines over three years, using their own criteria to reduce the number of lines. The farmers tagged favored varieties on the station with colored ribbons. A set of 20–25 lines was then taken to field trials on the farmers' plots. The farmers then chose the best performers and were responsible for multiplying and disbursing them to their neighbors.

Participation has long been a slogan of development. For the first time, effective techniques can make it a reality. Under the heading of participatory learning and action (PLA), there is a formidable array of methods that permit farmers to analyze their own situations and, most important, to engage in productive dialogue with research scientists and extension workers. PLA arose in the late 1980s out of earlier participatory approaches by combining semistructured interviewing and diagram making.[22] It enables rural people to take the lead, producing their own diagrams, undertaking their own analyses, and developing solutions to problems and recommendations for change and innovation. Maps are readily created by simply providing villagers with chalk and colored powder and no further instruction other than the request to produce a map of the village, watershed, or farm. People who are illiterate and barely numerate can construct seasonal calendars using pebbles or seeds. Pie diagrams—pieces of straw and colored powder lain out on an earthen floor—are used to indicate relative sources of income. Such diagrams not only reveal existing patterns but point to problems and opportunities and are seized on by rural people to make their needs felt.

PLA has now spread to most countries of the developing world and been adopted by government agencies, research centers, and university workers as well as by nongovernmental organizations (NGOs). In some ways, it has been a revolution—a set of methodologies, an attitude, and a way of working that has finally challenged the traditional top-down process that has characterized so much development work. Participants from outside find themselves, usually unexpectedly, listening as much as talking, experiencing close to firsthand the conditions of life in poor households, and changing their perceptions about the kinds of interventions and research required. In every exercise, the creation of productive dialogues replaced the traditional position of rural people as passive recipients of knowledge and instruction. A recent report by ActionAid describes a very sophisticated use of maps and preference rankings by the Sanaag people of Somaliland for their community-based livestock development.[23]

Conclusion

It is possible to provide food for all in the 21st century, but there is no simple or single answer. It is not just a matter of producing more or enough food. If hunger is to be banished, the rural poor either have to feed themselves or to earn the income to purchase the extra food they require. This means a new revolution in agricultural and natural resource production aimed at their needs, which cannot be achieved by ecology alone or by biotechnology alone or by a combination of the two. It requires participatory approaches as well—involving farmers as analysts, designers, and experimenters.

There are many hurdles to overcome. The developed world has to agree that there is a serious food problem and to vote the necessary aid monies. There has to be investment in training and research in the fields of agricultural ecology and agricultural biotechnology. Researchers and extension workers in government and the NGOs have to learn to facilitate genuinely participatory development. Governments have to be persuaded to adopt policies that do not discriminate against agriculture and encourage agricultural development and trade.

If all this can be done, then the world can be fed in a way that is not only equitable but also sustainable.

Gordon Conway is president of the Rockefeller Foundation in New York city. His expertise is in the field of agricultural ecology. This article is drawn from his most recent book, *The Doubly Green Revolution: Food for All in the 21st Century* (Ithaca, N.Y.: Cornell University Press, 1998). Conway is a former contributing editor of *Environment*. He can be contacted at the Rockefeller Foundation, 420 Fifth Avenue, New York, NY 10018.

NOTES

1. This article is largely based on G. R. Conway. *The Doubly Green Revolution: Food for All in the 21st Century* (Ithaca, N.Y.: Cornell University Press, 1998).
2. Food and Agriculture Organization, "Food Supplies and Prevalence of Chronic Undernutrition in Developing Regions as Assessed in 1992," Document ESS/MISC/1992 (Rome, 1992).
3. UNICEF, *The State of the World's Children 1998* (Oxford and New York: Oxford University Press for UNICEF, 1998).
4. H. J. Leonard, "Overview: Environment and the Poor," in H. J. Leonard, *Environment and the Poor: Development Strategies for a Common Agenda*, U.S.– Third World Policy Perspectives, no. 11 (Washington. D.C.: Overseas Development Council, 1989); M. Revallion and S. Chen, "What Can New Survey Data Tell Us about Recent Changes in Distribution and Poverty," *World Bank Economic Review* 11 (1997): 357–382; and Uvin, "Tragedy in Rwanda: The Political Ecology of Conflict," *Environment*, April 1996, 6–15, 29.
5. N. Alexandratos, ed., *World Agriculture: Towards 2010: An FAO Study* (Chichester, U.K.: Wiley and Sons, 1995); M. W. Rosengrant, M. Agcaoili-Sombilla, and N. D. Perez, "Global Food Projections to 2020: Implications for Investment, Food, Agriculture, and the Environment," Discussion Paper 5 (Washington, D.C.: International Food Policy Research Institute, 1995); D. O. Mitchell and M. D. Ingco, "Global and Regional Food Demand and Supply Prospects in N. Islam, ed., *Population and Food in the Early Twenty-First Century: Meeting Future Food Demands of an Increasing Population* (Washington, D.C.: International Food Policy Research Institute, 1995): 49–60; W. H. Bender, "How Much Food Will We Need in the 21st Century?," *Environment*, March 1997, 6–11, 27–28.
6. N. S. Raudhawa, "Some Concerns for Future of Punjab Agriculture," mimeo (New Delhi, India: no date).
7. P. L. Pingali and M. W. Rosengrant, "Intensive Food Systems in Asia: Can the Degradation Problems Be Reversed?" (Paper presented at the preconference workshop "Agricultural Intensification, Economic Development and the Environment" of the annual meeting of the American Agricultural Economics Association, Salt Lake City, Utah, 31 July–1 August 1998, in press).
8. C. C. Mann, "Crop Scientists Seek a New Revolution," *Science* 283 (1999): 311–314.
9. K. G. Cassman, "Ecological Intensification of Cereal Production Systems: Yield Potential, Soil Quality, and Precision Agriculture," *Proceedings of the National Academy of Sciences*, USA 96 (1999): 5952–59; and P. L. Pingali and P. W. Heisey, "Cereal Productivity in Developing Countries: Past Trends and Future Prospects," CIMMYT (Centro International de Mejoramiento de Maiz y Trigo) Economics Paper 99-03 (Mexico, 1999).
10. G. R. Conway and J. N. Pretty, *Unwelcome Harvest: Agriculture and Pollution* (London: Earthscan, 1991).
11. Ibid.
12. G. R. Conway, U. Lele, J. Peacock, and M. Pineirn, *Sustainable Agriculture for a Food Secure World* (Washington, D.C.: Consultative Group on Agricultural Research; and Stockholm, Sweden: Swedish Agency for Research Co-operation with Developing Countries, 1994).

13. J. Peng et al., " 'Green Revolution' Genes Encode Mutant Gibberellin Response Modulators," *Nature* 400 (1999): 256–261.
14. Mann, note 8 above and C. C. Mann, "Genetic Engineers Aim to Soup Up Crop Photosynthesis," *Science* 283 (1999): 314–316.
15. X. D. Ye et al., "Engineering the Complete Provitamin A (beta-carotene) Biosynthetic Pathway into (carotenoid-free) Rice Endosperm," *Science* (in press) (1999).
16. G. R. Conway and G. Toenniessen, "Feeding the World in the 21st Century," *Impacts of Foreseeable Science,* Supplement to *Nature* 402, C55–C58 (1999).
17. Royal Society, *Genetically Modified Plants for Food Use* (London: The Royal Society, 1998); Nuffield Council on Bioethics, *Genetically Modified Crops: The Ethical and Social Issues* (London, 1999); Food and Agriculture Organisation, *Biotechnology and Food Safety,* FAQ Food and Nutrition Paper 61, (Rome: World Health Organization and Food and Agriculture Organization, 1996); and L. Levidow, "Regulating Bt Maize in the United States and Europe: A Scientific-Cultural Comparison," *Environment,* December 1999, 10–22.; R. M. May, "Genetically Modified Foods: Facts, Worries, Policies, and Public Confidence," (www.2.dti.gov.uk/ost/ostbusiness/gen.html), 1999.
18. P. Kenmore, *How Rice Farmers Clean Up the Environment, Conserve Biodiversity, Raise More Food, Make Higher Profits: Indonesia's IPM—Model for Asia* (Manila, Philippines: Food and Agriculture Organisation, 1991); K. D. Gallagher, P. E. Kenmore, and K. Sogawa, "Judicial Use of Insecticides Deters Planthopper Outbreaks and Extends the Life of Resistant Varieties in Southeast Asian Rice," in R. F. Denno and T. J. Perfect, *Ecology and Management of Planthoppers* (1994): 599–614; and R. Stone, "Researchers Score Victory over Pesticides—and Pests—in Asia," *Science* 256 (1992): 5057.
19. International Livestock Research Institute, *Livestock and Nutrient Cycling* (Nairobi, Kenya, 1999).
20. International Livestock Research Institute, *Improving Smallholder Farming Through Animal Agriculture* (Nairobi, Kenya, 1999).
21. L. Sperling and U. Scheidegger, *Participatory Selection of Beans in Rwanda: Results, Methods, and Institutional Issues,* Gatekeeper Series, no. 51 (London: International Institute for Environment and Development, 1995).
22. Conway, note 1 above.
23. ActionAid, *Programme Review, June 1999, by Sanoag Community-Based Organisation* (London, 1999).

Escaping Hunger, Escaping Excess

The big myth of malnutrition is that it's a problem of poor countries. But in a world at once rich in food and filled with poverty, malnutrition now has many faces—all over the world.

by Gary Gardner and Brian Halweil

TODAY, ETHIOPIA AND ITS neighbors are once again in the grip of an unrelenting famine, which has left more than 16 million people on the brink of starvation. After a massive international mobilization to aid this region in the 1980s, the Horn of Africa has become synonymous with famine and malnutrition. But across the Atlantic Ocean, another country is currently facing an epidemic that has left not *tens* of millions, but more than *100* million people malnourished—a quarter of them morbidly so. This growing problem receives little attention as a public health disaster, despite warnings from health officials that malnourishment has reached epidemic levels and has left vast numbers of people sick, less productive, and far more likely to die prematurely.

In this country—the United States—55 percent of adults are overweight and 23 percent are obese. The medical expenses and lost wages caused by obesity cost the country an estimated $118 billion each year, the equivalent of 12 percent of the annual health budget. Being overweight and obese are major risk factors in coronary heart disease, cancer, stroke and diabetes. Together these diseases are the leading killers in the United States, accounting for half of all deaths.

Misconceptions of hunger and overeating abound worldwide. We tend to think of hunger as resulting from a desperate scarcity of food, and we imagine it occurring only in poor countries. However, in those nations in Africa and South Asia where hunger is most severe, there is often plenty of food to go around. And even food rich nations are home to many underfed people.

Meanwhile, as the concept of malnutrition stretches to encompass excess as well as deficiency, wealthy nations are seeing rates of malnourishment that rival those in desperately poor regions. And overeating is growing in poorer nations as well, even where hunger remains stubbornly high. In Colombia, for example, 41 percent of adults are overweight, a prevalence that rivals rates found in Europe. While hunger is a more acute problem and should be the highest nutritional concern, overeating is the fastest growing form of malnourishment in the world, according to the World Health Organization (WHO). For the first time in history, the number of overweight people rivals the number who are underweight, both estimated at 1.1 billion.

Because myth and misconception permeate the world's understanding of malnutrition, policy responses have been wildly off the mark in addressing the problem. Efforts to eliminate hunger often focus on technological quick fixes aimed at boosting crop yields and producing more food, for example, rather than addressing the socioeconomic causes of hunger, such as meager incomes, inequitable distribution of land, and the disenfranchisement of women. Efforts to reduce overeating single out affected individuals—through fad diets, diet drugs, or the like—while failing to promote prevention and education about healthy alternatives in a food environment full of heavily marketed, nutritionally suspect, "supersized" junk food. The result: half of humanity, in both rich and poor nations, is malnourished today, according to the WHO. And this is in spite of recent decades of global food surpluses.

Malnutrition has become a significant impediment to development in rich and poor countries alike. At the individual level, both hunger and obesity can reduce a person's physical fitness, increase susceptibility to illness, and shorten lifespan. In addition, children deprived of adequate nutrients during development can suffer from permanently reduced mental capacity. At the national level, poor eating hampers educational performance, curtails economic productivity, increases the burden of health care, and reduces general well-being. Confronting this epidemic of poor eating will have widespread benefits, but first the myths that obscure the causes of malnutrition must be dispelled.

The Scarcity Myth

IN THE EARLY 1980s, the world was flooded by news of hunger and death from the Horn of Africa. By 1985, nearly 300,000 people had died. But international observers paid little attention to the fact that in the midst of famine, these countries were exporting cotton, sugar cane, and other cash crops that had been grown on some of the country's best agricultural land. While only 30 percent of farmland in Ethiopia was affected by drought, ubiquitous images of emaciated people surrounded by parched land have served to reinforce the single largest myth about malnutrition: that hunger results from a national scarcity of food.

Indeed, for more than 40 years the world has produced regular and often bountiful food surpluses—large enough, in fact, to prompt major producing countries like the United States to pay farmers *not* to farm some of their land. Indeed, the Food and Agriculture Organization (FAO) estimates that 80 percent of hungry children in the developing world live in countries that produce food surpluses. And only about a quarter of the reduction in hunger between 1970 and 1995 could be attributed to increasing food availability per person, according to a study by the International Food Policy Research Institute (IFPRI).

This is not to say that scarcity might not one day become the principal source of hunger, as population growth and ongoing damage to farmland and water supplies shrink food availability per person in many countries. Countries like Nigeria and Pakistan, which are on track to double their populations in the next 50 years, have already seen stocks of surplus food erode steadily in the 1990s. And countries such as India, which overpump groundwater to prop up agricultural production, will be hard pressed to maintain self sufficiency once aquifers run dry or become uneconomical to pump. But for the billion or so people who are hungry *today*, the finger of blame points in other directions.

Hands down, the major cause of hunger is poverty—a lack of access to the goods and services essential for a healthy life. Where people are hungry, it's a good bet that they have little income, cannot gain title to land or qualify for credit, have poor access to health care, or have little or no education. Worldwide, 150 million people were unemployed at the end of 1998, and as many as 900 million had jobs that paid less than a living wage. These billion-plus people largely overlap with the 1.1 billion people who are underweight, and for whom hunger is a chronic experience. And nearly 2 billion more teeter at the edge of hunger, surviving on just 2 dollars or less per day, a large share of which is spent on food.

Hunger, like its main root, poverty, disproportionately affects females. Girls in India, for example, are four times as likely to be acutely malnourished as boys. And while 25 percent of men in developing countries suffer from anemia, a condition of iron deficiency, the rate is 45 percent for women—and 60 percent for those who are pregnant. This gender bias stems from cultural prejudices in households and in societies at large. Most directly, lean rations at home are often dished out to father and sons before mother and daughters, even though females in developing countries typically work longer hours than males do. Gender bias is also manifest in education. Inequitable schooling opportunities for girls lead to economic insecurity: women represent two-thirds of the world's illiterate people and three-fifths of its poor. With fewer educational and economic opportunities than men, women tend to be hungrier and suffer from more nutrient deficiencies.

Any serious attack on hunger, therefore, will aim to reduce poverty, and will give special attention to women.

Supersized

Food portions have steadily grown in recent decades. A standard serving of soda in the 1950s, for example, was a 6.5-ounce bottle. Today the industry standard is a 20-ounce bottle. "Supersizing" has evolved as a marketing strategy that costs food producers little, but appears to give significant added value to consumers. But this trend skews perceptions of normal servings: in the United States, one study found that participants consistently labeled as "medium" portions that were double or triple the size of recommended portions.

Food companies tend to push fatty or sweet foods for two reasons: they know we have an innate preference for them, and highly processed foods like white hamburger buns offer greater profits than more elemental products like fruits and vegetables. Adding more sugar, salt, fats, or oils (as typically concentrated in prepared mustard, ketchup, or pickles), can provide a tasty and profitable product that is often irresistible to consumers and companies alike.

The overeating phenomena is quickly spreading around the world, in part due to heavy advertising. Food companies spend more than any other industry on advertising in the United States. Coca-Cola and McDonald's are among the 10 largest advertisers in the world. This strategy has paid off: McDonald's opens five new restaurants every day, four of them outside the United States.

Americans consume 70 kilograms of caloric sweeteners per year—75 percent more than in 1909. That is nearly 200 grams or 53 teaspoons a day, the equivalent of a 5 pound bag of sugar every week and a half. In Europe and North America, fat and sugar count for more than half of all caloric intake, squeezing complex carbohydrates like grains and vegetables down to about one-third of total calories.

The IFPRI study on curbing malnutrition found that improving women's education and status together accounted for more than half of the reduction in malnutrition between 1970 and 1995. Such nutritional leverage stems from a woman's pivotal role in the family. A woman "eats for two" when she is pregnant and when she is nursing; pull her out of poverty, and improvements in her nutrition are passed on to her infant. But there's more: studies show that provided with an income, a woman will spend nearly all of it on household needs, especially food. The same money in a man's pocket is likely to be spent in part—up to 25 percent—on non-family items, such as cigarettes or alcohol.

From this perspective, microcredit initiatives, such as those of the Bangladesh-originated Grameen Bank, offer a promising means of combatting hunger. These unconventional programs provide small loans of tens or hundreds of dollars to help very poor women generate income through basket-weaving, chicken-raising, or other small projects. As the loans lift women out of poverty, they also yield nutritional benefits: a 10 percent increase in a woman's Grameen borrowing, for example, has been shown to produce a 6 percent increase in the arm circumference of her children (a measure of nutritional well-being). It also increases by 20 percent the likelihood that her daughter will be enrolled in school, which lowers the girl's risk of suffering malnutrition as an adult.

International support for such programs could expand them dramatically. One option is the nonprofit Microcredit Summit's campaign to raise $22 billion to increase the number of microcredit beneficiaries from 8 million in the late 1990s to 100 million by 2005. Such investments are a high-leverage option for a nation's foreign aid commitment, given all of the benefits—improved nutrition, better health, and slower rates of population growth—that come from reducing poverty, especially among women.

At a broader social level, the journey out of poverty and hunger can be expedited through better access to land and agricultural credit. These measures are especially important for women, since they produce more than half of the world's food, and a large share of what is consumed in rural households in developing countries. In India, Nepal, and Thailand, less than 10 percent of women own land, and those who do often have small, marginal tracts. For landless women, credit is next to impossible to obtain: in five African countries—Kenya, Malawi, Sierra Leone, Zambia, and Zimbabwe—where women constitute a large share of farmers, they receive less than one percent of the loans provided in agriculture. This despite their exceptional creditworthiness: women typically pay their debts more faithfully than men do.

Women also need access to sound nutritional information as a way to avoid nutritional impoverishment and unnecessary food expenditures. Breastfeeding campaigns, for example, can highlight the many advantages of this free and wholesome method of infant feeding. Baby formulas are often prepared in unsanitary conditions or watered down to reduce costs. Campaigns to promote breastfeeding and restrict sales of formula have been estimated to reduce illness from diarrhea—a condition that robs infants of needed vitamins and minerals—by 8 to 20 percent. They have reduced deaths from diarrhea by 24 to 27 percent. Breastfeeding also acts as a natural contraceptive following pregnancy, spacing births at greater intervals and thereby easing the pressure to feed everyone in poor families.

Nutritional education efforts are also essential to fighting hunger, and the most successful programs involve entire communities by enlisting affected people and local leaders. The BIDANI program in the Philippines, for example, provides orientation and training for villagers to participate in nutritional "interventions," which have worked to elevate 82 percent of enrolled children to a higher nutritional status. A similar program in Gambia substantially cut the death rate among women and children by working with the highly respected women elders of the matriarchal Kabilo tribe to educate community members about child-feeding practices, hygiene, and maternal health care.

Important as these social initiatives are for improving nutrition, more direct action is often required to meet the needs of those who suffer from hunger today. Even here, however, creative approaches can empower women and aid entire communities. In one simple case in Benin, food aid is dispensed not directly to families, but to girls at school, who bring it home to their parents. The practice combats the cultural bias against girls found in many countries, which often results in their removal from school at a young age to help at home or to allow a brother to get an education. It achieves two critical nutritional goals: it gets food to families that need it, and it increases girls' future employment prospects, which in turn reduces the likelihood of future malnutrition.

The Prone-to-Obesity Myth

FOR THOSE WHO HAVE ACCESS to enough food, eating habits around the world are in the midst of the most significant change since the development of agriculture thousands of years ago. Since the turn of the century, traditional diets featuring whole grains, vegetables, and fruits have been supplanted by diets rich in meat, dairy products, and highly processed items that are loaded with fat and sugar. This shift, already entrenched in industrial countries and now accelerating in developing nations as incomes rise, has created an epidemic of overeating and sparked a largely misunderstood public health crisis worldwide. In the United States, the leader in this global surge toward larger waist sizes, more than half of all adults are now overweight—a condition that, like hunger, increases susceptibility to disease and disability, reduces worker productivity, and cuts lives short.

The proliferation of high-calorie, high-fat foods that are widely available, heavily promoted, low in cost and nutrition, and served in huge portions has created what Yale psychologist Kelly Brownell calls a "toxic food environment." Sweets and fats increasingly crowd out nutritionally complete foods that provide essential micronutrients. For instance, one-fifth of the "vegetables" eaten today in the United States are servings of french fries and potato chips. Our propensity to eat sweet and fatty foods may have served our ancestors well for weathering seasonal lean times, but amidst unbridled abundance for many, it has become a handicap. When these eating habits are combined with increasingly urbanized, automated, and more sedentary lifestyles, it becomes clear why gaining weight is often difficult to avoid.

Failure to recognize the existence of this negative food environment has created the widespread misconception that individuals are entirely to blame for overeating. The reality is most countries embrace policies and practices that promote mass overconsumption of unhealthy foods, but abandon citizens when it comes to dealing with the health implications. Because individuals are stigmatized as weak-willed or prone to obesity, prevailing efforts to curb overeating have focused on techno-fixes and diets, not prevention and nutrition education.

This end-of-the-pipe mentality manifests itself in a variety of ways: liposuction is now the leading form of cosmetic surgery in the United States with 400,000 operations performed each year; fad diet books top the bestseller lists; designer "foods" such as olestra promise worry-free consumption of nutritionally empty snacks; and laboratories scurry to find the human "fat gene" in an effort to engineer our way out of obesity. While the U.S. Agriculture Department spends $333 million each year to educate the public about nutrition, the U.S. diet and weight-loss industry records annual revenues of $33 *billion*. And the highly lucrative weight-loss business feeds off of a global food industry that now has significant influence over food choices around the world.

Indeed, consumers get the majority of their dietary cues about food from food companies, who spend more on advertising—$30 billion each year in the United States alone—than any other industry. The most heavily advertised foods, unfortunately, tend to be of dubious nutritional value. And food advertisers disproportionately target children, the least savvy consumers, in order to shape lifelong habits. In fact, in the United States, the average child watches 10,000 commercials each year, more than any other segment of the population. And more than 90 percent of these ads are for sugary cereals, candy, soda, or other junk food, according to surveys by the Center for Science in the Public Interest.

Numerous studies show that these ads work. They prompt children to more frequently request, purchase, and consume advertised foods, even when they become adults. And as kids fill up on items loaded with empty calories like soda or candy, more nutritious items are squeezed out of the diet. Marketing to children has intensified in recent years as food companies have begun to target the school environment. More than 5,000 U.S. schools—13 percent of the country's total—now have contracts with fast-food establishments to provide either food service, vending machines, or both. Since 1990, soda companies have offered millions of dollars to cash-strapped school districts in the United States for exclusive rights to sell their products in schools.

With industrial country markets increasingly saturated, many food corporations are now looking to developing countries for greater profits. Mexico recently surpassed the United States as the top per capita consumer of Coca-Cola, for example. And that company's 1998 annual report notes that Africa's rapid population growth and low per capita consumption of carbonated beverages make that continent "a land of opportunity for us." The number of U.S. fast-food restaurants operating around the world is also growing rapidly: four of the five McDonald's restaurants that open every day are located outside the United States.

Overeating is also becoming a problem even in countries where hunger and poverty persist. In China, for example, consumption of high-fat foods such as pork and soy oil (which is used for frying) both soared after the economic boom of the 1980s, while consumption of rice and starchy roots dropped—changes that were most pronounced among wealthier households. The parallel trend of urbanization in the developing world also means ex-

---

Targeting Women

Hungry children are often scarred for life, suffering impaired immune systems, neurological damage, and retarded physical growth. Infants that are underweight *in utero* will be five centimeters shorter and five kilograms lighter as adults.

Chronic hunger leaves children and adults more susceptible to infectious diseases. Among the five leading causes of child death in the developing world, 54 percent of cases have malnutrition as an underlying cause.

Conflict and military spending exacerbate hunger directly by disrupting economies and food production, and indirectly by diverting funds away from poverty alleviation to militaries.

Where hunger exists, women are invariably more malnourished than men. In India, for example, girls are four times as likely to be hungry or suffer from micronutrient deficiencies as boys are. Hungry women bear and raise hungry children. Because impoverished families are less able to care for their offspring, hunger is perpetuated across generations.

Women produce more than half of the world's food, and in rural areas they provide the lion's share of food consumed in their own homes. Yet, note who's in control here. Women often cannot obtain access to land, credit, or the social and political support that men can.

posure to new foods and food advertising—particularly for highly processed and packaged items—and considerably more sedentary lifestyles. A recent study of 133 developing countries found that migration to the city—without any changes in income—can more than double per capita intake of sweeteners. Cash-squeezed households in Guayaquil, Ecuador, often spurn potatoes and fresh fruit juices in favor of fried plantains, potato chips, and soft drinks, replacing nutrient-dense foods with empty calories.

A world raised on Big Macs and soda isn't inevitable. But countering an increasingly ubiquitous toxic food environment will require dispelling the myths that surround overeating. Governments will have to recognize the existence of a health epidemic of overeating, and will have to work to counter the social pressures that promote poor eating habits. Empowering individuals through education about nutrition and healthy eating habits, particularly for children, is also essential.

If preventing overeating is the goal, rather than treating it after habits have been formed, then the school environment is an obvious place to start. In Singapore, for example, the nationwide Trim and Fit Scheme has reduced obesity among children by 33 to 50 percent, depending on the age group, by instituting changes in school catering and increasing nutrition and physical education for teachers and children. Similar programs in other countries have found comparable results, yet physical education programs in many nations are actually being scaled back.

Mass-media educational campaigns can also change long-standing nutritional habits in adults. Finland launched a campaign in the 1970s and 1980s to reduce the country's high incidence of coronary heart disease, which involved government-sponsored advertisements, national dietary guidelines, and regulations on food labeling. This broad, high-profile approach—it also advocated an end to smoking, and involved groups as diverse as farmers and the Finnish Heart Association—increased fruit and vegetable consumption per person two-fold and slashed mortality from coronary heart disease by 65 percent between 1969 and 1995. About half of the drop in mortality is credited to the lower levels of cholesterol induced by the nutrition education campaign.

A public health approach to overeating might also take some hints from successful campaigns against smoking, including warning labels and taxes to deter consumption. In Finland, the government now requires "heavily salted" to appear on foods high in sodium, while allowing low-sodium foods to bear the label "reduced salt content." A complement to the "low-fat" labels that grace so many new food products would be a more ominous "high-fat" or "high-sugar" label.

Consumption of nutrient-poor foods can be further reduced by fiscal tools. Yale's Kelly Brownell advocates adoption of a tax on food based on the nutrient value per calorie. Fatty and sugary foods low in nutrients and loaded with calories would be taxed the most, while fruits and vegetables might escape taxation entirely. The idea is to discourage consumption of unhealthy foods—and to raise revenue to promote healthier alternatives, nutrition education, or exercise programs, in essence to make it easier and cheaper to eat well. Large-scale cafeteria and vending machine studies show how powerful an influence price has on buying choices—reducing the price and increasing the selection of fruit, salad, and other healthy choices can often double or triple purchase of these items, even as total food purchases remain the same.

Such a tax is also justified as the cost of overeating to society grows. Graham Colditz at Harvard estimates the direct costs (hospital stays, medicine, treatment, and visits to the doctor) and indirect costs (reduced productivity, missed workdays, disability pensions) of obesity in the United States to be $118 billion annually. This sum, equal to nearly 12 percent of the U.S. annual health budget, is more than double the $47 billion in costs attributable to cigarette smoking—a better known and heavily taxed drag on public health. Fiscal measures to reduce overeating may be most attractive to developing countries, which must tackle growing caseloads of costly chronic diseases even as they struggle to eradicate infectious illness.

Putting the Pieces Together

THE EFFECTS OF POOR NUTRITION run deep into every aspect of a community, curtailing performance at school and work, increasing the cost of health care, and reducing health and well-being. By the same token, improving nutrition promises to have equally far-reaching, positive impacts on regions that choose to address the problem. Better eating can set into motion a host of other benefits, many in areas seemingly unrelated to food.

For this to occur, however, efforts to improve nutrition must be integrated into all aspects of a country's development decisions—from health care priorities to transportation funding to curricula planning for schools. A cleaner water supply, for example, would reduce the incidence of intestinal parasites that hamper the body's capacity to absorb micronutrients. Thus a ministry of public works dedicated to increasing access to clean water is a logical partner in a campaign to reduce micronutrient deficiencies. Similarly, transportation officials who promote bicycle commuting, ministers of culture who discourage TV watching, and an agriculture ministry that promotes nutritional education are all promoting lifestyles that, in conjunction with better eating, can reduce incidences of obesity.

There are numerous less obvious means, as well, by which nutritional improvement can be woven into daily life. To begin with, smart nutrition policies can be added to already-existing social programs. Health, education, and agricultural extension programs already reach deep into nutritionally vulnerable populations through existing networks of clinics, schools, and rural development offices. Nutrition is a natural outgrowth of their current responsibilities. Clinic staff, for example, could promote breast-feeding, and extension agents could encourage

2 ❖ WORLD'S POPULATION: PEOPLE AND HUNGER

Nutrition Split

Every region in the world now has large numbers of hungry or overweight people — or both — as affluence spreads and poverty persists.

Some of the clearest evidence that hunger is caused by poverty and not regional food scarcity is the presence of hunger in the **United States**. In 1998, 10 percent of U.S. households, home to nearly one in five American children, were "food insecure" — hungry, on the edge of hunger, or worried about being hungry.

From 1980 to 2000, the share of children who are underweight in **Latin America and the Caribbean** has dropped from 14 percent to 6 percent. But it seems this region has simply traded one form of poor eating for another: in most Latin American nations, the overweight population now exceeds the underweight population.

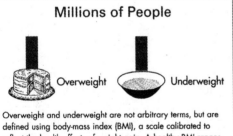

Millions of People

Overweight | Underweight

Overweight and underweight are not arbitrary terms, but are defined using body-mass index (BMI), a scale calibrated to reflect the health effects of weight gain. A healthy BMI ranges from 19 to 24; a BMI of 25 or above indicates "overweight" and brings increased risk of illnesses such as heart disease, stroke, diabetes, and cancer. A BMI above 30 signals "obesity" and even greater health risks. BMI is calculated as a person's weight in kilos divided by the square of height in meters.

home gardening. Such partnering is cost effective, not only because it uses existing infrastructure, but also because it often reduces the need for the original service. Women educated about breastfeeding on a prenatal visit, for example, are less likely to return months later with an infant suffering from diarrhea.

Programs intended to eradicate poverty, from microlending to employment creation, are most likely to raise nutritional levels when accompanied by education about health and nutrition. A "Credit with Education" program initiated in Ghana by the international group Freedom from Hunger coupled lending with education about breastfeeding, child feeding, diarrhea prevention, immunization, and family planning. A three-year follow-up study documented improved health and nutrition practices, fewer and shorter-lived episodes of food shortages, and dramatic improvements in childrens' nutrition among the participants compared with the control groups. Barbara MkNelly, the program's coordinator, warns that "simply improving a family's ability to buy food is no guarantee that poor baby-feeding practices, dietary choices or living conditions will not undercut nutritional gains."

8. Escaping Hunger, Escaping Excess

European levels of overeating are not far behind those of North America. The share of the adult population in Russia, Germany, and the United Kingdom that is overweight is roughly half, while the share in other European nations tends to be slightly lower.

Much of the **Middle East** faces an overeating crisis of North American proportions. But in poorer, war-torn nations, like Iraq and the Sudan, hunger reaches the desperate levels found in southern Africa.

Along with Sub-Saharan Africa, **South Asia** is home to a massive concentration of hungry people. Some 44 percent of the region's children are underweight, while the shares in India, Bangladesh, and Afghanistan are well above this average. At the same time, among the urban upper-class of this region, obesity is a growing problem.

Like Latin America, **East and Southeast Asia** have seen significant decreases in the share of the population that is hungry. Yet hunger remains stubbornly fixed in some countries, and overeating is spreading rapidly. The share of adults who are overweight in China jumped by more than half—from 9 percent to 15 percent—between 1989 and 1992.

The share of the world's population that is underweight is in decline, except in **sub-Saharan Africa,** where 36 percent of children are underweight due to poverty and other social factors.

The city of Curitiba, Brazil has even found links between nutrition and the city's waste flows. Concerned about the city's growing waste burden, and about malnutrition among the poorest sectors of the population, officials established a recycling program for organic waste that benefits farmers, the urban poor, and the city in general. City residents separate their organic waste from the rest of their garbage, bag it, then exchange it for fresh fruits and vegetables from local farmers at a city center. The city reduces its waste flow, farmers reduce their dependence on chemical fertilizer, and the urban poor get a steady supply of nutritious foods.

In any society, but especially where food cues come primarily from advertising, education is critical to making progress toward good nutrition. In the United States, the Berkeley Food System Project, for example, not only teaches kids about healthy eating, but promotes the use of vegetable gardens in school to help children learn about food at the source. The gardens also supply some of the food for school cafeterias, which were required in 1999 to begin serving all-organic lunches. The project encourages schools to incorporate this comprehensive view of food into their classwork. Janet Brown, who spearheaded the project for the Center for Ecoliteracy, explains that kids weaned on packaged and processed foods often shy away from fruits and veggies, because they have not been properly introduced. "But when a child pops a cherry tomato that she helped to grow into her mouth, then introducing a salad bar in the cafeteria is likely to be more successful."

Nutritional literacy is not just for kids, however. Doctors, nurses, and other health care professionals are well positioned to educate patients about the links between diet and health, and can be instrumental in improving eating habits. But modern medical systems often de-emphasize the role of nutrition: in the United States, only 23 percent of medical schools required that students take a separate course in nutrition in 1994. Doctors poorly trained in nutrition are less likely to take a preventive approach to health care, such as encouraging greater consumption of fruits and vegetables or increased physical activity, and are more likely to deal only with the consequences of poor eating—prescribing a cholesterol-lowering drug, for example, or scheduling bypass surgery. A recent U.S. survey by the Centers for Disease Control found "less than half of obese adults report being advised to lose weight by health care professionals."

Beyond educating medical professionals, health care as a whole could integrate nutrition by recognizing obesity as a disease and covering weight-loss programs and other nutritional interventions. Covering these expenses would not only reduce illness and patient suffering, but is likely to cut health care costs. An encouraging first step in this direction is Mutual of Omaha's decision to cover intensive dietary and lifestyle modification program of patients with heart disease, an initiative they hope will eliminate costly prescriptions and prevent surgeries months or years down the road. A local next step for the industry might be to cover regular nutrition checkups, akin to dental check-ups, as part of a basic insurance coverage.

Where communities have lost access to healthy food options, improving diets may require involving players throughout the food chain. Support of urban agriculture and urban farmers' markets has proven effective in getting good food to low-income urbanites. Urban gardens in Cuba, which meet 30 percent of the vegetable demand in some cities, have prospered under government nurturing. In the nutritionally impoverished inner cities of wealthier nations, farmers' markets are often the only source of fresh produce, as green grocers and supermarkets have left for the more affluent suburbs, and as fast food joints and convenience stores have replaced them. The Toronto Food Policy Council has used both farmers' markets and produce delivery schemes to connect local farmers and low-income urban residents, many of whom are single mothers. Some 70 percent of those buying food now eat more vegetables than they did when the program began in the early 1990s; 21 percent eat a greater variety;

and 16 percent now try new foods. More people also know about the recommended five or more servings a day of fruit and veggies.

Eliminating poor eating is the business of fiscal authorities as well. The food tax advocated by Kelly Brownell could raise funds for nutritional interventions. Michael Jacobsen, director of the Center for Science in the Public Interest, notes that even small taxes could generate sufficient revenues to fund "television advertisements, physical education teachers, bicycle paths, swimming pools, and other obesity prevention measures." In the United States, a $2/3$-cent tax per can of soda, a 5 percent tax on new televisions and video equipment, a $65 tax on each new motor vehicle, or an extra penny tax per gallon of gasoline would each raise roughly $1 billion each year.

Even without such a tax, authorities in some countries have begun to encourage lifestyle changes that are important complements to good nutrition. Australia's Department of Transport and Regional Services, Department of Health and Aged Care, and Department of Environment and Heritage teamed up in 1999 to promote the country's National Bicycling Strategy, which seeks to raise the level of cycling in the country. The involvement of this diverse set of government agencies demonstrates the broad impact that a commitment to good nutrition can have. More cycling means more exercise, an indispensable tool in the fight against overweight. But it can also mean cleaner air, less congested cities, and cheaper transportation infrastructure.

Where's the Nutrition?

Food advertisers disproportionately target children, the least savvy consumers. In the United States, the average child is bombarded with 10,000 commercials each year—90 percent of them for sugary cereals, candy, or other junk foods.

Junk foods often displace more nutritious foods, providing only "empty calories"—energy with little nutritional value. In the United Kingdom, per capita consumption of snack foods is up by nearly a quarter in the past five years—snack foods are now a $3.6 billion industry.

Eating in Industrial countries centers less than ever before on home and family. In 1998, just 38 percent of meals in U.S. homes were homemade, and one out of every three meals were eaten outside of the home.

Nutritionally poor foods are invading U.S. schools. Fast food companies have contracts, often worth millions of dollars, to provide food service or vending machines, at more than 5,000 U.S. schools. One deal prompted a Colorado school district to push Coca-Cola consumption, even in classrooms, when sales fell below contractual obligations.

A final part of reshaping the food environment is recultivating an appreciation of food as a cultural and nutritional treasure. The consumer culture, applied to eating, emphasizes brand allegiance and megameals, often at the expense of nutrition and health. Groups like the Slow Food Movement, based in Italy, and the Oldways Preservation and Exchange Trust in the United States, offer a postmodern critique of today's culinary norm by promoting a return to the art of cooking traditional foods and of socializing around food. Their work, which targets chefs as well as consumers, is the kind of cultural intervention that could help more people shift to a healthy diet, similar to the change in consciousness that encouraged a shift away from smoking in the United States. Government encouragement of these groups, perhaps through assistance with marketing and promotional activities, would insure that this important work benefits everyone, not just the affluent.

The experience of the Slow Food Movement and Oldways shows that as people care more about their food choices, their concerns are likely to evolve well beyond nutritional value. Health-conscious consumers often gravitate toward organic produce, in an effort to avoid agrochemical residues and to stop promoting farm practices that deplete the soil or pollute waterways. Many also reduce their consumption of animal products, which can reduce their intake of fat and cholesterol, but also eases the pressure on land and water resources. And these consumers are likely to seek out local food sources, which offer superior freshness and quality, as well as the opportunity to know the farmer and his methods.

The far-reaching effects of nutrition make it a central factor in personal and national development. Poor eating is as much a drag on national economic activity as it is on personal health. The reverse is also true: development choices, such as whether girls have as many years of schooling as boys, or whether food corporations are free to advertise without limit to young consumers, heavily influence what and how we eat.

Hidden Hunger

Hunger has been alleviated somewhat in the past 20 years, except in Africa.

Micronutrient deficiencies plague between 2 and 3.5 billion people around the world, including a considerable number of both the 1.1 billion who are hungry and the 1.1 billion who are overweight. Micronutrients—vitamins and minerals such as iron, calcium, and vitamins A through E—are crucial elements of a healthy diet.

Food aid is not the long-term answer for most of the world's hungry. Nearly 80 percent of all malnourished children in the developing world live in countries that have food surpluses. Today, hunger is the product of human decisions—people are denied access to food as a result of poverty and other social inequities, not as a result of net scarcity.

Deficiencies in nutrients such as iodine can stunt physical and mental growth. More than 740 million people—13 percent of the world—suffer from iodine deficiency, which is the most common preventable cause of mental retardation. Vitamin A deficiency is the world's leading cause of blindness. Iron deficiency, prevalent in 56 percent of women in developing countries who are pregnant, causes anemia, which can stunt the development of the fetus.

Gary Gardner is a senior researcher and Brian Halweil is a staff researcher at the Worldwatch Institute. They are co-authors of Worldwatch Paper 150, *Overfed and Underfed: The Global Epidemic of Malnutrition* (2000).

Unit 3

Unit Selections

9. **King Coal's Weakening Grip on Power,** Seth Dunn
10. **Oil, Profit$, and the Question of Alternative Energy,** Richard Rosentreter
11. **Here Comes the Sun: Whatever Happened to Solar Energy?** Eric Weltman
12. **The Hydrogen Experiment,** Seth Dunn
13. **Power Play,** Business Week
14. **Bull Market in Wind Energy,** Christopher Flavin

Key Points to Consider

❖ Why is coal such a costly fuel source once all the costs of its use are calculated? Describe some of the costs of coal use in terms of environmental quality, public health, and global climate.

❖ What is the relationship between the economic and political power of large energy corporations and the automotive industry and the weakening trend toward research and development of alternative energy sources? Are there situations in which it can be to the advantage of multinational corporations to encourage the use of solar power and other alternative energies?

❖ How can hydrogen power be created from the most abundant element in the universe without subsequent environmental damage? What are the prospects for success of the Icelandic experiment with hydrogen power and is there reason to believe that hydrogen as a fuel could also prove useful in large countries such as the United States?

❖ What are some of the major benefits of such alternate energy sources as solar power and wind power? Do these energy alternatives really have a chance at competing with fossil fuels for a share of the global energy market?

 **Links**   www.dushkin.com/online/

15. **Alternative Energy Institute, Inc.**
    http://www.altenergy.org
16. **Communications for a Sustainable Future**
    http://csf.colorado.edu
17. **Energy and the Environment: Resources for a Networked World**
    http://zebu.uoregon.edu/energy.html
18. **Institute for Global Communication/EcoNet**
    http://www.igc.org/igc/gateway/
19. **U.S. Department of Energy**
    http://www.energy.gov

These sites are annotated on pages 4 and 5.

Energy: Present and Future Problems

There has been a tendency, particularly in the developed nations of the world, to view the present high standards of living as exclusively the benefit of a high-technology society. In the "techno-optimism" of post–World War II years, prominent scientists described the technical-industrial civilization of the future as being limited only by a lack of enough trained engineers and scientists to build and maintain it. This euphoria reached its climax in July 1969 when American astronauts walked upon the surface of the Moon, an accomplishment brought about solely by American technology—or so it was supposed. It cannot be denied that technology has been important in raising standards of living and permitting Moon landings, but how much of the growth in living standards and how many outstanding and dramatic feats of space exploration have been the result of technology alone? The answer is few—for in many of humankind's recent successes, the contributions of technology to growth have been no more important than the availability of incredibly cheap energy resources, particularly petroleum, natural gas, and coal.

As the world's supply of recoverable (inexpensive) fossil fuels dwindles and becomes more important as an agency of international diplomacy, it becomes increasingly clear that the energy dilemma is the most serious economic and environmental threat facing the Western world and its high standard of living. With the exception of the population problem, the coming fossil fuel energy scarcity is probably the most serious threat facing the rest of the world as well. The economic dimensions of the energy problem are rooted in the instabilities of monetary systems produced by and dependent upon inexpensive energy. The environmental dimensions of the problem are even more complex, ranging from the hazards posed by the development of such alternative sources as nuclear power to the inability of developing world farmers to purchase necessary fertilizer produced from petroleum, which has suddenly become very costly, and to the enhanced greenhouse effect created by fossil fuel consumption. The only answers to the problems of dwindling and geographically vulnerable, inexpensive energy supplies are conservation and sustainable energy technology. Both require a massive readjustment of thinking, away from the exuberant notion that technology can solve any problem. The difficulty with conservation, of course, is a philosophical one that grows out of the still-prevailing optimism about high technology. Conservation is not as exciting as putting a man on the Moon. Its tactical applications—caulking windows and insulating attics—are dog-paddle technologies to people accustomed to the crawl stroke. Does a solution to this problem entail the technological fixes of which many are so enamored? Probably not, as it appears that the accelerating energy demands of the world's developing nations will most likely be first met by increased reliance on the traditional (and, in spite of recent price increases, still cheap) fossil fuels.

Although there is a need to reduce this reliance, there are few ready alternatives available to the poorer, but developing, countries. It would appear that conservation is the only option.

Indeed, it may be that the influence of at least one of the major fossil fuels is on the wane. In the first article in the unit, Seth Dunn of the Worldwatch Institute discusses "King Coal's Weakening Grip on Power." Beginning with a discussion of the recent Chinese decision to attempt to eliminate coal as the fuel of choice in Beijing, Dunn catalogues the social and environmental disadvantages of the world's most available fossil fuel. The benefits of a coal phase-out, Dunn notes, will be enormous. But reduction in coal use is not enough to solve the environmental problems related to fossil fuel use. In "Oil, Profit$, and the Question of Alternative Energy," columnist Richard Rosentreter notes that, even with the recent increases in oil and gas prices, the available alternative energy sources such as solar energy have not become a focal point for development. Rosentreter suggests that since the money and power of "Big Oil" is devoted to fossil fuel exploration and development, little is left over for alternative energy technologies, viewed as less profitable. The question of solar energy is also the topic of the third article in the unit. Environmental writer Eric Weltman, in "Here Comes the Sun: Whatever Happened to Solar Energy?" asks much the same question as Rosentreter. Rather than pointing to economic answers, however, Weltman concludes that the explanation for the decline in interest and development in solar energy is political. The real issue, he notes, is the unfavorable political environment for renewables.

The concluding three articles in the section discuss situations in which political relevance and economic incentives to the development of alternative energy are present. In "The Hydrogen Experiment" Seth Dunn describes one of the most promising of the relevant new technologies. In Iceland, long known for its willingness to develop such alternative sources as geothermal energy, a national experiment is under way. Scientists, politicians, and business leaders have committed to a grand experiment that may end the country's reliance on other forms of energy. They want nothing less than to become the world's first "hydrogen economy" over the next 30 years. And in "Bull Market in Wind Energy" Christopher Flavin, one of the world's foremost energy experts, suggests that the sustainable energy technology of wind power, once the mainstay of the United States' search for energy alternatives, has again surfaced as a powerful global alternative to fossil fuels. Answers to energy questions and issues are as diverse as the world's geography. But all the answers require a reorientation of thought and the action of committed groups of people who have the capacity to change the dominant direction of a culture.

King Coal's Weakening Grip on Power

The fuel that ushered in the Industrial Revolution still burns, but a new era beckons.

by Seth Dunn

Revolution was literally in the air on February 28, 1998, when officials in Beijing and 32 other Chinese cities—under pressure from the national environmental protection agency—began releasing pollution records that had been suppressed for 20 years. The weekly reports—intended to "enable the public to supervise the government's anti-pollution efforts"—revealed that the air outside Beijing's Gate of Heavenly Peace had become hellish. Prolonged exposure to the air posed serious health risks and had increased the city's death rate by 4 percent, according to research from Harvard and Beijing Medical Universities.

The news rocked Beijing, and media reports generated angry outcries from citizens who discovered that the haze hovering over their city—and its related health problems—were almost entirely the result of coal, which supplies 80 percent of the city's energy use for factories, power plants, ovens, and stoves. A few months later, in response to public pressure, city authorities announced a crackdown on coal burning, with the aim of banning it by the end of the century. Beginning with the city's 42-square-mile central limits, the government plans to establish coal-free zones, with local authorities helping residents switch from coal to cleaner-burning natural gas.

Beijing's move to banish what was known as "King Coal" in the nineteenth century in the United States and Europe illustrates how perceptions of this fossilized substance have changed over time. A thousand years ago, China fired coal in blast furnaces to produce the armor and arrowheads that defended its dynasties against outside invaders. But it was in the West that coal was first burned in massive amounts, beginning in the eighteenth century. If the Industrial Revolution was "Prometheus unbound," coal was the fire stolen from the gods that made it possible. With its production paralleling the rise of national powers, this fossil fuel became synonymous with wealth and modernity in the nineteenth century. In his classic 1865 work, *The Coal Question,* economist William Jevons went as far as to predict the collapse of the British Empire as its coal mines approached depletion.

But Prometheus paid dearly for his deed; chained to a mountaintop, he had his liver torn out daily by vultures. Likewise, the reign of King Coal has not been without heavy costs: its use has left a legacy of human and environmental damage that we have only begun to assess. At the close of the twentieth century, coal's smog-choked cityscapes are no longer the symbol of industrial opportunities and wealth that they were 100 years ago. Instead, coal is increasingly recognized as a leading threat to human health, and one of the most environmentally disruptive human activities.

Indeed, the sun may be setting on the empire of coal. Its share of world energy, which peaked at 62 percent in 1910, is now 23 percent and dropping. Although coal's market price has fallen 64 percent in the past 20 years to a historical low of $32 per ton, global use is at its lowest in a decade, having fallen 2.1 percent in 1998. One reason for this decline is that the price of dealing with coal's health and environmental toll—the "hidden cost"—is rising. And now King Coal's remaining colonies find them-

9. King Coal's Weakening Grip on Power

selves confronted with a concern of the sort that bedeviled Jevons. This time, however, it is coal dependence—not depletion—that is the potential threat to progress.

Even so, the mirage of coal as a source of cheap energy continues to be a powerful lure, and many countries have gone to great lengths to rationalize their reliance—suppressing information, compartmentalizing problems, or socializing costs. Until now, the problems of coal have been treated with an "emergency room" approach: ecological impacts have been addressed pollutant by pollutant, mine by mine; the health hazards, one urban crisis at a time. This narrow approach has been an expensive one, both economically and environmentally, and has had perverse, unforeseen consequences: each time one of coal's impacts is "mitigated," a more pervasive and chronic problem is created, exacerbating and spreading the fuel's negative effects out over space and time. For example, towering smokestacks, built to alleviate local air pollution, created the problem of acid rain. And efforts to curtail acid rain, in turn, are adding to greenhouse-gas emissions.

Increasingly, human health, ecological, climatic, and socioeconomic concerns are pushing us away from this piecemeal regulation—toward an end to the "end-of-pipe" approach. But for the world to judge whether continued dependence on coal is viable, a more comprehensive examination is in order. After centuries of treating coal like a first-time offender, there is a growing consensus that it is time to assess this fossil fuel in terms of its cumulative offenses and to seriously weigh the benefits of replacing it with cleaner, and ultimately cheaper, alternatives.

Exhibit A: Health Hazard

The solid blackish substance called coal is vegetation that has, over millions of years, accumulated in wetlands and been partially decomposed, suffocated, moisturized, compressed, and baked by the Earth's inner heat underground. During this process, unfathomable quantities of organic matter have been slowly broken down and stored. The act of extracting coal from the Earth's crust and burning it is an experiment without geological precedent, and it is altering the environment in profound, yet poorly understood, ways.

Coal has long been linked to air pollution and ill effects on health. In medieval London, an official proclamation banned coal burning as early as 1306 A.D. in an unsuccessful effort to curb the smog and sulfurous smell hanging over the city. Even today particulate matter (dust, soot, and other solid airborne pollutants) and sulfur are two of the most unhealthy by-products of coal combustion.

Particulates penetrate deep into lungs. Prolonged inhalation causes a range of respiratory and cardiovascular problems, such as emphysema, asthma, bronchitis, lung

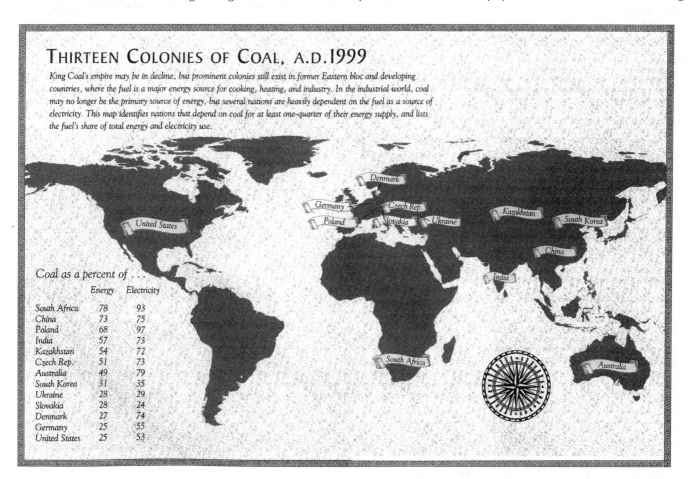

Thirteen Colonies of Coal, A.D. 1999

King Coal's empire may be in decline, but prominent colonies still exist in former Eastern bloc and developing countries, where the fuel is a major energy source for cooking, heating, and industry. In the industrial world, coal may no longer be the primary source of energy, but several nations are heavily dependent on the fuel as a source of electricity. This map identifies nations that depend on coal for at least one-quarter of their energy supply, and lists the fuel's share of total energy and electricity use.

Coal as a percent of . . .

	Energy	Electricity
South Africa	78	93
China	73	75
Poland	68	97
India	57	73
Kazakhstan	54	72
Czech Rep.	51	73
Australia	49	79
South Korea	31	35
Ukraine	28	29
Slovakia	28	24
Denmark	27	74
Germany	25	55
United States	25	53

cancer, and heart disease. It is also linked to higher infant mortality rates. The smallest particles can stay in an individual's lungs for a lifetime, potentially increasing the risk of cancer. Sulfur dioxide (SO_2) exposure is associated with increased hospitalization and death from pulmonary and heart disease, particularly among asthmatics and those with existing breathing problems.

These pollutants made up the "coal smogs" that killed 2,200 Londoners in 1880; the "killer fog" that caused 50 deaths in Donora, Pennsylvania in 1948; and the "London fog" that took 4,000 lives in 1952. Today, several coal-dependent cities—including Beijing and Delhi—are approaching the pollution levels of the Donora and London disasters, and the world's ten most air-polluted cities—nine in China, one in India—are all heavy coal users. Worldwide, particulate and SO_2 pollution cause at least 500,000 premature deaths, 4 to 5 million new cases of bronchitis, and millions of other respiratory illnesses per year. Such smogs have become transcontinental travelers: large dust clouds of particulates and sulfur from Asian coal now reach the U.S. West Coast.

Coal burning also releases nitrogen oxides, which react in sunlight to form ground-level ozone. In the United States and Europe, more than 100 cities are exposed to unhealthy ozone levels. Beijing, Calcutta, and Shanghai—all heavily coal dependent—expose millions of children to deadly mixes of particulates, sulfur dioxide, and nitrogen oxides.

Coal smoke contains potent carcinogens, affecting the more than 1 billion rural poor who rely on the fuel for cooking. Rural indoor air pollution from such cooking accounts for 1.8 to 2.7 million global annual deaths from air pollution, with women and children most at risk. In rural China, exposure to coal smoke increases lung cancer risks by a factor of nine or more.

Coal can also contain arsenic, lead, mercury, and fluorine—toxic heavy metals that can impair the development of fetuses and infants and cause open sores and bone decay. In rural China, where 800 million people use coal in their homes for cooking and heating, thousands of cases of arsenic poisoning, and millions of cases of fluorine poisoning have been reported. Millions of rural poor in other developing countries face similar risks.

Coal mining and extraction pose health hazards, as well. Explosions, falls, and hauling accidents injure or kill several thousand coal miners in China, Russia, and Ukraine each year. In China, more than five miners die for every million tons of coal mined. Perhaps the most serious and chronic threat to miners is pneumoconiosis, or "black lung"—a condition caused by continued inhalation of coal dust, which inflames, scars, and discolors lungs, and leads to a debilitating decline in lung function. In the United States, enough was known at the turn of the twentieth century about black lung to have spurred preventive action to remove or lessen the effects of the disorder, writes Alan Derickson, author of *Black Lung: Anatomy of a Public Health Disaster*. But company doctors misdiagnosed or concealed the illness for more than 50 years, until medical community mavericks and the largest strike in U.S. history forced lawmakers to enact compensatory and preventive measures. By then, the lives of hundreds of thousands of coal miners had been shortened. U.S. taxpayers have since paid more than $30 billion to compensate mining families.

Despite these advances, coal dust continues to plague miners. In Russia and Ukraine, official estimates range from 200 to 500 deaths per year. In China, where 2.5 million coal miners are exposed to dust diseases, the current annual death toll of 2,500 is expected to increase by 10 percent each year. Even in the United States, 1,500 miners died of black lung in 1994, and under-reporting is still prevalent.

Exhibit B: Environmental Damage

The coal smogs in Donora and London sparked public outrage, leading to the enactment of the first major clean-air laws. Setting local air quality standards, these acts prompted industries to install high smokestacks that would spread the pollutants over larger areas and to more distant regions. In parts of the United States, some smokestacks shot up higher than the top floor of the Empire State Building.

But this simple solution for local pollution had an unintended consequence. Carried aloft, nitrogen oxides and sulfur dioxide react in the atmosphere to form acids that

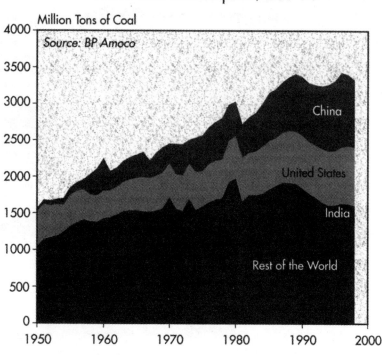

World Coal Consumption, 1950–98

fall as rain, snow, or fog or turn to acid on direct contact—corroding buildings and monuments and damaging vegetation, soils, rivers, lakes, and crops. The problems of acid rain and deposition surfaced first in Norwegian fish kills in the 1960s, and later in the "forest death" of Germany, the "Black Triangle" of dead trees in Central Europe, and the dying lakes and streams of the U.S. Adirondacks—all traced to coal burning hundreds of miles away.

Under pressure from environmental groups, industrial nations have addressed acid rain through an array of agreements focusing on sulfur emissions, which have been significantly reduced. But nitrogen emissions, which initially escaped regulation, have been slower to drop. In fact, in many regions they have risen, offsetting reductions made in sulfur emissions. In Europe, forest decline continues and hundreds of acid-stressed lakes face a long recovery time, as nitrogen persists well above tolerable levels. High-elevation forests in West Virginia, Tennessee, and Southern California are near saturation level for nitrogen, and high-elevation lakes in the Rocky Mountain, Cascade, and Sierra Nevada mountain ranges are on the verge of chronic acidity. In the Adirondacks, many waterways are becoming more acid even as sulfur deposits drop: by 2040, as many as half the region's 2,800 lakes and ponds may be too acid to support much life.

The West's acid deposition debacle is now replicating with potentially greater repercussions in Asia. A haze the size of the United States covers the Indian Ocean in winter, and in summer is blown inland and falls as acid rain, reportedly reducing Indian wheat yields. Acid rain falls on over 40 percent of China, and in 1995 caused $13 billion in damage to its forests and crops. Widening areas of China, India, South Korea, Thailand, Cambodia, and Vietnam are above critical levels of sulfur. Buildings, forests, and farmland close to or downwind from large urban and industrial centers are being hardest hit. Thousand-year-old sculptures from China's Song Dynasty have been corroded. And some scientists believe the Taj Mahal is in similar danger. A fifth of India's farmland faces acidification. China's sulfur emissions may overwhelm fertile soils across China, Japan, and South Korea by 2020.

Other types of ecosystem overload, too, are linked to coal. Nitrogen overfertilizes waterways, causing deadly algal blooms. Ground-level ozone damages forests and crops. Each year, ozone costs the United States between $5 and 10 billion in crop losses alone, and cuts wheat yields in parts of China by 10 percent. The formation and burning of massive slag heaps—piles of cinder left over from combustion—degrades land and emits carbon monoxide. Acidic or highly saline runoff from mines contaminate ground and surface water.

Air pollution regulations have prompted a hunt for low-sulfur coal, with companies turning from underground to surface—also known as strip, or open-pit—mining. In Canada, open-pit mines lie at the foot of Alberta's Jasper National Park, a World Heritage Site; in India's Bihar province, they endanger huge tracts of forest. These mines have uprooted hundreds of thousands of indigenous and poor people—aborigines in Australia, Native Americans in Arizona, villagers in northern Germany, tribals in Raniganj, India—from land they have inhabited for centuries, often with little advance notice or compensation. In West Virginia, huge machines engage in "mountain-top removal"—stripping away dozens of rolling hills, burying streams, and bulldozing mining communities.

As many developing countries follow the path of industrial nations, they too seem unable to steer clear of the pitfalls of a simplistic response to coal pollution. But the folly of focusing solely on coal's air pollutants proves most perverse in the developing world, where the added mining and processing requirements exacerbate severe land and water constraints. Chinese enterprises commonly violate emissions standards and burn high-sulfur coal rather than pay for precious water use to wash coal. In India, citizens' groups criticize the government's coal-washing mandate, arguing that it will waste energy, use up large quantities of scarce water and land, and increase pollution at mines.

Exhibit C: Shifting Climate

The second generation of coal-related pollution laws, motivated by public concern over acid rain, led companies to install another technological quick-fix. This time "clean-coal" technologies were the promised solution, namely flue-gas desulfurization and nitrogen-control equipment. While the equipment lowered emissions of the targeted pollutants, they, like higher smokestacks, had unforeseen side-effects. Clean coal creates added water demands, produces large amounts of sludge and other solid wastes, and decreases energy efficiently, thereby increasing emissions of other compounds—including carbon dioxide (CO_2).

Ranging from less than 20 to more than 98 percent in carbon content, coal is the most carbon-rich fossil fuel. The industrial era's heavy combustion of these fuels is short-circuiting the global carbon cycle, building up atmospheric CO_2 concentrations to their highest point in 420,000 years. The thickening blanket of these and other greenhouse gases has already trapped enough radiative heat to make the planet's surface its warmest in 1,200 years.

Many expected climatic dislocations are appearing: sea level rise; accelerating glacier retreat and ice shelf breakup; migrations and declines of forests, coral reefs, and other temperature-sensitive species; changes in the timing and duration of seasons; greater frequency and intensity of extreme weather events. Climate scenarios for the year 2050 from the Hadley Centre for Climate Prediction and Research show tropical forests turning to desert, adding more carbon to the atmosphere; malaria spreading to currently unaffected populations; an additional 30 million people at risk of hunger; another 66 million in danger of water stress; and 20 million more susceptible to flooding. Heat stress will have increased by 70 to 100 percent by then—adding several thousand deaths each year in large urban areas like New York, New Delhi, and Shanghai, according to Laurence Kalkstein of the University of Delaware.

Carbon emissions are not the only means by which coal changes climate: mining annually releases 25 million tons of methane, equal in warming potential to the United Kingdom's entire carbon output. But CO_2 is the most important contributor to climate change—and coal releases 29 percent more carbon per unit of energy than oil, and 80 percent more than natural gas. The climatic impact of coal burning is disproportionate to its importance as an energy source: with a 26 percent share of world energy, it accounts for 43 percent of annual global carbon emissions—approximately 2.7 billion tons. Climate instability also compounds other coal-related problems: heat stress exacerbates urban air pollution, and higher temperatures make natural systems more vulnerable to acid rain impacts.

Stabilizing atmospheric CO_2 levels at 450 parts per million during the next century, which some scientists believe necessary to avoid far more dangerous disruptions of climate, would constrain coal use to somewhere between 200 and 300 billion tons—less than 7 percent of the total resource base. Burning the entire coal resource, on the other hand, would release 3 trillion tons of carbon into the atmosphere, five times the safe limit. Thus, while energy analysts point to the apparent size of the fuel's reserves, the amount that could be safely used is far smaller. From their perspective of balancing the carbon budget, coal is a highly limited energy source.

Despite studies showing the economic feasibility of switching from coal, several governments and industries are pursuing another end-of-pipe solution: carbon sequestration. Firms and agencies in the United States, Norway, and elsewhere are devoting millions of dollars to test technologies for separating and capturing CO_2 from fossil fuels. The CO_2 would then be locked up by injecting it into oceans, terrestrial ecosystems, and geological formations. But the potential impacts on ocean chemistry and deep-sea ecosystems have not been explored, and injected emissions could be re-released due to geological activity. And if sites subject to slow release are used, carbon management could reduce atmospheric CO_2 concentrations in the near term but increase them in the long term—adding to the climate problem.

Meanwhile, some industrial nations seeking developing-country action on climate change are, contradictorily, redirecting clean coal programs overseas. In a novel form of trade "dumping," clean-coal equipment features prominently in bilateral energy missions, with firms and officials from the United States, Japan, and Australia proselytizing to poor nations that they "need clean-coal technologies." The World Bank and European Commission have aimed

ADDING UP THE COSTS OF COAL

While the market price for coal was $32 per ton in 1998, when environmental and health disruptions are factored into the equation, coal is not as cheap as it may seem.

	AIR	LAND	WATER	CLIMATE
MINING/ EXTRACTION	• Coal dust causes black lung and other respiratory diseases in miners. • Mining can result in explosions and fires. • Machinery causes dangerous fumes and disruptive noise.	• Mining causes soil degradation, erosion, and subsidence. • Farms and forests are destroyed and communities displaced by strip mining and mountain-top removal.	• Watersheds are degraded and streams filled in by mountain-top removal and strip mining. • Acid mine drainage from tailings as well as wastewater discharge pollute rivers and drinking water sources.	• Mining releases large quantities of methane, a potent greenhouse gas. • Greenhouse gases released by coal combustion play a significant role in destabilizing climate, contributing to sea-level rise, weather extremes, disease outbreaks, shifts in agriculture and water supply, extensive ecosystem damage, loss of species, and other serious disruptions.
TRANSPORTATION	• Coal may be shipped thousands of miles to power plants in open train cars and barges, producing "fugitive dust" that is blown into the air.	• A considerable amount of land has been developed for the rails and roads that transport coal.		• The engines and machines used to transport coal release CO_2, the most prevalent greenhouse gas. • Greenhouse gases released by coal combustion play a significant role in destabilizing climate, contributing to sea-level rise, weather extremes, disease outbreaks, shifts in agriculture and water supply, extensive ecosystem damage, loss of species, and other serious disruptions.
TREATMENT		• Smokestack scrubbers used to filter sulfur out of coal emissions produce large quantities of sludge and other wastes.	• Coal washing, used to strip sulfur from coal before it is burned, requires large quantities of water.	• Technologies used to trap sulfur and nitrogen emissions require more energy, which releases more CO_2. • Greenhouse gases released by coal combustion play a significant role in destabilizing climate, contributing to sea-level rise, weather extremes, disease outbreaks, shifts in agriculture and water supply, extensive ecosystem damage, loss of species, and other serious disruptions.
COMBUSTION/ CONVERSION	• Particulates, sulfur dioxide, ground-level ozone from nitrogen oxides, and toxic metals released by the burning of coal contribute to cancer risks, impair infant development, cause respiratory illness, and increase rates of morbidity and mortality.	• Acid deposition from sulfuric and nitric acid leaches nutrients from soils and damages forests, crops, and buildings. • Ozone impairs plant growth. • Power plants and massive "slag heaps"—piles of ashes from coal burning—take up land and cause degradation.	• Acid deposition and heavy metals poison rivers and lakes. • Nitrogen oxides cause eutrophication, where plant growth cuts off oxygen supplies to other species. • Cooling towers demand and heat up large amounts of water.	• Combustion of coal is the single largest source of CO_2 emissions. • Greenhouse gases released by coal combustion play a significant role in destabilizing climate, contributing to sea-level rise, weather extremes, disease outbreaks, shifts in agriculture and water supply, extensive ecosystem damage, loss of species, and other serious disruptions.

clean-coal technology initiatives at developing and former Eastern bloc nations, where the technologies remain unproven. Indeed, clean-coal equipment has failed to demonstrate financial viability in the West (its high capital investment costs make it less attractive than natural-gas-fired combined-cycle turbines), linking its peddling less to economics than to the political clout of the industry.

Exhibit D: Losing Labor

"The story of coal in America," writes Duane Lockard in *Coal: A Memoir and Critique*, "is the story of corporate successes and excesses generally." The same can now be said for the coal industry worldwide. Shrinking profits and growing deficits are leading to drastic cost-cutting practices that translate into lower prices but also major job losses, creating an employment crisis among coal miners round the globe. It is, however, both necessary and possible to reduce reliance on coal while minimizing the displacement of workers that inevitably accompanies the decline of an industry.

Worldwide, only about 10 million coal mining jobs remain, making up one-third of all mining jobs and accounting for one-third of 1 percent of the global workforce. In industrial nations, the coal-mining industry is no longer a major employer, and employment is falling even where production or exports are rising. In developing countries and transitional economies, where employment is still relatively high, pressures to reform the industry and cut costs are causing major job dislocations.

Like other sunset industries, the coal sector is increasingly characterized by bigger and fewer companies, more and larger equipment, and less labor-intensive operations. In the United States, the 10 largest firms account for 60 percent of output, up from 35 percent a decade ago. During coal's peak, in 1924, 705,000 miners toiled in U.S. mines; today there are fewer than 82,000. Thanks mostly to surface mining, employment has declined by two-thirds over the last 20 years and is expected to continue to fall; coal miners now count for less than 0.1 percent of the nation's workforce. Though domestic consumption continues to crawl upward, exports have dropped 25 percent since 1996, and experts agree that they will never return to pre-1998 levels.

The rate of contraction has as much to do with politics as with economic and environmental factors. Coal industries in both the United Kingdom and Germany have been weakened since the 1960s by environmental regulations and the switch to cleaner natural gas, now the fuel of choice for power generation in industrial nations. But while contraction in the United Kingdom has been rapid—only 13,000 union coal miners remain, out of 1.2 million in 1978—the decline in Germany has been more gradual, from 190,000 in 1982 to less than 90,000 today.

Similar struggles lie ahead for other coal-dependent nations. In Australia, 9,000 of the nation's 22,000 coal miners went on strike in 1997 when impending job cuts led Rio Tinto, the world's largest mining company, to try to deunionize the industry. In South Africa, coal production has risen 65 percent, but employment has fallen over 20 percent, since 1980. In India, where production has doubled since 1980, employment is still declining as a proportion of population. Poland's mines lose nearly $700 million each year. Russia has halted production in 90 mines and intends to have shut 130 of its 200 mines by 2000. Major future losses are expected in these countries as improved productivity and the shift to less energy-intensive service industries make more jobs redundant.

Cost cutting, mine closing, and job losses are greatest in China, where Li Yi, director of the Xishan coal mining bureau, summed up the industry's prevailing philosophy in a 1998 interview: "Our motto is: Cut people, improve efficiency." The world's leading coal producer and consumer, China has lost 870,000 workers over the last five years, and slashed production by 250 million tons in 1998 due to excess capacity and rail transport bottlenecks. (Like India, Australia, and South Africa, China faces a geographic mismatch between coal reserves and energy needs.) The government plans to close down 25,800 coal mines this year—most among the 75,000 mines in township and village areas—and shut off all small, unauthorized mines. In May 1999, the government halted the issuance of permits for new coal mining projects.

The United Kingdom and China highlight both the challenges of and chances for helping workers in the transition to the post-coal era. In both, thousands of laid-off workers have blocked traffic, stopped trains, and stormed official offices. But both governments recognize that coal's heyday is over: they are shifting from coal-reliant industries like steel works to more modern sectors—such as the high-tech and tourism industries—and both are planning solar-cell manufacturing sites in mining areas, to ease the transition for workers.

The Light at the End of the Tunnel

The current, emergency-room approach to coping with coal has proved so expensive, yielded such limited results, and contributed to so many environmental and health problems, that shifting to cleaner alternatives will help solve these problems at a much lower cost. Treating coal's symptoms in isolation has proved insufficient for improving human and planetary health. Fortunately, remedies are available that will allow the world to rapidly reduce the use of coal and accelerate the transition to cleaner energy sources.

Among the keys to cutting coal reliance are blocking mining and power projects through community activism, closing legislative loopholes, and reorienting coal-centric bilateral, multilateral, and multinational investment flows. But two policies are central to the "decoalonization" process: subsidy removal and energy taxation. Without them, the market will continue to deceive us into thinking coal is cheap, abundant, and irreplaceable, just when countries like China are beginning to realize how costly, limited, and unnecessary dependence on this fuel is.

Simply put, removing subsidies cuts coal consumption. Belgium, France, Japan, Spain, and the United Kingdom have collectively halved coal use since slashing or ending supports over the last fifteen years. Russia, India, and China have also made progress: China's coal subsidy rates have been more than halved since 1984, contributing to a slowing—and 5.2 percent drop in 1998—in consumption. Opportunities exist for further reductions. Total world coal subsidies are estimated to be $63 billion, including $30 billion in industrial nations, $27 billion in the former Eastern bloc, and $6 billion in China and India. In Germany, the total is $21 billion—including direct production supports of more than $70,000 per miner.

The experience of Germany highlights the opportunities for—and obstacles to—taxing coal. A European commission study shows that internalizing the external costs of coal from a German power plant would raise the price of power by 50 percent. Yet the government's 1998 ecological tax reform excluded coal due to industry opposition. As Ed Cohen-Rosenthal of Cornell University writes, "The question for coal miners is whether to dig in and fight or use the concern about global warming to negotiate the best deal for current members and retirees as one means of paving the way to a cleaner environment. This is a decision that only they can make and outsiders should respect their feelings. But their leverage for a negotiated outcome will never be higher than it is right now."

Digging in has predominated to date—coal labor groups underwrite skeptical "scientists" and oppose the Kyoto Protocol—though signs of reconciliation exist. In Australia, an Earthworker caucus of trade union and environmental groups is developing a plan for building solar and wind power industries. The AFL-CIO and U.S. environmental groups are crafting "worker-friendly" climate policies, like employing former miners in remediating abandoned mines. But while labor groups stress the need for "just transitions" to aid adversely affected workers, those representing coal miners appear less likely to become advocates of coal subsidy and tax reform, which could help fund such a transition, than to defend these endangered jobs to the bitter end—and at the expense of society at large.

Bold initiatives in coal taxation, meanwhile, can be found in China. The government has introduced a tax on high-sulfur coal to encourage a switch to plentiful natural gas and renewable-energy resources. Like cigarette taxes in the West, the coal levy may spread in the East; as with smoking in public places, coal use might also be banned outright where it is deemed too great a public burden to bear.

Back in Beijing, high-sulfur coal has been banned, 40 "coal-free zones" are planned, and natural-gas pipelines are under discussion. Hundreds of residents in Beijing are mobilizing through citizens' groups, such as the Global Village, to supervise implementation of the policies and raise public consciousness of the problem. The idea is catching on: four more Chinese cities—Shanghai, Lanzhou, Xian, and Shenyang—have followed suit with plans to phase out coal.

The challenge is to turn these local gains into a worldwide movement over the coming century, just as coal's negative consequences have risen from local to global during this one. A global coal phaseout has become as environmentally necessary and economically feasible as it might seem politically radical. Thirty years ago, few could have predicted the nascent anti-smoking effort would ever "go global," but it has. Coal now poses as serious a risk to our collective well-being, if not greater. If China's smoky cities can mobilize to begin eradicating the tobacco of our energy system, it is conceivable that the rest of the world's governments can as well.

Like sustainable development more broadly, achieving independence from King Coal will be no overnight coup, but a lengthy revolution. Yet the social, economic, and environmental rewards of a coal phaseout promise to be enormous. In the third millennium, societies will find themselves—to paraphrase Henry David Thoreau—rich in proportion to the coal they can afford to leave in the ground.

Seth Dunn is a research associate at the Worldwatch Institute.

Oil, Profit$, and the Question of Alternative Energy

As oil and gas prices continue to rise, the sun has apparently set on the development of solar power and other forms of alternative energy, despite official claims that the United States is committed to making them a success. The explosion in oil and gas prices has been attributed to numerous causes, but little attention has been given to the lackadaisical effort to develop alternative fuel sources and the continuous quest by the oil industry to discover more oil. Big oil has both money and power, and it shouldn't be any surprise how much can be accomplished, or prevented, with such a potent combination.

by Richard Rosentreter

The application of solar power is not a new idea. The ancient Greeks and Romans developed mirrors that would direct the sun's rays and cause a target to burst into flames within seconds. Nearly two centuries ago in 1839, Edmund Becqurel, a French experimental physicist, discovered that sunlight could produce electricity—almost fifty years before the first successful internal combustion engine was built.

During the late 1800s, harnessing the sun's rays to produce hot water was a booming business in the United States. Although the Industrial Revolution was in high gear and remarkable discoveries and inventions abounded, it took over 100 years for the first photovoltaic cell (a cell capable of producing wattage when exposed to radiant energy) to be developed by Bell Laboratories in 1954. Considering that photovoltaic cells have been the exclusive power source for satellites since the 1960s, and how rapidly television evolved during an era known as the Atomic Age, it is a wonder that solar technology hasn't advanced further.

The utilization of solar energy was briefly resurrected during the 1970s when the United States appeared to be committed to pursuing a technology that had the potential to reduce our dependency on fossil fuels. In April 1977, in the midst of an "energy crisis," President Jimmy Carter began a bold initiative to develop solar energy and other alternative fuels when he unveiled his National Energy Plan, which included setting an example by placing solar panels on the White House. Carter announced a "national goal of achieving 20 percent of the nation's energy from the sun and other renewable resources by the year 2000," and he introduced legislation that would provide homeowners with tax breaks for investing in this promising technology. In 1979, Congress followed Carter's lead and approved a $20 billion development fund for synthetic and alternative fuels. It appeared that the alternative energy industry was finally getting the financial backing it needed to have a profound impact on the nation's energy needs.

During the period in which financial support for solar energy was growing and a "windfall tax" on the profits of the oil industry was imposed, the proponents of big oil were gathering their own resources on Capitol Hill. Political action committees (PACs) that were affiliated with oil and gas interests began to sprout and, from 1977 to 1979, they contributed over $2.6 million to House and Senate candidates. A report by Alan Berlow and Laura Weiss in *Congressional Quarterly* concluded that most of the money went to candidates "with strong pro-industry voting." Support for alternative energy took a downward spiral when Ronald Reagan (a former spokesperson for General Electric) was elected U.S. president and became a staunch ally of corporate America.

By the late 1970s, oil companies had bought out many of the patents for photovoltaic cells, and corporate giants like Atlantic Richfield, Amoco, Exxon, and Mobil took control of solar power companies. This trend would lead Alfred Dougherty, former director of the Federal Trade Commission's bureau of competition to warn, "If the oil companies control substantial amounts of substitute fuels . . . they may slow the pace of production of alternative fuels in order to protect the value of their oil and gas reserves." Edwin Rothschild, a spokesperson for the Citizen Energy Labor Coalition, was concerned that the big oil companies "see solar power as a competing source of energy, and they want to control it and slow it down." However, ownership of solar technology by big oil was only the first step in the methodical dismantling of the alternative energy renaissance.

The Reagan administration would continue the squeeze on alternative energy as tax credits for residential investment in solar and wind power became "obsolete," as it was deemed to be "the responsibility of the private sector to develop and introduce new solar technologies." The $684 million requested by Carter for alternative energy in fiscal 1982 was slashed to $83 million in Reagan's 1983 proposal. What was transpiring in the realm of solar technology development didn't go unnoticed by the science community. In the November 1981 *New Scientist* article "Big Oil Reaches for the Sun," Ralph Flood reports that the "energy policy under the Reagan administration seems designed to accelerate the oil companies' control." The solar panels on the White House were discarded.

The lack of support for the development of alternative energy continued despite the findings of a study released in 1980 by the National Academy of Sciences at the request of the Energy Research and Development Administration. The study concluded that "a costly push by the government would lead to rapid growth in solar-related energy sources" and, if this occurred, "renewable energy sources could account for a quarter or more of the nation's energy needs within thirty years." Instead, the Reagan administration preferred to rapidly expand military aid and sent billions of dollars to foreign countries. It also greatly increased the federal budget for the research and production of atomic weapons.

Despite its promise, solar power faced several hardships and questions. Many fly-by-night companies popped up hoping to jump on the bandwagon and make a quick profit, which led to shoddy construction and customer dissatisfaction. The solar industry during this early stage lacked structure, and there was enormous confusion as industry representatives scrambled to establish guidelines for advertising claims, technical specifications, and warranty standards, which contributed to a higher cost.

Regardless of the problems the solar industry faced, greater financial support and stricter guidelines by the government could have offset many of the obstacles. There were some who argued that solar energy could only be harnessed during bright sunshine, which added to the skepticism surrounding it. However, in 1982 Charles Wurmfeld, former president of the New York State Professional Engineers Society, invented and patented an engine that drew energy from the ordinary atmosphere, not requiring bright sunshine. He believed his invention would save one-third of the oil burned in this country.

Two decades later, we are once again in the midst of an energy-related controversy. The dramatic rise in oil prices has been blamed on the Organization of Petroleum Exporting Countries (OPEC), pipeline problems, tougher standards imposed by the Environmental Protection Agency (EPA), and even consumer choice in vehicles. However, just as they did in 1979, oil companies are reporting large profit increases during this recent "energy crisis." Chevron Corporation announced that it earned a record $1.1 billion as surging oil and gas prices boosted first-quarter profits in 2000 nearly fourfold. BP Amoco said its profits more than doubled because of rising oil prices and cost-cutting measures. Exxon Mobil, Texaco, and Conoco also reported between two and four times more profits than in 1999.

During a House hearing in June, oil company profits were updated: Texaco profits were up 473 percent; Conoco, 371 percent; BP Amoco, 290 percent; and Chevron, 271 percent. The list goes on, and it is predicted that the oil business will continue to be lucrative. According to Sam Fletcher in the February 21 *Oil and Gas Journal*, the rise in oil prices is "fueling a recovery among producers . . . that is expected to propel the oil and gas industry through 2000 and beyond."

Analysts across the country agree with this prediction. According to Deutsche Banc Alex Brown, a major investment banking firm in Baltimore, Maryland, there is a "bullish outlook for the commodity." David Garcia, a financial analyst, says, "These are the best industry conditions for at least fifteen years." And Fadel Gheit of Fahnestock and Company says, "The industry has never looked better."

Motorists paying high prices at the pump would beg to differ. According to Steve Leisman in his April 26 *Wall Street Journal* article, "High Prices Give Oil Companies a First-Quarter Boost," analysts describe the stocks of several big oil companies as "increasingly attractive." This attractiveness relies on profitability, which relies on finding more oil reserves. According to a June 1 article in the *Houston Chronicle*, shareholders of Exxon Mobil recently rejected overwhelmingly proposals to invest in renewable energy and study potential harm from oil drilling in Alaska.

While the development of alternative energy sources continues to lag, supporters of the oil industry continue to promote the use of fossil fuels. During a recent House hearing on high gas prices, representatives from the oil industry argued that a possible solution would be to begin drilling in environmentally sensitive regions, such as the Arctic and Rocky Mountains. According to Red Cavaney, president and chief executive officer of the American Petroleum Institute (API), the nation's energy woes could be resolved if the oil companies were allowed to drill in areas that have been safeguarded by environmental protection legislation.

Already there is congressional support to begin such action. Senate Energy and Natural Resources Chair Frank Murkowski (Republican–Alaska) proposed legislation that

Running on Empty

Phillip M. Morse

With the flap this summer over gas prices in the United States, I feel impelled to ask why anyone is actually surprised. Petroleum is a finite resource and hence, before it runs out, it's only natural that the price should go up. Current estimates, based upon today's rate of consumption, indicate that cheap oil will be gone within fifty years. Therefore, as that time approaches, local conditions and price gouging can be expected to cause painful spikes in what U.S. consumers pay at the pump.

We must adjust to reality by reducing demand and converting to alternative sources. Higher oil prices today are a wakeup call to the fact that infinite growth is impossible and population growth is approaching the limits of what the environment can acceptably support.

The standard of living of individual societies results from conversion of available natural resources into useful products through technology. These natural resources have limits that can be temporarily exceeded by importing them from regions with a surplus. But when such reserves have also been consumed, only the development of new technology to utilize renewable resources will preserve or improve a traditional standard of living.

Unfortunately for the people with underdeveloped economies, resources that are necessary to raise their living standard are being consumed by the developed economies. For the people living in the underdeveloped economies to raise their standard of living now, new technological advances are required on a magnitude comparable to those that made fools of earlier predictors.

World population was about one billion in 1800 when Thomas Malthus predicted that starvation would halt population growth. He was wrong because he didn't anticipate the technological advancements of the Industrial Revolution, which raised Western living standards and increased life expectancy. World population was about four billion in 1960 when Paul Ehrlich wrote his book *The Population Bomb,* in which he also predicted that starvation would stop population growth. He, too, was wrong because he didn't anticipate the technology that became known as the green revolution.

But both Malthus and Ehrlich were right in terms of the underdeveloped economies, which are still very nearly in the state they envisioned. Technology gave Western societies the two revolutions that developed their economies, raised their middle-class standards of living, and reduced their death rates, resulting in rapid population growth. World population has now passed six billion and is expected to peak at ten billion within the next 100 years. But there is a tradeoff between population size and the living standard. The higher the standard of living, the less people the environment can sustain with its finite resources. Some estimate that two billion is the maximum world population sustainable with an acceptable standard of living. At two billion, poverty and starvation might be ended throughout the world. But at over ten billion, virtually all will be at a starvation level of existence.

In the United States, we behave as though there are no limits. An automobile driver who runs out of gas in the middle of nowhere is thought to be irresponsible, but such common sense isn't applied to our civilization. Politicians like to give the impression that they can influence the price of fuel by government action. But in the long run it is a supply-and-demand problem, and when the demand exceeds supply the price defies controls. Rationing will occur either through higher prices or long lines at gas stations.

At present, alternatives to oil aren't cost-effective, but rising oil prices will eventually resolve that problem. We are entering a time of transition to the post-oil era. There is a whole new growth industry waiting in the wings based upon energy alternatives stimulated by higher oil prices. For example, hydrogen fuel cells and photovoltaic cells have yet to be applied to their full potential. These alternative energy sources and more must be applied now by new technology to transportation, heating, food production, plastic products, and related industries before the price of petroleum becomes prohibitive.

Change is coming whether we like it or not, forced on us by the natural increase in the price of petroleum. But if the producing countries are persuaded to temporarily increase their production to keep prices low, they will run out sooner rather than later. If the price increase is gradual, the transition could be smooth; but if it is manipulated, when change does come it could be catastrophic for Western societies. And while it might have less effect on the underdeveloped economies, it will make it harder for them to catch up with the developed world.

Such fundamental changes should take many years, but we have less than fifty to complete the transition. The price of petroleum will probably go ballistic over the next ten years—and this seems to have already begun. We must learn how to function on the energy we receive daily from the sun. We must allow and encourage our population to shrink back to a sustainable size while we preserve a tolerable standard of living and the freedom of individual choice.

As nations become increasingly dependent upon oil controlled by unsympathetic suppliers, national security crises are imminent—perhaps even wars will be fought to control the remaining supply. The Gulf War may be a preview of things to come. Saddam Hussein's goal was to control the world's largest supply of oil. He was stopped in time. But the next Gulf War may not be resolved so easily.

At some point an epiphany will occur, and the sooner it does the better. But as Mark Twain said, "People refuse to see the handwriting on the wall until their backs are against it."

Philip M. Morse can be reached by e-mail at pmmorse@worldpath.net.

would allow oil companies to drill in the Arctic National Wildlife Preserve. A Senate budget measure has already projected $1.2 billion in royalties from the Alaska refuge in 2005, and it recently voted in favor of opening the region to commercial drilling.

Currently, American oil corporations are sinking millions of dollars into exploration and gaining access to large oil deposits in the Caspian Sea region. According to Jeanne Whalen in the March 4, 2000, *Wall Street Journal,* the Caspian holds as much as 2.2 billion barrels, and Michael Davis reports in the April 22 *Houston Chronicle* that Conoco and Exxon Mobil have received the green light to proceed with an estimated $5 billion oil development project in that region and will pay $75 million for the right to develop in the Azerbaijan fields of the Caspian. Secretary of Energy Bill Richardson, who recently addressed the House of Representatives, summed up the rationale for the movement toward searching for oil when he said that the "world's thirst for oil is steadily rising" and "demand will continue to grow."

Imagine if the money spent by the U.S. government on foreign aid and by oil corporations on fossil fuel exploration were invested in the development of alternative fuels. Apparently loyalty to fossil fuels is too deep. According to a recent analysis by the Congressional Research Service, reported in the March/April issue of *Mother Jones,* seventy-seven cents of every energy research dollar from 1973 to 1997 went to nuclear and fossil fuels; just fourteen cents went to alternative energy.

The adverse effects that burning fossil fuels have had on our environment and their contribution to global warming are well documented. Yet, according to Exxon Mobil's 1999 annual report, the company acknowledged the public's concern over "climate changes" due to the use of fossil fuels but said that the projected serious effects "rely on speculative assumptions and results from unproven models." Exxon Mobil doesn't believe that "the current scientific understanding justifies mandatory restrictions on the use of fossil fuels" as it is certain "that significant economic harm would result from restricting fuel availability to consumers." The fact is that the company would lose profits—just as any other oil company would if alternative energy sources were developed. Oil companies are searching for more oil reserves, not alternative sources for fossil fuels. More oil means more profit.

While there is no concrete evidence that proves oil companies purposely sabotaged the progress of alternative energy, there are clues that point to political favoritism on their behalf. The U.S. Senate recently passed an amendment by Senator Kay Bailey Hutchinson (Republican–Texas) that saves oil companies from paying $66 million a year in royalties to the government. Opposing the amendment was Senator Russ Feingold (Democrat–Wisconsin), who said, "I am very concerned that Congress is abdicating its responsibility." Feingold cited soft money political contributions from oil giants—including Exxon, Chevron, BP Oil, and Amoco—that totaled over $2 million during a two-year span. In all likelihood, the oil industry will use the latest "oil crisis" as leverage to promote continued legislation in its favor.

As for future legislation, the current poll leader for the presidency, Republican George W. Bush—a former oil company executive who has amassed "substantial financial contributions" from the industry—has also been criticized for being a "tool of oil interests." It is highly unlikely that alternative fuels have a chance for further development if Bush—who, according to Joby Warrick in the April 4 *Washington Post,* "supports oil exploration in the Arctic National Wildlife Preserve while opposing the United Nation's 1997 Kyoto protocol that requires industrialized countries to cut emissions of greenhouse gases blamed for global warming"—lands in the White House in 2001.

Nearly two decades ago, a group of students from Crowder College in Neosho, Montana, built a car at the cost of about $5,000 that traveled across the continental United States powered only by the sun. Just last year, students from the University of Oklahoma built a solar-powered vehicle that won a biennial intercollegiate competition which provides them an opportunity to "design, build, and race solar-powered cars." Modern science technology has given humanity the ability to access hundreds of channels on cable television, develop computers to communicate on a global scale, clone animals, and produce state-of-the-art weapons of mass destruction. Imagine if the inventors and scientists of the world focused their minds and energy on developing alternative energy sources for the public good.

Time magazine recently published a special issue entitled "The Future of Technology." In a section depicting the future of the automobile, surprisingly, there was no mention of cars powered by solar technology or any other alternative fuels. Could it be that the fate of alternative energy has been sealed by an industry that would crumble if it were to face competition from other sources?

At the end of the movie *Back to the Future,* a brilliant scientist called "Doc" refueled a DeLorean with a handful of household trash. It is a fantastic concept that falls into the same category as solar energy because it is nonprofitable. According to the July 7 *Fortune 500,* Exxon Mobil, Ford Motors, and General Motors are some of the top profit-making corporations in the United States, and they wield a great amount of economic and political clout. Although there are alternative energy sources and related technologies available for development to meet our growing energy needs, there is currently not enough profit in them to be an attractive alternative for corporations. Perhaps renewable energy, too, is destined to become fossilized.

Richard Rosentreter holds an associate's degree in journalism from Suffolk County Community College in New York, where he has been cited for his investigative journalism, and a bachelor's degree in English from Southampton College at Long Island University, New York, where he regularly writes a column on social and political issues. He can be reached by e-mail at richinsight@hotmail.com.

Here Comes the Sun
Whatever happened to solar energy?

By Eric Weltman

Looking back, the worst of the '70s—polyester, Nixon, disco—is remembered, even celebrated. But it's forgotten that in the same decade, amidst the oil shocks and nuclear debacles, the future seemed to belong to solar energy. Indeed, with gas-guzzling SUVs clogging our roads, it's difficult to remember the urgency of that time.

There even once was a day called Sun Day. The idea, recalls organizer Denis Hayes, was "to convey to the American public that there are options, that it is possible to run a modern industrial state on sunshine." On May 3, 1978, Sun Day began with a sunrise ceremony at the United Nations led by Ambassador Andrew Young and continued with hundreds of events across the country. President Carter used the occasion to announce an additional $100 million in federal solar spending and the installation of a solar water heater on the White House roof. The White House Council on Environmental Quality ambitiously declared, "A national goal of providing significantly more than half of our energy by solar sources by the year 2020 should be achievable."

But then the '80s happened. With the election of Ronald Reagan, solar energy entered a dark age of malign neglect. Reagan eliminated tax credits for solar energy and removed the solar panels from the White House roof. Federal research-and-development funding for solar power fell from $557 million in 1980 to $81 million in 1990. At the same time, oil prices plummeted, diminishing demand for alternatives and taking energy off the agenda of the nation and much of the environmental movement. "If oil had remained expensive," Hayes says, "everything would have fallen into place."

Consequently, things now look a lot different than the sunny optimists of the '70s predicted. Consumers pay more for a gallon of bottled water than they do for a gallon of gas, while, at $20 a ton, coal is cheaper than topsoil. The universe of renewable energy sources—including solar, wind and geothermal power (but not hydroelectric)—provides only 2.1 percent of the nation's electricity. The future isn't much brighter: Absent any new policies, according to federal projections, by 2020 renewable energy is expected to provide just 3 percent of the nation's electricity.

Of course, the problems of fossil fuels—toxic spills, mining waste, acid rain, smog, etc.—haven't gone away. Meanwhile, a new problem has emerged: global climate change, with its multiple threats of rising sea levels, disrupting agriculture, increasing weather-related disasters and spreading infectious diseases. The scientific consensus is that climate change is happening, and its chief source is carbon dioxide released by the combustion of fossil fuels. The United States accounts for about a quarter of the world's energy consumption, so it's no surprise that this country also is responsible for 24 percent of carbon dioxide emissions—the largest source of which is power plants.

In the past, the world would have turned to the United States for renewable energy solutions. After all, the United States invented photovoltaic (PV) panels, devices that turn sunlight into electricity, and, in the '80s, California produced more than 90 percent of the world's wind energy. But the torch has been seized by Europe and Japan, which support renewable energy with a range of tax benefits, mandates and pricing programs. In fact, the European Union prohibits subsidies for fuels other than renewables. Meanwhile, according to the Worldwatch Institute, wind power is the most rapidly growing source of energy in the world, increasing 20 percent per year since 1990. The Danes have captured half of the market for wind technologies.

But the potential for renewables remains great in the United States. The Solar Energy Industries Association (SEIA) claims that PV panels covering 0.3 percent of the country, a quarter of the land occupied by railroads, could provide all of the nation's electricity. Likewise, the 11 states stretching between North Dakota and Texas have been dubbed the "Saudi Arabia of wind energy," with enough gusts to supply more than the nation's electricity consumption.

The American public continues to support renewable energy in survey after survey. Yet renewables face the continued obstacle of the political power of the utility, nuclear and fossil fuels industries. That clout has translated, among other things, into billions of dollars in subsidies in the form of federal research and development, tax benefits and ratepayer bailouts. However, there are several reasons that renewables may finally earn their day in the sun. One is the rising concern over the global climate change, which has spurred the interest of environmental organizations and the foundations that fund them. Another is the restructuring of the electric utility industry.

> **Absent any new policies, by 2020 renewable energy is expected to provide just 3 percent of the nation's electricity.**

A growing number of states are dismantling the monopolies that have controlled the entire process of electricity generation and distribution, and Congress is contemplating national restructuring legislation. Large industrial customers tout the lower energy prices they say will result from opening up the market to competition. Environmentalists see opportunities, perils and problems in the rush to deregulate.

Deregulation has allowed some consumers to choose "green energy." Since electricity from all sources is mixed up in the utility grid, no one can guarantee that a particular home will receive "green electrons." However, consumers can sign up to pay their bills to companies that generate their electricity from renewable sources. In California, the municipal governments of San Diego, Santa Monica and San Jose, as well as the Los Angeles Dodgers, have opted for green power.

The vision of true energy independence—households generating their own energy—has been advanced by developments in PV "solar roofing shingles." A top item on the solar industry's wish list is federal legislation requiring "net metering," which would allow solar-powered homes to cut their bills by sending extra electricity to utilities and running their meters backward. Meanwhile, David Morris of the Institute for Local Self-Reliance advocates more research and development in batteries that could store excess "home-grown" energy, which could "potentially make obsolete a trillion dollars in transmission and distribution lines."

But environmentalists are quick to point out the limits of green consumerism. First, utilities are insisting that ratepayers bail them out for billions of dollars invested in nuclear boondoggles that would otherwise die a quick death in a competitive market. Furthermore, since California's markets opened up in 1998, only 1 percent of the state's consumers have chosen green energy—and that's with a subsidy scheduled to end in 2001. The most optimistic marketers expect that 20 percent of residential customers and 10 percent of commercial customers will choose green power. As Rob Sargent of the Massachusetts Public Interest Research Group (PIRG) notes, "The utilities would like nothing better than to use consumer choice as an argument against policies requiring renewable energy."

The utilities argue that in a competitive marketplace there should be no restrictions on giving customers what they want: cheap power. But environmentalists counter that the price of this power doesn't include the environmental costs born by society, including dirty air, dangerous wastes and climate change. Their concern that a focus on narrow, short-term costs could have dire social consequences has already been realized: utilities have slashed investments in energy efficiency by half since the mid-'90s.

Environmentalists have fought some tenacious state-by-state battles to incorporate green energy policies into utility restructuring, with mixed results. Fourteen states have established "public benefits trusts," which tax electricity use to fund renewable energy, energy efficiency and low-income energy programs. Eleven states require that a certain percentage of their electricity be generated by renewables, but these "Renewable Portfolio Standards" (RPS) are largely unambitious. For example, Arizona requires that solar energy supply just 1 percent of the state's power by 2002.

In Congress, environmentalists are supporting a bill introduced by Vermont Republican Sen. Jim Jeffords to add a shade of green to federal restructuring legislation. The bill would establish an RPS of 20 percent renewable energy use (excluding hydropower) by 2020, create a public benefits trust, require utilities to tell consumers how much pollution they produce and place a cap on emissions of carbon dioxide and other pollutants. It's a much more ambitious bill than the Clinton administration's, which would establish an RPS of only 7.5 percent by 2010. But the Jeffords bill doesn't have universal support among environmentalists, some of whom criticize its failure to prohibit nuclear bailouts and its allowance of emissions trading.

The biggest problem, though, is that it doesn't stand a chance of passing. The bill doesn't even have the full support of the 151-member House Renewable Energy Caucus or the newly formed 24-member Senate caucus. "If you took a vote today," says Ken Bossong of the Sun Day Campaign, "it would go down in flames."

The bill's poor prospects, Bossong says, stem from the lack of grassroots momentum behind it. Scott Denman of the Safe Energy Communication Council adds, "The movement needs to develop a political base and be much more politically aggressive."

11. Here Comes the Sun: Whatever Happened to Solar Energy?

Energy advocacy lost its populist edge in the '80s, Sargent says, when environmentalists and utilities began collaborating, chiefly to promote investments in energy efficiency. Getting a "seat at the table" was a positive thing, he says, but it encouraged environmentalists to forget that "our power is derived from the size of our constituency, not from our access."

Environmentalists trying to overcome this mistake face several key challenges. Today's movement lacks the built-in activist base of opponents to nuclear power that existed in the '70s, when more than 50 nuclear power plants were under construction. Now energy is so far off society's radar screen, most people don't even know where their power comes from. "Most people we talk to think their electricity comes from hydropower," says Andrea Kavanagh of the National Environmental Trust (NET).

The most ambitious effort to re-energize the movement is Earth Day 2000, chaired by Hayes, an original Earth Day and Sun Day organizer. The focus of Earth Day 2000 is energy and Hayes hopes that the month-long series of events will provide a spark missing from the issue. Lacking the immediate context of an energy crisis or the Three-Mile Island disaster, he says, "we kind of have to create the timeliness of the issue ourselves."

His Earth Day Network claims to have nearly 3,000 groups in 163 countries involved thus far and boasts a flagship event on the mall in Washington on April 22 featuring actor Leonardo DiCaprio. The Earth Day agenda—endorsed by about 500 organizations, including the Natural Resources Defense Council (NRDC) and U.S. PIRG—calls for quadrupling federal investments in renewable energy and efficiency in five years and halting subsidies for fossil fuels and nuclear power, with a goal of producing at least one-third of the nation's energy from renewables by 2020. In the "changed environment" after Earth Day 2000, Hayes hopes, there will [be] an opportunity for the environmental movement to achieve such goals.

Not everyone shares his optimism, however. Kalee Kreider of NET acknowledges the tremendous boost recycling received from Earth Day 1980's focus on solid waste, but says she's "not going to plan on a similar bounce for energy." Citing the lack of infrastructure to build on any momentum, Bossong predicts, "A lot of money will be spent, a modest amount of media coverage will be generated and probably nothing will happen."

That said, Bossong himself maintains a database of about 1,000 organizations across the country that have some level of involvement in clean energy issues, including several national outfits with extensive field operations. PIRG, with 27 state-based organizations and six U.S. PIRG field offices, has campaigns to clean up dirty power plants and promote clean energy. The Sierra Club, with 65 chapters, is focusing its energy program on transportation. Ozone Action has paired up with the International Council on Local Environmental Initiatives to help municipal officials take action against global warming.

The new kid on the block is NET, a group started by the Pew Charitable Trusts in 1994, which organized a "Pollution Solutions" bus tour of 36 cities this fall to demonstrate how people can consume less energy. "You have to build people toward political action," Kreider says. "Most people, as a first dipping of their toes in the energy issue, are not prepared to slam their senator or take on a multinational corporation."

In addition to national organizations, Bossong's database includes approximately 800 state and regional organizations, from the Northeast Sustainable Energy Association to the Northwest Energy Coalition. Local activists—whether they're campaigning to clean up dirty power plants in Massachusetts or to stop nuclear waste storage in Minnesota—are ready to be plugged into national clean energy initiatives. "These battles are creating a constituency for clean energy in a way that I've not seen in a long time," Sargent says.

Of course, some groups differ on how the movement should proceed. The NRDC's Ralph Cavanaugh cautions that the fossil fuel industry is "not monolithic," pointing to the key support of some utilities in winning a recent extension of wind energy tax credits and British Petroleum's ownership of one of the nation's largest PV manufacturers. While acknowledging that the coal mining industry has been "unsupportive," Cavanaugh notes that it "is a declining force both economically and politically." He says, "I don't find a lot of organized opposition to renewable energy in general."

But don't tell that to U.S. PIRG, Friends of the Earth or Taxpayers for Common Sense, who have been battling to reduce federal subsidies for fossil fuels. The fiscal year 2000 budget contains $1.5 million more for coal research and development than last year, a total of $124 million. Likewise, Rebecca Stanfield of U.S. PIRG maintains that the utility industry's opposition to a national renewables mandate has been "relentless." Sargent adds, "There are some in the environmental community who place too much faith in the goodwill and enlightenment of corporate leaders and are unwilling to point the finger at our enemies."

Another question is what place renewable energy has in the environmental movement's clean air agenda. Many environmentalists, some reluctantly, acknowledge that relying more on natural gas, which contains less carbon and other pollutants than coal and oil, is essential to combat global climate change. But a NET fact sheet—to the chagrin of other advocates—goes so far as to declare natural gas a "solution for today," while renewable energy is a "solution for tomorrow."

There are also different views on how fast the renewable energy industry can increase production to meet clean air needs. "You can only ramp up new technologies and industries so fast without creating bottlenecks, increasing costs and creating a backlash," argues Alan Nogee of the Union of Concerned Scientists, who supports the RPS standard in the Jeffords bill. Ironically, he points to nuclear power "as a really good example of an industry that created a lot of its own problems by growing too fast." Hayes disagrees: "Technically, we can do pretty much what we want to pay to do."

The real issue is political will: The political environment for renewables is not good. Clinton's hallmark has been programs that are big on pronouncements and goals but lack the cash to translate them into action. Members of Congress can greenwash themselves by joining their green energy caucus and then, as Denman says, "stab sustainable energy in the back." The fiscal year 2000 budget of $247 million for renewables research and development is a decrease of nearly $20 million from last year, which Scott Sklar of SEIA blames on pre-election shenanigans by congressional Republicans trying to embarrass Vice President Al Gore. However, the environmentalists' own budget recommendations also were down from the previous year. "We slightly tempered our request to make sure we were politically relevant," Sklar says. "It didn't work."

The problem with what is politically relevant is that it may not be enough to save the planet. The Jeffords bill, for example, would freeze utility carbon dioxide emissions at 2000 levels by 2020, while scientists say that emissions reductions of more than 60 percent are necessary to stabilize carbon levels in the atmosphere—and the sooner those cuts are made, the better. Ross Gelbspan, author of *The Heat Is On,* a best-selling book on global warming, charges that environmental organizations involved in the climate change negotiations "are more concerned with their access to government officials than solving the problems with global warming."

To counter this troubling disconnect, Gelbspan and a group of energy experts have proposed their own "World Energy Modernization Plan." The plan calls for the creation of a 0.25 percent tax on international currency transactions, yielding $150 to $200 billion for a fund to promote the global adoption of renewable and energy efficient technologies. "Even if people reject the details of the plan," Gelbspan says, "our hope is that it communicates the scope and scale of what's needed to deal with this crisis. The science on what needs to be done is unambiguous."

Eric Weltman *is a writer and activist in Cambridge, Massachusetts.*

The Hydrogen Experiment

In Reykjavík, Iceland, scientists, politicians, and business leaders have conspired to put into motion a grand experiment that may end the country's—and the world's—reliance on fossil fuels forever. The island has committed to becoming the world's first hydrogen economy over the next 30 years.

by Seth Dunn

Riding from the airport to Iceland's capital, Reykjavík, gives one the sensation of having landed on the moon. Black lava rocks cover the mostly barren landscape, which is articulated by craters, hills, and mountains. Other parts of the island are covered by a thin layer of green moss. American astronauts traveled here in the 1960s to practice walking the lunar surface, defining rock types, and taking specimens.

I, too, have traveled here on a journey of sorts to a new world—a world that is powered not by oil, coal, and other polluting fossil fuels, but one that relies primarily on renewable resources for energy and on hydrogen as an energy carrier, producing electricity with only water and heat as byproducts. My quest has brought me to the cluttered office of Bragi Árnason, a chemistry professor at the University of Iceland whose 30-year-old plan to run his country on hydrogen energy has recently become an official objective of his government, to be achieved over the next 30 years. "I think we could be a pilot country, giving a vision of the world to come," he says to me with a quiet conviction and a deep, blue-eyed stare that reminds me of this country's hardy Viking past.

When he first proposed this hydrogen economy decades ago, many thought he was crazy. But today, "Professor Hydrogen," as he has been nicknamed, is something of a national hero. And Iceland is now his 39,000-square-mile lab space for at long last conducting his ambitious experiment. Already, his scientific research has led to a multi-million-dollar hydrogen venture between his university, his government, other Iceland institutions, and a number of major multinational corporations.

I am not alone in my expedition to ground zero of the hydrogen economy: hundreds of scientists, politicians, investors, and journalists have visited over the past year to learn more about Iceland's plans. My journey is also an echo of what happened in the 18th century, when

3 ❖ ENERGY: PRESENT AND FUTURE PROBLEMS

Iceland
Near the Arctic Circle, Iceland is home to about 270,000 people—90 percent of whom live in cities. While the island has some of the most active volcanoes in the world, nearly 11 percent of the land mass is buried under massive glaciers.

"Yes, my friends, I believe that water will one day be employed as fuel, that hydrogen and oxygen which constitute it, used singly or together, will furnish an inexhaustible source of heat and light, of an intensity of which coal is not capable . . . There is, therefore, nothing to fear . . . Water will be the coal of the future."

—Jules Verne
The Mysterious Island, 1874

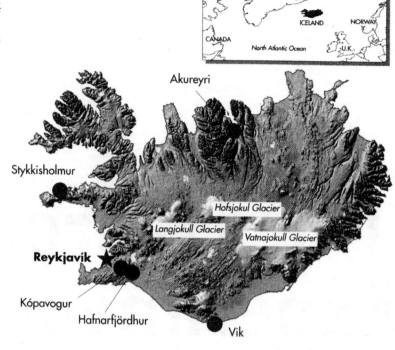

merchants and officials flocked to another North Atlantic island—Great Britain—to witness the harnessing of coal.

Today, many experts are watching Iceland closely as a "planetary laboratory" for the anticipated global energy transition from an economy based predominantly on finite fossil fuels to one fueled by virtually unlimited renewable resources and hydrogen, the most abundant element in the universe. The way this energy transition unfolds over the coming decades will be greatly influenced by choices made today. How will the hydrogen be produced? How will it be transported? How will it be stored and used? Iceland is facing these choices right now, and in plotting its course has reached a fork in the road. It must choose between developing an interim system that produces and delivers methanol, from which hydrogen can be later extracted, or developing a full infrastructure for directly transporting and using hydrogen. Whether the country tests incremental improvements or more ambitious steps will have important economic and environmental implications, not only for Iceland but for other countries hoping to draw conclusions from its experiment.

Iceland is not undertaking this experiment in isolation. Its hydrogen strategy is tied to three major global trends. The first of these is growing concern over the future supply and price of oil—already a heavy burden on the Icelandic economy. The second is the recent revolution in bringing hydrogen-powered fuel cells—used for decades in space travel—down to earth, making Árnason's vision far more economically feasible than it was just ten years ago. The third is the accelerating worldwide movement to combat climate change by reducing carbon emissions from fossil fuel burning, which in its current configuration places constraints on Iceland that make a hydrogen transition particularly palatable. How the island's plans proceed will both help to shape and be shaped by these broader international developments.

A Head Start

Straddling the Mid-Atlantic Ridge, Iceland is a geologist's dream. Providing inspiration for Jules Verne's *Journey to the Center of the Earth,* the island's volcanoes have accounted for an estimated one-third of Earth's lava output since 1500 A.D. Eruptions have featured prominently in Icelandic religion and history, at times wiping out large parts of the population. Reykjavík is the only city I know that has a museum devoted solely to volcanoes. There, one can find out the latest about the 150 volcanoes that remain active today.

Iceland's volcanic activity is accompanied by other geological processes. Earthquakes are frequent, though usually mild, which has made natives rather blasé about them. Also common are volcanically heated regions of hot water and steam, most visible in the hot springs and geysers scattered across the island. In fact, the word "geyser" originated here, derived from *geysir,* and Reykjavík translates to "smoky bay." During my visit, the well-known Geysir, which erupts higher than the United States' Old Faithful, was reemerging from years of dormancy, to the delight of Icelanders everywhere. The country first began to tap its geothermal energy for heating homes and other buildings (also called district heating) in the 1940s. Today, 90 percent of the country's buildings—and all of the capital's—are heated with geothermal water. Several towns in the countryside use geothermal heat to run greenhouses for horticulture, and geothermal steam is also widely harnessed for power generation. One tourist hotspot, the Blue Lagoon bathing resort, is supplied by the warm, silicate-rich excess water from the nearby Reykjanes geothermal power station. Yet it is estimated that only 1 percent of the country's geothermal energy potential has been utilized.

Falling water is another abundant energy source here. Although it was floating ice floes that inspired an early (but departing) settler to christen the island Iceland, the country's high latitude has exposed it to a series of ice ages. This icy legacy lingers today in the form of sizable glaciers, including Europe's largest, which have carved deep valleys with breathtaking waterfalls and powerful rivers. The first stream was harnessed for hydroelectricity in the 1900s. The country aggressively expanded its hydro capacity after declaring independence from Denmark in the 1940s, beginning an era of economic growth that elevated it from Third World status to one of the world's most wealthy nations today. Hydroelectricity currently provides 19 percent of Iceland's energy—and that share could be significantly increased, as the country has harnessed only 15 percent of potential resources (though many regions are unlikely to be tapped, due to their natural beauty, ecological fragility, and historical significance).

Iceland is unique among modern nations in having an electricity system that is already 99.9 percent reliant on indigenous renewable energy—geothermal and hydroelectric. The overall energy system, including transportation, is roughly 58 percent dependent on renewable sources. This, some experts believe, prepares the country well to make the transition from internal combustion engines to fuel cells, and from hydrocarbon to hydrogen energy. With its extensive renewable energy grid, Iceland has a headstart on the rest of the world, and is positioned to blaze the path to an economy free of fossil fuels.

Peat and Petroleum

When Vikings first permanently settled Iceland in the 9th century A.D., they used bushy birchwood and peat reservoirs to make fires for cooking and heating, and to fuel iron forges to craft weapons. But deforestation soon led to the end of wood supplies, and the cold climate would freeze the peat bogs, limiting their use as fuel.

Beyond its peat supplies, Iceland has virtually no indigenous fossil fuel resources. As the Industrial Revolution gathered momentum, the nation began to import coal and coke for heating purposes; coal would remain the primary heating source until the development of geothermal energy. In the late 1800s, as petroleum emerged as a fuel, Iceland turned to importing oil. Today, imported oil—

> **But does the Iceland experiment really apply to the rest of the world?**
>
> Other countries don't have the abundant renewable (geothermal and hydroelectric) resources Iceland has, but they do have other renewable energy sources—solar, wind, biomass, and ocean tides—that can also be used to produce hydrogen. The United States has enough wind to meet its entire electricity demand, if only it can be delivered where and when it is needed. The Middle East and the tropics, with their abundant sunshine, could one day become major hydrogen producers and exporters. By allowing intermittent energy to be cheaply stored and transported, hydrogen technology will turn renewables from marginal to mainstream sources.
>
> Because they have not developed their renewable energy resources as fully as Iceland has, other countries may first use natural gas as a "bridge" to hydrogen. This would mean using existing natural gas pipelines to transport the gas to fueling stations, reforming it to hydrogen at the stations, and using the hydrogen directly in vehicles. The ultimate step, though, is producing the hydrogen from renewable energy and delivering it through a hydrogen infrastructure—which is exactly the challenge Iceland is facing.

about 850,000 tons per year—accounts for 38 percent of national energy use, 57 percent of this used to run its motor vehicles and the boats of its relatively large fishing industry, the nation's leading source of exports (see "Energy Use In Island, by Source"). Dependence on oil imports costs the nation $150 million annually, and explains why transport and fishing each account for one-third of its carbon emissions.

The final third of Iceland's greenhouse emissions is found in other industries—primarily the production, or smelting, of metals like aluminum. The availability of low-cost electricity—at $.02 per kilowatt, it is the world's cheapest—has made Iceland a welcome haven for these energy-intensive industries. Metals production, along with transport and fishing, makes the island one of the world's top per-capita emitters of carbon dioxide, and offsets much of the greenhouse gas savings Iceland has achieved in space heating and electricity.

These features of Iceland's energy economy—a carbon-free power sector, costly dependence on oil for fishing and transportation, rising emissions from the metals industry—have placed the nation in a difficult situation with regard to complying with international climate change commitments. The 1997 Kyoto Protocol's guidelines for reducing greenhouse gas emissions in industrial nations are based on emission levels from the year 1990, which prevents Iceland from taking credit for its previously completed transition to greenhouse gas-free space heating and electricity generation. Although the government, arguing its special situation, negotiated a 10 percent reprieve between 1990 and 2010 under the Protocol, officials esti-

3 ❖ ENERGY: PRESENT AND FUTURE PROBLEMS

THE HYDROGEN CYCLE
A blueprint for the post-fossil fuel energy economy

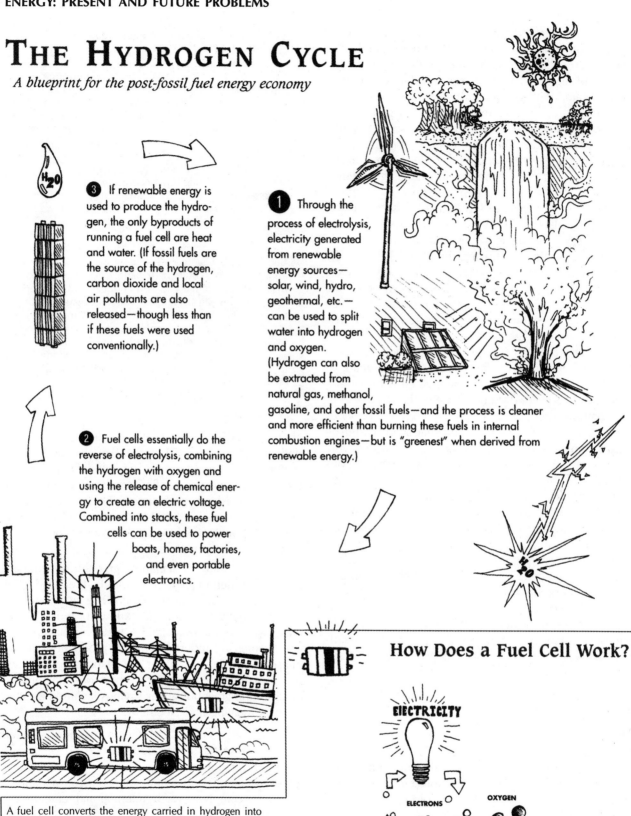

❸ If renewable energy is used to produce the hydrogen, the only byproducts of running a fuel cell are heat and water. (If fossil fuels are the source of the hydrogen, carbon dioxide and local air pollutants are also released—though less than if these fuels were used conventionally.)

❶ Through the process of electrolysis, electricity generated from renewable energy sources—solar, wind, hydro, geothermal, etc.—can be used to split water into hydrogen and oxygen. (Hydrogen can also be extracted from natural gas, methanol, gasoline, and other fossil fuels—and the process is cleaner and more efficient than burning these fuels in internal combustion engines—but is "greenest" when derived from renewable energy.)

❷ Fuel cells essentially do the reverse of electrolysis, combining the hydrogen with oxygen and using the release of chemical energy to create an electric voltage. Combined into stacks, these fuel cells can be used to power boats, homes, factories, and even portable electronics.

How Does a Fuel Cell Work?

A fuel cell converts the energy carried in hydrogen into electricity. One of the most promising types of fuel cells uses a "proton exchange membrane." When hydrogen molecules are injected, the cell's membrane splits them into protons and electrons. The protons pass through the membrane and react with oxygen, forming water and releasing heat. The electrons, which cannot pass through the membrane, travel along a circuit and create electricity.

ILLUSTRATIONS BY JONATHAN GUZMAN

12. Hydrogen Experiment

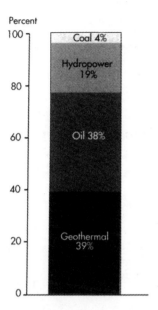

Energy Use In Iceland, by Source

mate that plans to build new aluminum smelters will cause it to exceed this target. Because of this so-called "Kyoto dilemma," Iceland is among only a few remaining industrial nations that have not signed the agreement.

In 1997, as the Protocol talks gathered momentum and the nation's dilemma was becoming apparent, a recently elected Parliamentarian named Hjalmar Árnason submitted a resolution to the Parliament, or Althing, demanding that the government begin to explore its energy alternatives. Árnason, a former elementary-school teacher who says he was "raised by an environmental extremist" father (he is not related to the scientist Bragi Árnason), soon found himself chairing a government committee on alternative energy, which was commissioned to submit a report. One of the first people he tapped for the committee was Professor Hydrogen.

Science Meets Politics

Bragi Árnason began studying Iceland's geothermal resource "as a hobby," he tells me, while a graduate student pursuing doctoral research in the 1970s. His deep knowledge of the island's circulatory system of hot water flows enables him to explain, for example, why the water you shower with in Reykjavík probably last fell as rain back in 1000 A.D. As he came to grasp the size of the resource, he began to consider ways in which this untapped potential might be used. At the time, the climbing cost of oil imports was beginning to hit the fishing fleet, prompting discussion of alternative fuels—including hydrogen.

Iceland has been producing hydrogen since 1958, when it opened a state fertilizer plant on the outskirts of Reykjavík under the post-war Marshall Plan. The production process uses hydro-generated electricity to split water into hydrogen and oxygen molecules—a process called electrolysis (see "The Hydrogen Cycle"). The fertilizer plant uses about 13 megawatts of power annually to produce about 2,000 tons of liquid hydrogen, which is then used to make ammonia for the fertilizer industry. In 1980, Bragi Árnason and colleagues completed a lengthy study on the cost of electrolyzing much larger amounts of hydrogen, using not only hydroelectricity but geothermal steam as well—which can speed up what is a very high-temperature process. Their paper found that this approach would be cheaper than importing hydrogen or making it by conventional electrolysis, but it did not find a receptive audience as oil prices plummeted during the 1980s.

The early 1990s saw a reemergence of Icelandic interest in producing hydrogen, both for powering the fishing fleet and for export as a fuel to the European market. In a 1993 paper, Dr. Árnason argued that a transition in fuels from oil to hydrogen may be "a feasible future option for Iceland and a testing ground for changing fuel technology." He also contended that the country could benefit from using hydrogen sooner than other countries. Some of his reasons included Iceland's small population and high levels of technology; its abundance of hydropower and geothermal energy; and its absence of fossil fuel supplies. Another was the relatively simple infrastructural change involved in converting the fishing fleet from oil to hydrogen, by locating small production plants in major harbor areas and adapting the boats for liquid hydrogen.

Early on, the plan was to use liquid hydrogen to fuel the boats' existing internal combustion engines. But "then came the fuel cell revolution," as Dr. Árnason puts it. By the late 1990s, the fuel cell, an electrochemical device that combines hydrogen and oxygen to produce electricity and water, had achieved dramatic cost reductions over the previous two decades. The technology had become the focus of engineers aiming to make fuel cells a viable replacement not only for the internal combustion engine, but for batteries in portable electronics and for power plants as well. Demonstrations of fuel cell-powered buses in Vancouver and Chicago, and their growing use in hundreds of locations in the United States, Europe, and Japan, caught the attention of governments and major automobile manufacturers. The fuel cell was increasingly viewed as the "enabling technology" for a hydrogen economy.

One Icelander particularly taken with these developments was a young man named Jón Björn Skúlason, who while attending the University of British Columbia in Vancouver became familiar with Ballard Power Systems, a leading fuel cell manufacturer headquartered just outside the city. Upon returning home, Skúlason encouraged the politician Hjalmar Árnason in his promotion of energy alternatives and hydrogen; his enthusiasm earned him a position on the expert committee. In 1998, the panel formally recommended that the nation consider converting fully to a hydrogen economy within 30 years.

> **Iceland's "Hydrogen Society" Strategy, 2000–2030**
>
> 2000   2020
>
> ▶ Test three hydrogen fuel cell buses in Reykjavík.
>
> ▶ Gradually replace the entire Reykjavík city bus fleet, and possibly other bus fleets, with hydrogen fuel cell buses.
>
> ▶ Introduce hydrogen-based fuel cell cars for private transportation.
>
> ▶ Test hydrogen fuel cell fishing vessels.
>
> ▶ Gradually replace fishing fleet with hydrogen fuel cell fishing vessels.

By then, Hjalmar Árnason had already given the process a push. During a phone interview with a reporter from the *Economist,* he floated the year 2030 as a target date for the government's evolving hydrogen plans. The resulting article, published in August 1997, created a buzz abroad, and the parliamentarian received hundreds of phone calls from around the world. That fall, Iceland's prime minister released a statement announcing that the government was officially moving the country toward a hydrogen economy. The ministers of energy and industry, commerce, and environment signed on, as well as both sides of the two-party Althing. And Árnason obtained permission to start negotiating with interested members of industry.

A Piece of the Action

Iceland has a tradition of "stock companies," or business cooperatives that evolved in the eighteenth century to help domestic farmers and fishers compete with the formidable Danish trading companies that at the time controlled fishing and goods manufacturing. The first of these, granted royal support in 1752, brought in weavers from Germany, farmers from Norway, and other overseas experts to teach the Icelanders the best methods of agriculture, boat-building, and the manufacture of woolen goods. Over the years, these long-lasting business associations helped the nation's enterprises survive and sometimes thrive.

The formation of the Icelandic Hydrogen and Fuel Cell Company (now Icelandic New Energy) can be seen as the latest example of this stock company tradition—but with a contemporary twist: German carmakers instead of weavers, Norwegian power companies rather than farmers. The first to contact Hjalmar Árnason after publication of the *Economist* article was DaimlerChrysler. Its roots traceable back to Otto Benz, designer of the first internal-combustion engine car, DaimlerChrysler now aspires to be the first maker of fuel cell-powered cars. The firm has entered into a $800 million partnership with Ballard Power Systems and Ford to produce fuel-cell cars, and plans to have the first buses and cars on European roads in 2002 and 2004, respectively—making Iceland a potentially valuable training ground, especially for testing fuel cell vehicles in a cold climate.

The second company to touch base with the Iceland government was Royal Dutch Shell, the Netherlands-based energy company that, among those now in the oil business, has perhaps the most advanced post-petroleum plans. Birthplace of the "scenario planning" technique that prepared it for the oil shocks of the 1970s better than most businesses, Shell has posited an Iceland-like future for the rest of the world, with 50 percent of energy coming from renewable sources by 2050. The firm surprised its colleagues in mid-1998 by creating a formal Shell Hydrogen division, and then sending its representatives to the World Hydrogen Energy Conference in Buenos Aires.

The third group to establish communications with island officials was Norsk Hydro, a Norwegian energy and industry conglomerate. The company is involved in a trial run of a hydrogen fuel cell bus in Oslo, and has considerable experience in hydrogen production: it has its own fertilizer business, and Norsk Hydro electrolyzers run Iceland's hydrogen-producing fertilizer plant. Norsk Hydro is also involved in the politically sensitive issue of Iceland's planned aluminum smelters, having signed commitments with the national power company and the ministries of energy and industry and commerce to construct a new smelter on the island's east coast.

Negotiations among these companies and the Icelandic government culminated in February 1999 with the creation of the Icelandic Hydrogen and Fuel Cell Company. Shell, DaimlerChrysler, and Norsk Hydro each hold shares of the company. The majority partner, *Vistorka* (which means "eco-energy"), is a holding company owned by a diverse array of Icelandic institutions and enterprises: the New Business Venture Fund, the University of Iceland, the National Fertilizer Plant, the Reykjanes Geothermal Power Plant, the Icelandic Technological Institute, and the Reykjavík Municipal Power Company. Also indirectly involved with the holding company is the Reykjavík City-Bus Company.

The stated purpose of the new joint venture is to "investigate the potential for replacing the use of fossil fuels in Iceland with hydrogen and creating the world's first hydrogen economy." On the day of its announcement, Iceland's environment minister stated: "The Government of Iceland welcomes the establishment of this company by these parties and considers that the choice of location for this project is an acknowledgement of Iceland's distinctive status and long-term potential." Like the *Economist* article, the announcement attracted industry attention. But for some companies, it was too late to climb on the bandwagon. Toyota officials reportedly attempted, to no avail, to take over the project by offering to foot its entire bill and supply all the needed engineers.

Buses, Cars, and Boats

Bragi Árnason and a colleague, Thorsteinn Sigfússon, have outlined a gradual, five-phase scenario for the hydrogen transformation. (See timeline.) In phase one (an estimated $8 million project that has received $1 million from the government), hydrogen fuel cells are to be demonstrated in Reykjavík's 100 municipal public transit buses. The current plan is to have three buses on the streets by 2002. The fertilizer plant will serve as the filling station for the buses, its hydrogen pressurized as a gas and stored on the roofs of the vehicles. Because enough hydrogen can be stored onboard to run a bus for 250 kilometers, the average daily distance traveled by a Reykjavík bus, there is no need for a complicated infrastructure for distributing the fuel.

In phase two, the entire city bus fleet—and possibly those in other parts of the island—will be replaced by fuel cell buses. The Reykjavík bus fleet program has a price tag estimated at $50 million and this spring received $3.5 million from the European Community. Phase three involves the introduction of private fuel cell passenger cars—which requires a more complicated infrastructure. At present, storing pressurized hydrogen gas onboard a large number of smaller vehicles, with more geographically dispersed refueling requirements, is too expensive to be considered a realistic option. The first fuel cell cars are therefore expected to run not on hydrogen directly, but rather on liquid methanol—which contains bound hydrogen but must be reformed, or heated, onboard the vehicle to produce the hydrogen to power the fuel cell.

Methanol is also, at the moment, the preferred fuel for the final two phases: the testing of a fuel cell-powered fishing vessel, followed by the replacement of the entire boating fleet. These trawlers use electric motors that are in the range of one to two megawatts—larger than those for cars and buses, but close to the size of the fuel cells that are now starting to be commercialized for stationary use in homes and buildings. Several European vessel manufacturers have already expressed their interest in becoming involved in this phase, and Dr. Árnason would like to see a fuel cell boat demonstrated no later than 2006.

But using methanol as an intermediate step to hydrogen is not without its problems. Skúlason, who is now president of Icelandic New Energy, notes that Shell is concerned about the use of methanol, particularly its toxicity. And since methanol reforming releases carbon dioxide, the environmental benefit is much less than if a way can be found to store the direct hydrogen onboard, which in Iceland's case would mean complete elimination of greenhouse gas emissions. It's a difficult decision, notes Skúlason: "We must deal with the technologies we are given by the global companies."

Iceland will have to choose between two options: producing and distributing pure hydrogen and storing it onboard vehicles (the "direct hydrogen" option); or producing hydrogen onboard vehicles from other fuels—natural gas, methanol, ethanol, or gasoline—using a reformer (the "onboard reformer" option). In general, the automobile industry strongly favors the onboard option, using methanol and gasoline, because most existing service stations already handle these fuels. A third path, reforming natural gas at hydrogen refueling stations, is under consideration in countries like the United States, that already have an extensive natural gas network, but is not practical in Iceland.

The up-front costs of direct hydrogen will be high because such a change requires a new infrastructure for transporting hydrogen, handling it at fueling stations, and storing the fuel onboard as a compressed gas or liquid. According to DaimlerChrysler's Ferdinand Panik, retrofitting 30 percent of service stations in the U.S. states of New York, Massachusetts, and California for methanol distribution would cost about $400 million. Supplying hydrogen to these stations would cost about $1.4 billion.

But in terms of long-term societal benefits, direct hydrogen is the clear winner. Using hydrogen directly is more efficient, because of the extra weight of the processor and lower hydrogen content of the methanol or gasoline. It is also less complex than having a reformer onboard each vehicle—which adds $1,500 to the cost of a new car, takes time to warm up, and creates maintenance problems. As the vehicle population grows large enough to cover the capital costs of providing refueling facilities, the costs of direct hydrogen will become comparable to the onboard option. Once the infrastructure and vehicles are put in place, using hydrogen fuel will be more cost-effective than having cars with reformers—even excluding the environmental gains. If Iceland, with its heavy renewable energy reliance, were to switch directly to hydrogen, the country would have *no* greenhouse emissions. And in fact, it is much easier to produce hydrogen than methanol from renewable energy through electrolysis. Thus, as renewables become more prominent around the world, a hydrogen infrastructure will emerge as the most practical option. In Iceland, rather than require that hydrogen first be used to create, and then be reformed from, methanol, the simplest approach would be to use geothermal power and hydropower, augmented by geothermal steam, to electrolyze water, creating pure hydrogen to drive cars and boats. But behind the seeming solidarity of the public/private venture, a fateful struggle may be emerging.

A Fork in the Road

In spite of the long-term economic and environmental advantages of the direct hydrogen approach, industry and government—both in Iceland and worldwide—have devoted substantially greater attention and financial support to the intermediate approach of using methanol and onboard reformers. Car companies are hesitant to mass-produce a car that cannot be easily refueled at many locations. Energy companies, similarly, are loathe to invest in pipelines and fuel stations for vehicles that have yet to hit the market. This is a classic case of what some engineers call the chicken-and-egg dilemma of creating a fueling infrastructure. But the potential public benefits—especially for addressing climate change—give governments around the world incentive to steer the private sector toward the optimal long-term solution of a hydrogen infrastructure, by supporting additional research into hydrogen storage and by collaborating with industry.

In Iceland's case, producing pure hydrogen through electrolysis by hydropower is at the moment three times as expensive as importing gasoline. But the fuel cells now being readied for the transportation market are three times as efficient as an internal combustion engine. In other words, running the island's transport and fishing sectors off pure hydrogen from hydropower is becoming economically competitive with operating conventional gasoline-run cars and diesel-run boats.

> "The transition is messy. We have one leg in the old world, and one in the new."
>
> Hjalmar Árnason,
> member of Iceland's Parliament

Since the methanol reformers these fuel cells will presumably use are still several years away from mass production, some scientists see the next few years as an important window of opportunity to prove the viability of direct hydrogen technology. But the history of technology is littered with examples of inferior technologies "locking out" rivals: witness VHS versus Beta in the videocassette recorder market. If methanol does gain market dominance, and locks out the direct hydrogen approach, it may be decades before real hydrogen cars become widespread—a wrong turn that could take the Icelandic venture kilometers from its destination. By the time a full-blown methanol infrastructure were put in place, it would probably no longer be the preferred fuel—committing the country to a fleet of obsolete cars and causing the consortium to strand millions of *kronur* in financial assets.

Yet some outside developments are pointing in the direction of direct hydrogen. In California, where legislation requires that 10 percent of new cars sold in 2003 must produce "zero-emissions," a consortium called the California Fuel Cell Partnership is planning to test out 50 fuel cell vehicles and build two hydrogen fueling stations that will pump hydrogen gas into onboard fuel tanks. Hydrogen fueling stations have already been built in Sacramento (California's capital), Dearborn, Michigan (home to Ford headquarters), and the airport at Frankfurt, Germany—the last of which expects to eventually import hydrogen from Iceland. The prospect of Iceland becoming a major hydrogen exporter, perhaps the new energy era's "Kuwait of the North," surfaces several times during my interviews—and is no doubt a good selling point for the strategy to officials inclined to think more in narrow economic terms.

Skúlason assures me that there is a "very open discussion" underway within the consortium, and says "we have to take steps slowly because there might be a shift." He admits that he would prefer to see compressed hydrogen gas used, noting the advantages of having direct hydrogen fuel infrastructure and vehicles. Shell and DaimlerChrysler themselves seem to recognize the potential competitive advantage of putting up hydrogen filling stations and reformer-free cars right from the beginning, giving them a headstart in preparing for a world fueled by hydrogen. At a June 2000 conference in Washington, DC, Shell Hydrogen CEO Don Huberts asserted that direct hydrogen was the best fuel for fuel cells, and suggested that geothermal energy converted to hydrogen would be the main means

12. Hydrogen Experiment

> "Many people ask me how soon this will happen. I tell them, 'We are living at the beginning of the transition. You will see the end of it. And your children, they will live in this world.'"
>
> Bragi Árnason,
> University of Iceland

for converting the Icelandic economy. DaimlerChrysler representatives have admitted that their methanol reformers are relatively expensive and large—they take up the entire back seat—and the company has recently rolled out "next generation" prototype cars that run on liquid and compressed hydrogen—prime candidates for the Iceland strategy.

"The transition is messy," the politician Hjalmar Árnason tells me. "We have one leg in the old world, and one in the new." It's an apt metaphor, given Iceland's geography. But the question is whether the Icelandic venture will, in rather un-Viking fashion, cautiously creep ahead—sticking to the onboard methanol approach—or, brashly set *both* feet in the new world, voyaging straight to direct hydrogen. As a world leader in utilizing renewable energy sources, if Iceland does not take the "newest" path, governments and businesses elsewhere may extract the wrong conclusion from its experiment and give short shrift to the direct hydrogen option. Skúlason nails the conundrum: "How many times will we shift? Will it be cheaper for society to pay a little more now and not have to rebuild? This argument doesn't always work with government or the consumer."

Professor Árnason is quick to note that, whichever short-term infrastructural path the country takes, "the final destination is the same:" pure hydrogen, derived from renewable energy and used directly in fuel cells. But he acknowledges that there may be significant costs in taking the gradual approach. And he agrees that the assumption on which his scenario is based—that methanol is the most economical option—is "subject to revision." The cost and efficiency of fuel cells will continue to improve, and advances in carbon nanotubes, metal hydrides, and other storage technologies are making it more feasible to store hydrogen onboard. The high cost of electrolysis is likely to decline sharply with technical improvements, while other sources of hydrogen—tapping solar, wind, and tidal power, splitting water with direct sunlight, playing with the metabolism of photosynthetic algae—are on the horizon. And new climate policies or fluctuating fuel prices from volatile oil markets "would change the whole picture."

Why Iceland?

When he first met with his prospective joint venture partners, Bragi Árnason posed this query: "Why are you interested in coming to Iceland?" He asked the question because "we were quite surprised to learn about the strong interest of these companies in participating in a joint venture with little Iceland." Their answers shed light on some of the elements that may be useful for developing a hydrogen economy elsewhere in the world.

Without a doubt, the most critical element of getting the Iceland experiment underway has been the government's clearly stated commitment to transforming itself into a hydrogen economy within a set timeframe. A similar dynamic is at work in California, where the zero-emission mandate has forced energy and transport companies to join forces with the public sector to seriously explore hydrogen. For Dr. Árnason, the lesson is clear: a strong public commitment can attract and encourage the participation of private sector leaders, resulting in partnerships that provide the financial and technical support needed to move toward environmental solutions. "You *must* have the politicians," he says.

In addition, companies have shown interest in the Iceland experiment because the results will be applicable around the world. While the country's hydrogen can be produced completely by renewable energy, its car and bus system and heavy reliance on petroleum—amplified by its island setting—are common characteristics of industrial nations, making the result somewhat adaptable. The island's head start in transitioning to renewable energy also makes it a good place to test out this larger shift.

Iceland may also have something more to tell us about the more general cultural building blocks that can enable the evolution of a hydrogen society. Icelanders treasure their hard-won independence, and the prospect of energy self-reliance is attractive. Hjalmar Árnason likes to emphasize his homeland's "free, open society," which he believes has maintained a political process more conducive to bold proposals and less subject to special-interest influence and partisan gridlock. He points, too, to the country's openness to new technology—to its willingness to take part in international scientific endeavors such as global research in human genetics. He hopes Iceland will become a training ground for hydrogen scientists from around the world, cooperating internationally to convert its NATO base to hydrogen. Skúlason cites a poll of Reykjavík citizens indicating that 60 percent of the citizens were familiar with and supportive of the hydrogen strategy—though some ask about the safety of the fuel (it is as safe as gasoline), pointing to the need for public education campaigns before people will be persuaded to buy fuel cell cars.

Another important cultural factor has been what Árni Finnsson, of the Icelandic Nature Conservation, describes as his nation's relatively recent but increasing "encounter with the globalization of environmental issues." This encounter originated with the emotional whaling disputes of the 1970s and 80s, and today includes debates about persistent organic pollutants and climate change. As Icelanders seek to become more a part of global society, so too do they seek legitimacy on global issues, forcing their government to sensitize itself to emerging cross-border debates—a process that has sometimes created Iceland's political equivalent of volcanic eruptions.

Finnsson points out that, thanks to the "Kyoto dilemma," Icelandic climate policy is not terribly progressive, consisting mainly of efforts to create loopholes that would allow additional greenhouse gas emissions from its new aluminum smelters. But there is little doubt that this dilemma has also unwittingly helped encourage the hydrogen strategy, by forcing the nation to explore deep changes in its energy system. In a land that, even as it becomes wired to the information age, routinely blocks new road projects due to age-old superstitions of upsetting elves and other "hidden people," it's a contradiction that somehow seems appropriate. A country that has stubbornly refused to sign the Kyoto Protocol provides the most compelling evidence to date that climate change concerns—and commitments—will increasingly drive the great hydrogen transformation.

But my favorite, if least provable, theory for "Why Iceland?" comes from the heroic ideals of its sagas. One of the recurring themes of these remarkable literary works is that a person's true value lies in renown after death, in becoming a force in the lives of later generations through one's deeds. Listening to Bragi Árnason, who is now 65, one cannot help but wonder whether this cultural concern for renown is playing a part in the saga now unfolding: how Iceland became the world's first hydrogen society, inspiring the rest of the globe to follow its lead. "Many people ask me how soon this will happen. I tell them, 'We are living at the beginning of the transition. You will see the end of it. And your children, they will live in this world.'"

Seth Dunn is a research associate at the Worldwatch Institute. He is the author of Worldwatch Paper 151, *Micropower: The Next Electrical Era.*

POWER PLAY

Enron, the nation's largest energy merchant, won't let California stand in its way

On this cold January day in Houston, Enron President Jeffrey K. Skilling could easily play the pirate that California consumer groups are casting him as these days. After two weeks of sailing with his three children in the Virgin Islands, Skilling's face is slightly sunburned, and he sports a rakish post-vacation beard. But the CEO-elect isn't buying the buccaneer image that some have slapped on his company. He clearly thinks Californians should be thanking Enron, not castigating it, for its role in trying to push open the state's power markets. "We're on the side of angels," he says. "We're taking on the entrenched monopolies. In every business we've been in, we're the good guys."

Alas, the nation's largest energy merchant is garnering no such accolades from California's great deregulation experiment. Soaring power prices have pushed the state's utilities to the brink of bankruptcy and forced Third World-style blackouts across the world's sixth-largest economy. Enron and other electricity marketers and generators are being investigated by the state attorney general and sued by consumers amid accusations of profiteering and market manipulation. "Every trading company in the country has been feasting on California, and Enron is the shrewdest of them all. They are like sharks in a feeding frenzy," says Michael Shames, executive director of the Utility Consumers' Action Network in San Diego. Enron, an early critic of California's deregulation plan, hotly denies those charges.

> Enron hotly **DENIES** accusations of market manipulation from consumers. Meanwhile, an **INVESTIGATION** of electricity marketers is under way

FRICTIONLESS? The glaring spotlight on California's botched attempt at deregulation casts Enron in the uncomfortable role of defending its radical business model. Though often grouped with utilities, Enron produces little power itself and owns relatively little in the way of hard assets. Instead it has pioneered the financialization of energy; making the company more akin to Goldman Sachs than Consolidated Edison. Its impressive profits stream is squeezed out of a torrent of often low-margin trades, in which it buys and sells a dazzling variety of contracts. The more buyers and sellers, the better for Enron, which is now twice the size of its nearest competitor.

Enron executives paint this as nothing less than a holy war on inefficiency. When the power giants are busted up, scrappy Enron believes it can thrive by delivering just the products and services particular customers want most. And they're thinking way beyond just energy. Skilling, a former McKinsey & Co. director, is now applying the model to a slew of new markets, including data storage, steel, even advertising space. He believes the company can thrive as a market maker. "Whether it's Enron or not is almost irrelevant," says Skilling. "It's going to happen."

For now, though, California's crisis has abruptly slammed the brakes on efforts to deregulate U.S. retail electricity sales, which are under way in nearly 25 other states. Moreover, that mess may prove to be the death knell for industry hopes that retail deregulation would spread far beyond the 20% of U.S. businesses and homeowners who now can name their own electricity supplier. "Deregulation has lost a lot of momentum," says Kemm C. Farney, vice-president for electric power at consulting firm WEFA Inc. "It's hard to argue that this has resulted in lower bills." Even in overseas markets, where Enron is counting on huge future gains, supporters are watching nervously. Big utilities, particularly in the south of Europe, "want to use California to slow down liberalization," says Jan van Aken, secretary general of the European Federation of Energy Traders.

It's not hard to see why. Sure, the old regulated system was expensive, but at least we knew the lights would turn on when we flipped the switch. However, Skilling and others argue that the battle in California

THE ENRON WAY

Enron makes money by trading commodities and other contracts. Here's how it works:

ENERGY DEALS

1 Enron has a contract to provide an aluminum smelter with electricity at a fixed price for three years. Enron buys aluminum from the smelter.

2 Prices move down for aluminum and up for electricity. Enron pays the smelter to shut down temporarily, buying cheaper aluminum in the market to meet the smelter's commitments.

3 Enron then sells the electricity elsewhere at a higher market price.

RISK MANAGEMENT

1 A small clothing cataloger asks Enron to create a derivative to protect against hot weather that might hurt sales of winter clothes.

2 For every degree above normal over some period, Enron must pay the retailer a certain amount; for every degree below, the retailer pays Enron.

3 Enron lays off its risk through a deal with a soft drink maker who benefits from hot weather.

---

has obscured the reasons for pushing deregulation in the first place. The old utilities, he says, "were incredibly expensive and provided horrible service to their customers." Under the regulated model where utilities could simply pass on costs to customers, they lacked incentives to utilize capacity more efficiently or to offer innovative services. In California, for instance, business leaders pushed for deregulation because they were paying 50% more for power than their counterparts in other parts of the country. At the same time, the utilities themselves were operating high-cost plants with little access to capital to fix their problems, says Farney. The upside of deregulation? "We have a tremendous amount of new construction. It's a direct result of the industry having new access to capital markets," he says.

One of the biggest problems now is that the nation is in a muddled state between regulation and deregulation. For instance, freezing retail prices for consumers and businesses in California while deregulating the wholesale market where utilities shop left the power companies unable to cover their skyrocketing energy costs and pushed them near the brink of financial collapse. Paul W. MacAvoy, a professor at the Yale School of Management, compares that to the blunder that regulators made in the 1970s with the savings and loan industry. The S&Ls were restricted to investing in long-term assets, like mortgages, while paying volatile market rates on shorter-term deposits. "There's a widespread opinion in the academic world that regulation is bad, but partial and phased-in deregulation is much worse," says MacAvoy.

Skilling argues that Enron's exposure to fallout from California is minimal. Its biggest business is in the wholesale market, serving utilities and big industrial customers who under federal law already have the right to choose their electricity suppliers. And wholesale markets in

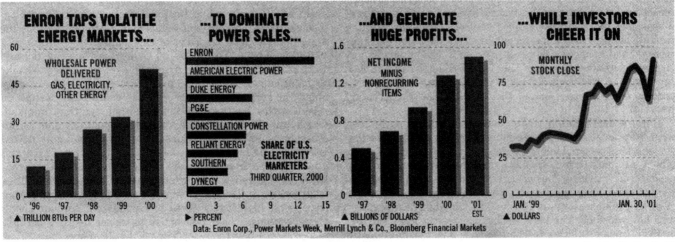

Charts by Alan Baseden/BW

Europe and Japan are rapidly opening their doors. That explains why most observers believe that Enron will emerge with its earnings engine intact. "We do not expect the California situation to have any significant impact on Enron's financial outlook," says Skilling, who is set to take over the CEO title from Enron Chairman and CEO Kenneth L. Lay on Feb. 12. In the fourth quarter, Enron's core wholesale trading and services business reported income of $777 million before interest and taxes, nearly triple that of a year earlier. Enron's total net income excluding nonrecurring items shot up 32% for all of last year, to $1.27 billion, on sales of $100.8 billion, up 150%. Physical volumes of gas and electricity delivered jumped 59%, boosted in part by EnronOnline, the company's year-old Web-based trading operation.

POWERFUL ALLIES. Under a new Republican Administration and a President with strong ties to Lay ("Enrons's Big Wheel Has a Heavy Tread"), Washington isn't likely to derail deregulation. If anything, Enron will be using the crisis to push for more favorable federal rules for moving electricity from state to state, which could boost its flexibility and trading options. Certainly, Curt L. Hebert, the new chairman of the Federal Energy Regulatory Commission (FERC), is a free-market advocate friendly to Enron's views. "I don't think California is going to be any lesson that stops America from wanting choice. It's going to be a lesson that allows America to learn what not to do," he says.

Still, Skilling and Lay will have to spend precious time and energy on a deregulation battle that Enron seemed well on its way to winning. Already, Oregon, Arkansas, and Nevada are considering a slowdown in their deregulation plans, while northeastern states have put in place wholesale price caps. Some Western governors want Congress to force regional price caps on the wholesale electricity market. And Representative Peter A. DeFazio (D-Ore.) is ready to introduce an "energy re-regulation" bill that would overturn the 1992 U.S. Energy Policy Act. "Electricity is not a commodity fit for the competitive marketplace. Private investors don't have public safety in mind," says Doug Heller, consumer advocate with the Foundation for Taxpayer & Consumer Rights in Santa Monica, Calif., which is pushing for a referendum to have California take over much of the power system.

If that kind of radical sentiment spreads beyond the California border, Enron could find its growth prospects, even in the $220 billion wholesale market, sharply curtailed. And it would surely be bad news for Enron's latest effort to crack the retail market: a separately owned company formed last May with then-America Online Inc. and IBM called the New Power Co. Enron tried four years ago to enter the residential market in California but quickly backed out when it found the state's new rules wouldn't let it turn a profit.

The hyperconfident Skilling is unrepentant about his company's long

---

IS THERE ANYTHING THEY CAN'T TRADE?

Enron still makes most of its money trading energy. Its earnings from sales to utilities and other wholesale customers should grow at a 35% to 40% annual pace for at least several years, it says. However, the company is also branching out into entirely new markets:

RETAIL ENERGY

Enron signed new contracts worth $16 billion last year to manage energy needs for big business customers.

BROADBAND

It projects a $450 billion worldwide market for communications bandwidth trading and services by 2005. It has begun distributing video-on-demand and other content over its own high-speed fiber network to customers like Blockbuster.

PAPER, PULP, AND LUMBER

Enron launched a specialized Web site, Clickpaper.com, tailored to the $330 billion wood-products market, and bought newsprint maker Garden State Paper Co. for $72 million last October to ensure access to supplies.

MEDIA

The plan is to create a market for trading advertising time and space in TV, radio, and print. That includes offering advertisers, TV networks, and publishers ways to hedge risks, such as long-term fixed-priced contracts.

STEEL

The company's first target is a $22 billion segment of the commodity steel business. Enron plans to lease a network of regional distribution centers to ensure reliable deliveries.

role as a deregulation crusader. Enron benefits from volatility, not high prices, he says. In California, where Enron doesn't own power plants, the company insists it is dealing in thin margins. "Depending how markets are swinging around day to day, we're either making a little money or losing a little money on that," says Lay, although he won't break out specific figures for California. If anything, says analyst Donato J. Eassey of Merrill Lynch & Co., operating margins are getting thinner—in the entire wholesale business they were 2.4% last year, vs. 3.7% in '99 and 3.6% in '98. That's O.K., says Eassey: "I'll take that margin pressure with a 72% earnings growth rate on that [wholesale] business."

In its gleaming Houston office tower, Enron has seven floors of some 1,500 traders making markets in gas, electricity, metals, bandwidth, and other products. Enron, for example, posts prices for an array of energy contracts. With the click of a mouse, utilities caught short on supply can make a deal. Likewise, generators with excess capacity can find prices Enron is willing to pay. Enron pockets a spread on the deal. The traders are supported by a back-office team that schedules pipeline and transmission capacity to actually deliver gas and electricity, checks credit, and handles billing. Each desk of two or three traders used to handle about 100 transactions a day when doing their business by phone. Now a desk handles 800 to 900, mostly on EnronOnline.

Enron's success has not gone unnoticed, but it has a few advantages that competitors would be hard pressed to match. First is its expertise; it pioneered the model. It has dozens of PhDs in mathematics, physics, and other disciplines, and even hired a former shuttle astronaut to schedule satellite time. Enron also has built a huge network of buyers and sellers with a wide range of commodities to trade. With its deep pockets Enron has plenty of capital to keep its markets liquid.

> The company's recent **SUCCESS** is sure to make its competition more aggressive

A big part of the value Enron provides is helping companies manage risk, especially the risk of big price swings or delivery snafus. One example: In 1999, Peoples Gas, Light & Coke Co. of Chicago, a local distribution company, signed a five-year gas procurement deal with Enron. Enron took over Peoples' scheduling of gas pipelines and storage, assets that Enron then could use to meet commitments for a broad array of other customers. In the meantime, it procures gas daily as Peoples needs it, provides working capital, handles accounts receivable and payable, manages storage, and finances the gas in storage. "This is our distinctive competence—bringing the whole thing to the table," says Skilling. "No one's our equal in that." And Enron has pioneered contracts to manage other kinds of risk too, including weather.

Enron's deregulation roots reach back to the early '70s, when Lay, an economist by training, began promoting opening gas markets as an Under Secretary of the Interior in the Nixon Administration. In 1986, Lay took over Enron, the product of a merger between two pipeline companies. He warned federal regulators that pipelines like his were in danger of sinking under rules that crippled their ability to compete with low-priced oil. The FERC finally changed the rules starting in 1985, freeing utilities to shop for gas and the pipelines to search for customers. Enron embraced the changes with gusto, rapidly becoming the largest buyer and seller of gas in North America It then pushed just as aggressively to open wholesale and retail electricity markets, to the chagrin of the nation's entrenched utilities. In Ohio, it offered consumers coupons for free electricity to build political support for deregulation. In California, it fielded an army of regulatory attorneys to battle the state's utilities over the rules of engagement.

SERIOUS FLAWS. The seeds of California's energy debacle were planted long before deregulation set in, though. Thanks to environmental concerns, poor planning, and uncertainty about the rules of deregulation, California's power supply in recent years has been growing far slower than demand, which was driven by an expanding population and its increasing use of energy-sucking technology. The badly flawed 1996 deregulation scheme—some terms of which Enron and others fought from the start—only exacerbated the problems. It included a pricing system that prevented utilities from locking in long-term supplies at fixed prices and from passing on higher fuel costs to customers. Without rising prices, consumers had no incentive to reduce demand and utilities couldn't pay their soaring wholesale power bills. That scared away some power suppliers and created the financial crisis from which the state is still trying to dig out.

By now there is nearly universal agreement that the biggest flaw in California's deregulation plan was the decision to force utilities to buy all of their power needs one day in advance from a newly formed entity called the California Power Exchange. The theory was that the exchange would provide the most transparent prices, since every buyer and seller had to operate through it. Whatever power didn't get bought through the exchange would be purchased on a last-minute basis the following day by another entity called the Independent System Operator (ISO).

But many observers believe this two-step setup encouraged generators to offer less power to the ex-

13. Power Play

ENRON'S BIG WHEEL HAS A HEAVY TREAD

Texas business has always prided itself on its "can-do" attitude. But when it comes to prodding government policymakers to action, nobody comes close to Enron Corp. CEO Kenneth L. Lay. "In recent years, that has become Kendo," says Chase Untermeyer, director of government affairs at Compaq Computer Corp. and chairman of the Texas State Board of Education.

So what has Ken done? In Houston, Lay led the fight for a new baseball stadium, which opened last April under the Enron name and with $100 million of its funding. As chairman of the Governor's Business Council under former Governor George W. Bush, Lay pushed for and won state education reform, litigation curbs, and tax cuts. In Washington, Lay's army of lobbyists has sought everything from electricity deregulation to tax breaks.

Now, with Bush in the White House, Lay stands to become one of the most influential corporate players in the country. With the public furor over California's power crisis, he'll work to stave off cries for reregulation of electricity. Says Craig McDonald, executive director of Texans for Public Justice, a liberal consumer advocacy group: "Enron is at the top of the top group of corporations that has the ear of George W. Bush."

"**PIONEER.**" Indeed, the ties between the Texans run deep. They share a background in the energy business and a free-market worldview. Lay was a big backer of the first President Bush, whose Secretary of State James A. Baker III and Commerce Secretary Robert A. Moshbacher were hired as Enron consultants after Bush's '92 defeat. Enron was the younger Bush's leading patron in Austin, donating more than $550,000 to the governor, according to the Center for Public Integrity, a Washington watchdog group. When Bush set his sights on the White House, Lay was one of 212 "pioneers" who raised more than $100,000, and Enron was a campaign's top supplier of corporate jets. Enron donated $250,000 for the Republican national convention. In addition, Lay sent the maximum $5,000 to the Florida recount legal fund, and gave $100,000 to the inaugural committee.

> Lay was **PRESIDENT BUSH'S** leading patron during his years in Austin

Lay, who steps down as Enron's CEO this month but will remain chairman, intends to stay plugged in. During the Presidential campaign, he was a key Bush adviser on energy. He has since been named a "transition adviser" to the Energy Dept. High on Enron's list of priorities is federal authority to approve interstate electricity transmission lines. Lay says he was willing to serve in a key Bush Cabinet post but didn't lobby for a job, and no offer came: "I've got a job, and I can support President Bush in a lot of other ways, just like I did his father."

Beyond the White House, Enron has a vast lobbying operation that pushed issues such as deregulation, tax breaks, telecom, trade, and environmental quality. While Enron tilts toward Republicans, it gives generously to both parties. In the 2000 election, the company and its top execs funneled $1,095,350 in soft money to Republican Party committees and $519,565 to Democrats, says the nonpartisan Center for Responsive Politics.

Those who have been subjected to Enron's lobbying efforts say the company is persistent but not overbearing in pushing its agenda. "They are forceful and they are persuasive," says former Hill staffer David M. Nemtzow, now president of the Alliance to Save Energy, a nonprofit coalition of companies and government officials. With friends in high places, the next few months will determine just how much Houston's Ken-do can do.

By Richard S. Dunham and Laura Cohn in Washington

---

change and instead to wait until the last minute to sell power to the ISO, which out of desperation would have to pay higher prices. In three lawsuits, consumer groups, municipal water districts, and the City of San Francisco allege that Enron and other electricity marketers engaged in unlawful market manipulation. Several investigations so far have failed to prove collusion by Enron or others. But one study said that the soaring prices couldn't be explained by such factors as high demand and rising fuel and environmental costs. "We concluded that power was being withheld inexplicably, at the exact time at which prices were most vulnerable to manipulation," says Edward Kahn of the Analysis Group/Economics, a co-author of one study backed by the parent of utility Southern California Edison. Adds coauthor Paul L. Joskow, a professor of economics at Massachusetts Institute of Technology: "It was bad regulation, bad market design, bad luck, and greed."

NOT ROLLING OVER. Enron dismisses the study as flawed and slanted to favor the utility that paid for it. Meanwhile, the attorney general's office, state Public Utility Commission, and FERC are all continuing to investigate. California's current plans to fix the problem involve the state becoming the lead buyer, through long-term contracts that take the uncertainty out of pricing. The Power Exchange is being phased out. Enron applauds that move. And even some of the hardest-hit power users in California haven't given up on deregulation. California Steel Industries Inc., a Fontana (Calif.)-based steel producer, lost power 14 times in January, costing it millions of dollars. Yet Lourenco Goncalves, its president and CEO, is not advocating a return to regulation: "Deregulation is a good idea," he says. "We just need to treat this like a business and fix what went wrong."

The energy crisis comes just as Enron is vigorously pushing its promise of efficiency and lower prices into a slew of new commodities. "Look at pulp and paper. Look at all those salesmen who play golf all day. For a commodity, for crying out loud. You don't need to play golf to sell a commodity," blurts Skilling in his typically blunt style. In these new arenas, Enron won't face the kind of regulatory obstacles it must eliminate in energy. But it's no wonder that many players will resist what they see as an arrogant and menacing middleman. The California experience gives them new ammunition. Enron and other traders "sure as hell haven't done a very good job of protecting the American public from the rising price of natural gas or electricity. I don't see that they bring any real value to the steel marketplace," says Daniel R. DiMicco, CEO of Nucor Corp., one of the nation's leading steelmakers.

In one of his first big bets outside of the energy patch, Skilling is attacking the telecommunications industry. In the past year, Enron has spent nearly $500 million to create its own high-speed fiber network. But the goal is not to be a traditional telecom player. Instead, Enron wants to create a vast spot market for high-speed communications capacity, and ultimately a futures market. Consider its deal with Blockbuster Inc. The two signed a 20-year pact for Enron to deliver movies-on-demand to Blockbuster customers, a service that's now in four test markets. Instead of having to build its own network and coordinate a multitude of agreements with local high-speed communications providers, Blockbuster turned the whole thing over to Enron. It pays Enron per movie, getting only the capacity it needs, when it needs it. For now, those movies are moving mostly on Enron's 13,500-mile fiber network. But as a robust market for trading bandwidth develops, Enron could pluck the capacity it needs from other communications companies. It may even sell its fiber network when it no longer needs the physical capacity to ensure delivery to its customers. "The minute I don't need it, it's gone," says Skilling.

Indeed, Enron's new CEO has already demonstrated that he doesn't linger over troubled assets. To help fund its vast ambitions, Enron expects to shed more than $2 billion in plants and other properties around the world this year, including a gas production plant in India and a wind-power company in California. In the fourth quarter, the company took a $326 million aftertax charge to cover problems at Azurix Corp., its failed water business that's likely to be busted up. Under Skilling, Enron is essentially abandoning its once-ambitious plan to build power plants, pipelines, and other facilities around the globe.

Now, Enron will build assets where they clearly support the trading operations, including places like Spain and Japan. Rival energy marketer and producer Dynegy Inc. of Houston questions just how competitive Enron can be without a significant physical presence in its new commodities "to give you that incremental intelligence," says CEO Chuck Watson. Without that, "you don't really have a competitive advantage." But investors don't seem worried. Indeed, Enron's stock is up about 175% in the past two years. "One thing about Enron is they're incredibly well managed, very smart, and entrepreneurial," says portfolio manager Robert L. Shoss of shareholder AIM Capital Management Inc. "They've proven they deserve the benefit of the doubt."

As Enron gets bigger, maintaining its entrepreneurial culture and stiff risk-management controls will surely prove a bigger challenge. And more competition from strong rivals, like Dynegy, means "a lot of the margins are going to go away," says one utility analyst. "Early entrants get generous margins that attract other competitors." One of the biggest risks is that Enron simply can't create the open markets it needs. California stands as a stark reminder of that. If the state goes back to a "cost-of-service" utility model, where utilities simply pass through their costs to consumers, "that would cause us trouble," says Skilling.

But Skilling and other industry players don't think that's likely to happen, as other states such as Pennsylvania prove their deregulation models can work. Indeed, some believe that the nation's muddled effort to move toward deregulation is the best of all worlds for Enron. "If the market were really open and very, very efficient, there isn't a lot of need for these trading intermediaries," says Lawrence J. Makovich, a senior director of electric power research for Cambridge Energy Research Associates. Luckily for Skilling and Enron, that's one vision that's likely to remain a pipe dream for some time.

By Wendy Zellner in Houston, with Christopher Palmeri in Los Angeles, Peter Coy in New York, and Laura Cohn in Washington, D.C.

Bull Market in Wind Energy

Many countries may soon find that the cheapest way to produce electricity is to pull it out of the air

By Christopher Flavin

The Spanish city of Pamplona has long been known for its annual running of the bulls. But this mid-sized industrial center, the capital of the state of Navarra in the rugged Pyrenees region, is quickly gaining another distinction: it has the world's fastest growing wind energy industry. Starting from scratch just three years earlier, Navarra was obtaining 23 percent of its electricity from the wind by the end of 1998.

With a population of 180,000, Pamplona has an economy based heavily on manufacturing, including a sizeable car industry. But along with much of the rest of Spain, the city has had a relatively stagnant economy and a high rate of unemployment in recent years. In an effort to deal with that problem—and replace the coal and nuclear energy it imports from other parts of Spain with local power—Navarra recently introduced a set of tax incentives and other inducements for harnessing wind energy using locally manufactured turbines.

These policies paid off well beyond the dreams of the government officials who crafted them. Several wind-energy companies were quickly established in Navarra, most of them joint ventures owned in part by the Danish firms that supplied the technology. And much of the investment is coming from Energia Hidroelectrica de Navarra, the regional electric utility. These firms have provided a strong political base for the region's burgeoning wind power industry. Navarra's wind companies are already looking to expand their horizons to even larger potential markets in areas where Spain has strong historic ties, such as North Africa and South America.

The sudden transformation of Navarra's energy mix may turn out to be foreshadowing something much bigger. During the 1990s, wind power has already become the world's fastest growing energy source. Propelled by supportive new government policies—most of them motivated by environmental concerns—some 2,100 megawatts of new wind generating capacity were added in 1998, according to our preliminary estimate. That's not only a new record, but 35 percent more than the previous record set the year before.

Global wind generating capacity now stands at 9,600 megawatts—a 26 percent increase from a year earlier. (See figure) Wind turbines will generate a projected 21 billion kilowatt-hours of electricity in 1999—enough power for 3.5 million suburban homes. And though it now provides less than 1 percent of the world's electricity, these double-digit growth rates could make the wind a major power supplier soon.

Wind power is also one of the world's most rapidly expanding industries. Valued at roughly $2 billion in 1998, the wind industry is creating thousands of jobs at a time when manufacturing employment is falling in many nations. And as a booming new industry, wind energy has become a major investment opportunity—comparable perhaps to some of the internet stocks that are now so hot on Wall Street.

Ancient Ideas, Brand New Markets

Although it has had a recent rebirth, wind power is actually an ancient source of energy. The first windmills for grinding grain appeared in Persia just over 1,000 years ago, and later spread to China, throughout the Mediterranean, and to northern Europe, where the Dutch developed the massive machines for which the country is still known. In the Middle Ages, windmills allowed peasants to grind grain without depending on watermills controlled by feudal lords.

As the fossil-fuel age emerged in the early 20th century, wind power seemed to have become a permanent footnote in the history of energy technology. But in the 1970s, Danish companies invented a machine composed of three propeller-like fiberglass blades that pointed upwind of a steel tower on which they were mounted. The latest versions, which are also manufactured by companies based in Germany, India, Spain, and the United States, have variable pitch blades whose angle of attack varies depending on the wind speed. The blades, which can be as long as 40 meters, will spin in winds of little more than 15 kilometers per hour. They maintain a relatively slow and constant speed, though a new generation of electronic variable-speed drives allow the blade speed to vary, increasing the machines' efficiency. The generator—similar in design to those connected to diesel engines—sits atop the tower, along with the transmission, brakes, and sophisticated microprocessors that coordinate all of the equipment.

The 1998 wind energy boom was led by Germany, which added 790 megawatts, pushing its capacity to 2,875 megawatts, nearly double the total capacity in the United States. Germany's wind industry is only seven years old; it grew out of a 1991 electricity reform law that was motivated in part by the Chernobyl nuclear disaster. But already, wind generators are producing as much power as two of Germany's large coal-fired power plants, or a little more than 1 percent of the country's electricity. In the northern state of Schleswig Holstein, wind now provides 15 percent of the electricity, and is on course to supplant nuclear energy as the state's leading power source. The new German government plans to shut down the nuclear plants that supply 30 percent of the country's electricity—a move that may give wind power another substantial boost.

One of the most notable developments in 1998 was the emergence of Spain as the number-two player in the industry. Spain added an estimated 395 megawatts of wind power last year. That increase pushed the country's total wind capacity up 86 percent, to 850 megawatts. Robust wind energy industries have sprung up not only in Navarra, but also in the northwest state of Galicia, and in the south near Gibraltar. With development in all of these regions accelerating steadily, Spain could soon surpass Germany as the world's leading wind energy producer.

Wind generation also expanded in the United States in 1998. Some 230 megawatts of new capacity were added, to make up the largest increase in wind power that the country has seen since 1986. The new installations are spread across 10 different states, and were spurred by the desire to take advantage of a wind energy tax credit that is currently scheduled to expire in June 1999. They include a 107-megawatt wind farm in Minnesota, a 42-megawatt farm in Wyoming, a 25-megawatt farm in Oregon, and many small projects, ranging from Maine to New Mexico.

In contrast, to the erratic ups and downs of the U.S. industry, Denmark maintained its moderate, steady pace in 1998. The 235 megawatts of capacity added in the last year took Denmark's total wind capacity to 1,350 megawatts. Wind now generates over 8 percent of the country's power. Most notably, Danish wind companies utterly dominate the global export market; more than half the new wind turbines installed worldwide in 1998 were made in Denmark. The Danish companies are also involved in joint-venture manufacturing in India and Spain—an arrangement that has allowed for the rapid transfer of wind technology to the host countries. Altogether, the Danish wind industry had a turnover of just under $1 billion last year. That's roughly equal to the combined sales value of the nation's natural gas and fishing industries—two leading Danish sectors.

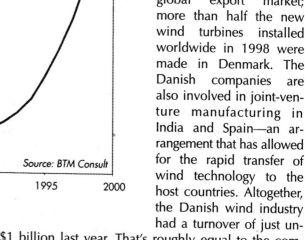

World Wind Energy Generating Capacity, 1980–98

Source: BTM Consult

A Technology That Has Come of Age

The nations that could benefit most from wind power are in the developing world, where power demand is growing rapidly and where most countries lack adequate local supplies of fossil fuels. The developing world's largest wind industry is in India, which has over 900 megawatts of wind power in place. But expansion has slowed there in the last two years, due to a suspension of the generous tax breaks that were in effect in the mid-1990s. Observers expect the new government to restore some of these incentives, which could give the industry a boost in 1999. Although wind power now provides less than 1 percent of India's electricity, its share could one day rise to 20 or 30 percent.

The wind potential of China is even greater. China has not yet established a solid legal basis for wind power, although several companies have installed small projects there in the last few years, with the help of foreign aid. But China could become a wind superpower. It has vast wind resources in several regions, including a huge stretch of Inner Mongolia that by itself could provide most of the power needed in Beijing and the rest of northern China.

14. Bull Market in Wind Energy

China's wind potential is estimated to exceed its total current electricity use. That fact has enormous international implications, since China's coal-based economy exacts a heavy environmental toll. A Chinese wind industry could allow a significant reduction in global greenhouse gas emissions.

The dramatic growth of wind power in the 1990s stems primarily from laws that guarantee access to the grid for wind generators at a fixed price. (The price offered is usually a bit higher than the cost of fossil-fuel power—a recognition that the environmental benefits of renewable energy technologies are worth paying for.) These laws have established a stable market for the new industry, and have overcome resistance from the coal- and nuclear-dependent utilities that monopolize the market today. Some 70 percent of the global wind power market in recent years has been centered in just three countries, Germany, Spain, and Denmark—a distribution that reflects both the success of such laws where they exist, and the failure to adopt them broadly.

Wind energy is also being spurred by steady advances in technology. Larger turbines, more efficient manufacturing, and more careful siting of the machines are among the improvements that have pushed wind power costs down precipitously—from $2,600 per kilowatt in 1981 to $800 in 1998. A typical wind turbine today produces 700 kilowatts of electricity, costs $700,000 (including installation), and provides enough power annually for 200 homes.

State-of-the-art wind turbines are highly automated and reliable; their downtime for maintenance, of less than 5 percent, is less than for fossil fuel plants, and maintenance costs are minimal. The "footprint" of a wind farm is also very small, since wind turbines blend readily with traditional uses of the rural landscape. (Farmers can either install their own wind turbines or lease the land to wind companies. The bulk of the land can still be used for grazing animals or raising crops.)

In many areas, wind power is already less expensive than electricity from coal-fired power plants. And as the technology continues to improve, further cost declines are projected. In many countries, the wind could become the most economical source of power in the next decade.

Over the past few years, wind seems to have achieved the kind of "critical mass" necessary to attract serious corporate interest. In 1996 and 1997, the largest U.S. natural gas company, Enron Corporation, purchased two wind manufacturing companies and is now developing projects around the world. Japanese trading companies have announced plans to build large wind projects, as have the German power giant Siemens and Florida Power and Light, a major U.S. electric utility. Companies in Denmark and the Netherlands are making plans for even larger offshore wind farms in the North Sea.

Continued and perhaps accelerated growth of the wind industry is likely in 1999 and beyond. Spain and the United States are projected to have particularly good years, probably exceeding 500 megawatts of new turbines each. Other countries where growth is likely include Canada, Italy, Japan, Norway, and the United Kingdom. The strongest potential developing-country markets include Argentina, Brazil, Costa Rica, Egypt, Morocco, and Turkey.

Ultimately, wind power could be a major force for transforming the global energy economy. In the United States, the states of North Dakota, South Dakota, and Texas have sufficient wind capacity to provide all U.S. electricity. In windy regions such as Patagonia, the American Great Plains, or the steppes of Central Asia, wind farms could churn out vast amounts of electricity. A study by Danish researchers in 1998 laid out a strategy for providing 10 percent of the world's electricity from the wind within the next few decades. In the longer run, wind power—both onshore and off—could easily exceed hydropower—which now supplies 20 percent of the world's electricity—as an energy source.

Wind power is still considered a laughing matter by many energy industry executives. But soon, such people may look almost as silly as those who once called the airplane an absurd idea. If that happens, Navarra may enter the history books alongside Kitty Hawk, North Carolina as the proving ground for a technology that changed the way the world lives.

Christopher Flavin is senior vice president at Worldwatch Institute, and co-author with Nicholas Lenssen of Power Surge: Guide to the Coming Energy Revolution *(W.W. Norton, 1994).*

Unit 4

Unit Selections

15. **Planet of Weeds,** David Quammen
16. **Invasive Species: Pathogens of Globalization,** Christopher Bright
17. **Mass Extinction,** David Hosansky
18. **Watching vs. Taking,** Howard Youth

Key Points to Consider

❖ How and why are alien plants and animals spreading so rapidly and what threats do they pose to native species in nearly all parts of the world? What gives the alien species advantages that native plants do not have?

❖ Why do biologists have such a difficult time in being precise about the number of plants and animals that are going extinct? How does the current rate of mass extinction compare with past periods of species loss in terms of numbers and in terms of time?

❖ How can newer forms of ecotourism such as wildlife watching protect species against extinction? Does wildlife watching, as contrasted with hunting and poaching, pose similar kinds of problems for wild animals?

 **Links** www.dushkin.com/online/

20. **Friends of the Earth**
    http://www.foe.co.uk/index.html
21. **GORP: Great Outdoor Recreation Pages**
    http://www.gorp.com/gor/resource/Us_National_Park/AK/wild_den.htm
22. **Smithsonian Institution Web Site**
    http://www.si.edu
23. **Tennessee Green**
    http://korrnet.org/tngreen/
24. **World Wildlife Federation**
    http://www.wwf.org

These sites are annotated on pages 4 and 5.

Biosphere: Endangered Species

Tragically, the modern conservation movement began too late to save many species of plants and animals from extinction. In fact, even after concern for the biosphere developed among resource managers, their effectiveness in halting the decline of herds and flocks, packs and schools, or groves and grasslands has been limited by the ruthlessness and efficiency of the competition. Wild plants and animals compete directly with human beings and their domesticated livestock and crop plants for living space and for other resources such as sunlight, air, water, and soil. As the historical record of this competition in North America and other areas attests, since the seventeenth century human settlement has been responsible—either directly or indirectly—for the demise of many plant and wildlife species. It should be noted that extinction is a natural process—part of the evolutionary cycle—and not always created by human activity, but human actions have the capacity to accelerate a natural process that might otherwise take millennia.

In the opening article of this unit, the losses of Earth's plants and animals from human impact are tallied. Author David Quammen makes the point in "Planet of Weeds" that the nature of human impact on the biosphere reduces biodiversity and encourages the dominance of plant and animal species that can be termed "weeds": hardy, resilient, able to live almost anywhere, and capable of choking out native species almost everywhere. Among the weedy species, human beings are paramount and Quammen's pessimistic view of a future world is of humans and a few other survivors "picking through the rubble" of a once fertile planet.

This theme is continued in the article in the subsection on plants. The most important component of the biosphere is the primary production of living vegetation, and, in the article in this subsection, the issue of human impact on vegetation is central. Environmental researcher Christopher Bright describes one of the least anticipated results of the growing global economy: the spread of "invasive species." In "Invasive Species: Pathogens of Globalization," Bright notes that thousands of alien species are traveling aboard ships, planes, and railroad cars from one continent to another, often carried in commodities themselves. This consequence of the worldwide global trading network poses enormous hazards for native species in affected areas and confronting the problem may be as important a challenge to environmental quality as global carbon emissions.

Prospects for the future of wildlife are not appreciably better than they are for many of the world's plant species. Land developers destroy animal habitats as cities encroach upon the countryside. Living space for all wildlife species is destroyed as river valleys are transformed into reservoirs for the generation of hydropower and as forests are removed for construction materials and for paper. Toxic wastes from urban areas work their way into the food chain, annually killing thousands of animals in the United States alone. Rural lands are sprayed with herbicides and pesticides, which also kill bird life and small mammals. Most important, wild animals are placed in jeopardy for the very simple reason that they compete with domestic stock and humans for the same resources. And in any instance in which the protection of wildlife comes at the expense of livestock or humans, the wildlife is going to lose.

One of the clearest and most recent examples of heightened competition between wildlife and animals is currently being played out in virtually all of the world's environments where thousands of species are becoming extinct, most of them because of competition or competition-related processes. The rate of extinction produced through human activities is apparently between 100 to 10,000 times the normal biological rate of change. In "Mass Extinction," science writer David Hosansky catalogues this global process and warns that we can't even imagine the long-term consequences of the loss of biodiversity that extinctions pose. One hopeful sign in the ongoing territorial battle between humans and other animals is the subject of the concluding article in the section. In "Watching vs. Taking," writer and ecotourist Howard Youth describes the shift in the relationship between humans and wildlife. He explains that millions of people are turning away from the hunting of species for food, pelts, or sport and turning to the simple pleasure of watching. The trend began with the "birders" who turned the hobby of bird-watching into an important economic enterprise in many areas. It continues with the increasing number of people willing to engage in this new kind of hunt.

PLANET OF WEEDS

Tallying the losses of Earth's animals and plants

By David Quammen

Hope is a duty from which paleontologists are exempt. Their job is to take the long view, the cold and stony view, of triumphs and catastrophes in the history of life. They study the fossil record, that erratic selection of petrified shells, carapaces, bones, teeth, tree trunks, leaves, pollen, and other biological relics, and from it they attempt to discern the lost secrets of time, the big patterns of stasis and change, the trends of innovation and adaptation and refinement and decline that have blown like sea winds among ancient creatures in ancient ecosystems. Although life is their subject, death and burial supply all their data. They're the coroners of biology. This gives to paleontologists a certain distance, a hyperopic perspective beyond the reach of anxiety over outcomes of the struggles they chronicle. If hope is the thing with feathers, as Emily Dickinson said, then it's good to remember that feathers don't generally fossilize well. In lieu of hope and despair, paleontologists have a highly developed sense of cyclicity. That's why I recently went to Chicago, with a handful of urgently grim

15. Planet of Weeds

questions, and called on a paleontologist named David Jablonski. I wanted answers unvarnished with obligatory hope.

Jablonski is a big-pattern man, a macroevolutionist, who works fastidiously from the particular to the very broad. He's an expert on the morphology and distribution of marine bivalves and gastropods—or clams and snails, as he calls them when speaking casually. He sifts through the record of those mollusk lineages, preserved in rock and later harvested into museum drawers, to extract ideas about the origin of novelty. His attention roams back through 600 million years of time. His special skill involves framing large, resonant questions that can be answered with small, lithified clamshells. For instance: By what combinations of causal factor and sheer chance have the great evolutionary innovations arisen? How quickly have those innovations taken hold? How long have they abided? He's also interested in extinction, the converse of abidance, the yang to evolution's yin. Why do some species survive for a long time, he wonders, whereas others die out much sooner? And why has the rate of extinction—low throughout most of Earth's history—spiked upward cataclysmically on just a few occasions? How do these cataclysmic episodes, known in the trade as mass extinctions, differ in kind as well as degree from the gradual process of species extinction during the millions of years between? Can what struck in the past strike again?

The concept of mass extinction implies a biological crisis that spanned large parts of the planet and, in a relatively short time, eradicated a sizable number of species from a variety of groups. There's no absolute threshold of magnitude, and dozens of different episodes in geologic history might qualify, but five big ones stand out: Ordovician, Devonian, Permian, Triassic, Cretaceous. The Ordovician extinction, 439 million years ago, entailed the disappearance of roughly 85 percent of marine animal species—and that was before there were any animals *on land*. The Devonian extinction, 367 million years ago, seems to have been almost as severe. About 245 million years ago came the Permian extinction, the worst ever, claiming 95 percent of all known animal species and therefore almost wiping out the animal kingdom altogether. The Triassic, 208 million years ago, was bad again, though not nearly so bad as the Permian. The most recent was the Cretaceous extinction (sometimes called the K-T event because it defines the boundary between two geologic periods, with K for Cretaceous, never mind why, and T for Tertiary), familiar even to schoolchildren because it ended the age of dinosaurs. Less familiarly, the K-T event also brought extinction of the marine reptiles and the ammonites, as well as major losses of species among fish, mammals, amphibians, sea urchins, and other groups, totaling

THE EARTH HAS UNDERGONE FIVE MAJOR EXTINCTION PERIODS, EACH REQUIRING MILLIONS OF YEARS OF RECOVERY

76 percent of all species. In between these five episodes occurred some lesser mass extinctions, and throughout the intervening lulls extinction continued, too—but at a much slower pace, known as the background rate, claiming only about one species in any major group every million years. At the background rate, extinction is infrequent enough to be counterbalanced by the evolution of new species. Each of the five major episodes, in contrast, represents a drastic net loss of species diversity, a deep trough of biological impoverishment from which Earth only slowly recovered. How slowly? How long is the lag between a nadir of impoverishment and a recovery to ecological fullness? That's another of Jablonski's research interests. His rough estimates run to 5 or 10 million years. What drew me to this man's work, and then to his doorstep, were his special competence on mass extinctions and his willingness to discuss the notion that a sixth one is in progress now.

Some people will tell you that we as a species, *Homo sapiens,* the savvy ape, all 5.9 billion of us in our collective impact, are destroying the world. Me, I won't tell you that, because "the world" is so vague, whereas what we are or aren't destroying is quite specific. Some people will tell you that we are rampaging suicidally toward a degree of global wreckage that will result in our own extinction. I won't tell you that either. Some people say that the environment will be the paramount political and social concern of the twenty-first century, but what they mean by "the environment" is anyone's guess. Polluted air? Polluted water? Acid rain? A frayed skein of ozone over Antarctica? Greenhouse gases emitted by smokestacks and cars? Toxic wastes? None of these concerns is the big one, paleontological in scope, though some are more closely entangled with it than others. If the world's air is clean for humans to breathe but supports no birds or butterflies, if the world's waters are pure for humans to drink but contain no fish or crustaceans or diatoms, have we solved our environmental problems? Well, I suppose so, at least as environmentalism is commonly construed. That clumsy, confused, and presumptuous formulation "the environment" implies viewing air, water, soil, forests, rivers, swamps, deserts, and oceans as merely a milieu within which something important is set: human life, human history. But what's at issue in fact is not an environment; it's a living world.

Here instead is what I'd like to tell you: The consensus among conscientious biologists is that we're headed into another mass extinction, a vale of biological impoverishment commensurate with the big five. Many experts remain hopeful that we can brake that descent, but my own view is that we're likely to go all the way down. I visited David Jablonski to ask what we might see at the bottom.

On a hot summer morning, Jablonski is busy in his office on the second floor of the Hinds Geophysical Laboratory at the University of Chicago. It's a large open room furnished in tall bookshelves, tables piled high with books, stacks of paper standing knee-high off the floor. The walls are mostly bare, aside from a chart of the geologic time scale, a clipped cartoon of dancing tyrannosaurs in red sneakers, and a poster from a Rodin exhibition, quietly appropriate to the overall theme of eloquent stone. Jablonski is a lean forty-five-year-old man with a dark full beard. Educated at Columbia and Yale, he came to Chicago in 1985 and has helped make its paleontology program perhaps the country's best. Although in not many hours he'll be leaving on a trip to Alaska, he has been cordial about agreeing to this chat. Stepping carefully, we move among the piled journals, reprints, and photocopies. Every pile represents a different research question, he tells me. "I juggle a lot of these things all at once because they feed into one another." That's exactly why I've come: for a little rigorous intellectual synergy.

Let's talk about mass extinctions, I say. When did someone first realize that the concept might apply to current events, not just to the Permian or the Cretaceous?

He begins sorting through memory, back to the early 1970s, when the full scope of the current extinction problem was barely recognized. Before then, some writers warned about "vanishing wildlife" and "endangered species," but generally the warnings were framed around individual species with popular appeal, such as the whooping crane, the tiger, the blue whale, the peregrine falcon. During the 1970s a new form of concern broke forth—call it wholesale concern—from the awareness that unnumbered millions of narrowly endemic (that is, unique

> BIOLOGISTS BELIEVE THAT WE ARE ENTERING ANOTHER MASS EXTINCTION, A VALE OF BIOLOGICAL IMPOVERISHMENT

and localized) species inhabit the tropical forests and that those forests were quickly being cut. In 1976, a Nairobi-based biologist named Norman Myers published a paper in *Science* on that subject; in passing, he also compared current extinctions with the rate during what he loosely called "the 'great dying' of the dinosaurs." David Jablonski, then a graduate student, read Myers's paper and tucked a copy into his files. This was the first time, as Jablonski recalls, that anyone tried to quantify the rate of present-day extinctions. "Norman was a pretty lonely guy, for a long time, on that," he says. In 1979, Myers published *The Sinking Ark,* explaining the problem and offering some rough projections. Between the years 1600 and 1900, by his tally, humanity had caused the extinction of about 75 known species, almost all of them mammals and birds. Between 1900 and 1979, humans had extinguished about another 75 known species, representing a rate well above the rate of known losses during the Cretaceous extinction. But even more worrisome was the inferable rate of unrecorded extinctions, recent and now impending, among plants and animals still unidentified by science. Myers guessed that 25,000 plant species presently stood jeopardized, and maybe hundreds of thousands of insects. "By the time human communities establish ecologically sound life-styles, the fallout of species could total several million." Rereading that sentence now, I'm struck by the reckless optimism of his assumption that human communities eventually will establish "ecologically sound life-styles."

Although this early stab at quantification helped to galvanize public concern, it also became a target for a handful of critics, who used the inexactitude of the numbers to cast doubt on the reality of the problem. Most conspicuous of the naysayers was Julian Simon, an economist at the University of Maryland, who argued bullishly that human resourcefulness would solve all problems worth solving, of which a decline in diversity of tropical insects wasn't one.

In a 1986 issue of *New Scientist,* Simon rebutted Norman Myers, arguing from his own construal of select data that there was "no obvious recent downward trend in world forests—no obvious 'losses' at all, and certainly no 'near catastrophic' loss." He later co-authored an op-ed piece in the *New York Times* under the headline "Facts, Not Species, Are Periled." Again he went after Myers, asserting a "complete absence of evidence for the claim that the extinction of species is going up rapidly—or even going up at all." Simon's worst disservice to logic in that statement and others was the denial that *inferential* evidence of wholesale extinction counts for anything. Of inferential evidence there was an abundance—for example, from the Centinela Ridge in a cloud-forest zone of western Ecuador, where in 1978 the botanist Alwyn Gentry and a colleague found thirty-eight species of narrowly endemic plants; including several with mysteriously black leaves. Before Gentry could get back, Centinela Ridge had been completely deforested, the native plants replaced by cacao and other crops. As for inferential evidence generally, we might do well to remember what it contributes to our conviction that approximately 105,000 Japanese civilians died in the atomic bombing of Hiroshima. The city's population fell abruptly on August 6, 1945, but there was no one-by-one identification of 105,000 bodies.

Nowadays a few younger writers have taken Simon's line, pooh-poohing the concern over extinction. As for Simon himself, who died earlier this year, perhaps the truest sentence he left behind was, "We must also try to get more reliable information about the number of species that might be lost with various changes in the forests." No one could argue.

But is isn't easy to get such information. Field biologists tend to avoid investing their precious research time in doomed tracts of forest. Beyond that, our culture offers little institutional support for the study of narrowly endemic species in order to register their existence *before* their habitats are destroyed. Despite these obstacles, recent efforts to quantify rates of extinction have supplanted the old warnings. These new estimates use satellite imaging and improved on-the-ground data about deforestation, records of the many human-caused extinctions on islands, and a branch of ecological theory called island biogeography, which connects documented island cases with the mainland problem of forest fragmentation. These efforts differ in particulars, reflecting how much uncertainty is still involved, but their varied tones form a chorus of consensus. I'll mention three of the most credible.

W. V. Reid, of the World Resources Institute, in 1992 gathered numbers on the average annual deforestation in each of sixty-three tropical countries during the 1980s and from them charted three different scenarios (low, middle, high) of presumable forest loss by the year 2040. He chose a standard mathematical model of the relationship between decreasing habitat area and decreasing species diversity, made conservative assumptions about the crucial constant, and ran his various deforestation estimates through the model. Reid's calculations suggest that by the year 2040, between 17 and 35 percent of tropical forest species will be extinct or doomed to be. Either at the high or the low end of this range, it would amount to a bad loss, though not as bad as the K-T event. Then again 2040 won't mark the end of human pressures on biological diversity or landscape.

Robert M. May, an ecologist at Oxford, co-authored a similar effort in 1995. May and his colleagues noted the five causal factors that account for most extinctions: habitat destruction, habitat fragmentation, overkill, invasive species, and secondary effects cascading through an ecosystem from other extinctions. Each of those five is more intricate than it sounds. For instance, habitat fragmentation dooms species by consigning them to small, island-like parcels of habitat surrounded by an ocean of human impact and by then subjecting them to the same jeopardies (small population size, acted upon by environmental fluctuation, catastrophe, inbreeding, bad luck, and cascading effects) that make island species especially vulnerable to extinction. May's team concluded that most extant bird and mammal species can expect average life spans of between 200 and 400 years. That's equivalent to saying that about a third of one percent will go extinct each year until some unimaginable end point is reached. "Much of the diversity we inherited," May and his co-authors wrote, "will be gone before humanity sorts itself out."

The most recent estimate comes from Stuart L. Pimm and Thomas M. Brooks, ecologists at the University of Tennessee. Using a combination of published data on bird species lost from forest fragments and field data gathered themselves, Pimm and Brooks concluded that 50 percent of the world's forest-bird species will be doomed to extinction by deforestation occurring over the next half century. And birds won't be the

> EVEN BY CONSERVATIVE ESTIMATES, HUGE PERCENTAGES OF EARTH'S ANIMALS AND PLANTS WILL SIMPLY DISAPPEAR

15. Planet of Weeds

IN THE NEXT FIFTY YEARS, DEFORESTATION WILL DOOM ONE HALF OF THE WORLD'S FOREST-BIRD SPECIES

sole victims. "How many species will be lost if current trends continue?" the two scientists asked. "Somewhere between one third and two thirds of all species—easily making this event as large as the previous five mass extinctions the planet has experienced."

Jablonski, who started down this line of thought in 1978, offers me a reminder about the conceptual machinery behind such estimates. "All mathematical models," he says cheerily, "are wrong. They are approximations. And the question is: Are they usefully wrong, or are they meaninglessly wrong?" Models projecting present and future species loss are useful, he suggests, if they help people realize that *Homo sapiens* is perturbing Earth's biosphere to a degree it hasn't often been perturbed before. In other words, that this is a drastic experiment in biological drawdown we're engaged in, not a continuation of routine.

Behind the projections of species loss lurk a number of crucial but hard-to-plot variables, among which two are especially weighty: continuing landscape conversion and the growth curve of human population.

Landscape conversion can mean many things: draining wetlands to build roads and airports, turning tallgrass prairies under the plow, fencing savanna and overgrazing it with domestic stock, cutting second-growth forest in Vermont and consigning the land to ski resorts or vacation suburbs, slash-and-burn clearing of Madagascar's rain forest to grow rice on wet hillsides, industrial logging in Borneo to meet Japanese plywood demands. The ecologist John Terborgh and a colleague, Carel P. van Schaik, have described a four-stage process of landscape conversion that they call the land-use cascade. The successive stages are: 1) *wildlands,* encompassing native floral and faunal communities altered little or not at all by human impact; 2) *extensively used areas,* such as natural grasslands lightly grazed, savanna kept open for prey animals by infrequent human-set fires, or forests sparsely worked by slash-and-burn farmers at low density; 3) *intensively used areas,* meaning crop fields, plantations, village commons, travel corridors, urban and industrial zones; and finally 4) *degraded land,* formerly useful but now abused beyond value to anybody. Madagascar, again, would be a good place to see all four stages, especially the terminal one. Along a thin road that leads inland from a town called Mahajanga, on the west coast, you can gaze out over a vista of degraded land—chalky red hills and gullies, bare of forest, burned too often by graziers wanting a short-term burst of pasturage, sparsely covered in dry grass and scrubby fan palms, eroded starkly, draining red mud into the Betsiboka River, supporting almost no human presence. Another showcase of degraded land—attributable to fuelwood gathering, overgrazing, population density, and decades of apartheid—is the Ciskei homeland in South Africa. Or you might look at overirrigated crop fields left ruinously salinized in the Central Valley of California.

Among all forms of landscape conversion, pushing tropical forest from the *wildlands* category to the *intensively used* category has the greatest impact on biological diversity. You can see it in western India, where a spectacular deciduous ecosystem known as the Gir forest (home to the last surviving population of the Asiatic lion, *Panthera leo persica*) is yielding along its ragged edges to new mango orchards, peanut fields, and lime quarries for cement. You can see it in the central Amazon, where big tracts of rain forest have been felled and burned, in a largely futile attempt (encouraged by misguided government incentives, now revoked) to pasture cattle on sun-hardened clay. According to the United Nations Food and Agriculture Organization, the rate of deforestation in tropical countries has increased (contrary to Julian Simon's claim) since the 1970s, when Myers made his estimates. During the 1980s, as the FAO reported in 1993, that rate reached 15.4 million hectares (a hectare being the metric equivalent of 2.5 acres) annually. South America was losing 6.2 million hectares a year. Southeast Asia was losing less in area but more proportionally: 1.6 percent of its forests yearly. In terms of cumulative loss, as reported by other observers, the Atlantic coastal forest of Brazil is at least 95 percent gone. The Philippines, once nearly covered with rain forest, has lost 92 percent. Costa Rica has continued to lose forest, despite that country's famous concern for its biological resources. The richest of old-growth lowland forests in West Africa, India, the Greater Antilles, Madagascar, and elsewhere have been reduced to less than a tenth of their original areas. By the middle of the next century, if those trends continue, tropical forest will exist virtually nowhere outside of protected areas—that is, national parks, wildlife refuges, and other official reserves.

How many protected areas will there be? The present worldwide total is about 9,800, encompassing 6.3 percent of the planet's land area. Will those parks and reserves retain their full biological diversity? No. Species with large territorial needs will be unable to maintain viable population levels within small reserves, and as those species die away their absence will affect others. The disappearance of big predators, for instance, can release limits on medium-size predators and scavengers, whose overabundance can drive still other species (such as ground-nesting birds) to extinction. This has already happened in some habitat fragments, such as Panama's Barro Colorado Island, and been well documented in the literature of island biogeography. The lesson of fragmented habitats is Yeatsian: Things fall apart.

Human population growth will make a bad situation worse by putting ever more pressure on all available land.

Population growth rates have declined in many countries within the past several decades, it's true. But world population is still increasing, and even if average fertility suddenly, magically, dropped to 2.0 children per female, population would continue to increase (on the momentum of birth rate exceeding death rate among a generally younger and healthier populace) for some time. The annual increase is now 80 million people, with most of that increment coming in less-developed countries. The latest long-range projections from the Population Division of the United Nations, released earlier this year, are slightly down from previous long-term projections in 1992 but still point toward a problematic future. According to the U.N.'s middle estimate (and most probable? hard to know) among seven fertility scenarios, human population will rise from the present 5.9 billion to 9.4 billion by the year 2050, then to 10.8 billion by 2150, before leveling off there at the end of the twenty-second century. If it happens that way, about 9.7 billion people will inhabit the countries included within Africa, Latin America, the Caribbean, and Asia. The total population of those countries—most of which are in the low latitudes, many of which are less developed, and which together encompass a large portion of Earth's remaining tropical forest—will be more than

THE LESSON TO BE LEARNED FROM FRAGMENTED, ISOLATED HABITATS IS YEATSIAN: THINGS FALL APART

> WE CONFRONT THE VISION OF A HUMAN POPULATION PRESSING SNUGLY AROUND WHATEVER NATURAL LANDSCAPE REMAINS

twice what it is today. Those 9.7 billion people, crowded together in hot places, forming the ocean within which tropical nature reserves are insularized, will constitute 90 percent of humanity. Anyone interested in the future of biological diversity needs to think about the pressures these people will face, and the pressures they will exert in return.

We also need to remember that the impact of *Homo sapiens* on the biosphere can't be measured simply in population figures. As the population expert Paul Harrison pointed out in his book *The Third Revolution,* that impact is a product of three variables: population size, consumption level, and technology. Although population growth is highest in less-developed countries, consumption levels are generally far higher in the developed world (for instance, the average American consumes about ten times as much energy as the average Chilean, and about a hundred times as much as the average Angolan), and also higher among the affluent minority in any country than among the rural poor. High consumption exacerbates the impact of a given population, whereas technological developments may either exacerbate it further (think of the automobile, the air conditioner, the chainsaw) or mitigate it (as when a technological innovation improves efficiency for an established function). All three variables play a role in every case, but a directional change in one form of human impact—upon air pollution from fossil-fuel burning, say, or fish harvest from the seas—can be mainly attributable to a change in one variable, with only minor influence from the other two. Sulfur-dioxide emissions in developed countries fell dramatically during the 1970s and '80s, due to technological improvements in papermaking and other industrial processes; those emissions would have fallen still farther if not for increased population (accounting for 25 percent of the upward vector) and increased consumption (accounting for 75 percent). Deforestation, in contrast, is a directional change that *has* been mostly attributable to population growth.

According to Harrison's calculations, population growth accounted for 79 percent of the deforestation in less-developed countries between 1973 and 1988. Some experts would argue with those calculations, no doubt, and insist on redirecting our concern toward the role that distant consumers, wood-products buyers among slow-growing but affluent populations of the developed nations, play in driving the destruction of Borneo's dipterocarp forests or the hardwoods of West Africa. Still, Harrison's figures point toward an undeniable reality; more total people will need more total land. By his estimate, the minimum land necessary for food growing and other human needs (such as water supply and waste dumping) amounts to one fifth of a hectare per person. Given the U.N.'s projected increase of 4.9 billion souls before the human population finally levels off, that comes to another billion hectares of human-claimed landscape, a billion hectares less forest—even without allowing for any further deforestation by the current human population, or for any further loss of agricultural land to degradation. A billion hectares—in other words, 10 million square kilometers—is, by a conservative estimate, well more than half the remaining forest area in Africa, Latin America, and Asia. This raises the vision of a very exigent human population pressing snugly around whatever patches of natural landscape remain.

Add to that vision the extra, incendiary aggravation of poverty. According to a recent World Bank estimate, about 30 percent of the total population of less-developed countries lives in poverty. Alan Durning, in his 1992 book *How Much Is Enough? The Consumer Society and the Fate of the Earth,* puts it in a broader perspective when he says that the world's human population is divided among three "ecological classes": the consumers, the middle-income, and the poor. His consumer class includes those 1.1 billion fortunate people whose annual income per family member is more than $7,500. At the other extreme, the world's poor also number about 1.1 billion people—all from households with less than $700 annually per member. "They are mostly rural Africans, Indians, and other South Asians," Durning writes. "They eat almost exclusively grains, root crops, beans, and other legumes, and they drink mostly unclean water. They live in huts and shanties, they travel by foot, and most of their possessions are constructed of stone, wood, and other substances available from the local environment." He calls them the "absolute poor." It's only reasonable to assume that another billion people will be added to that class, mostly in what are now the less-developed countries, before population growth stabilizes. How will those additional billion, deprived of education and other advantages, interact with the tropical landscape? Not likely by entering information-intensive jobs in the service sector of the new global economy. Julian Simon argued that human ingenuity—and by extension, human population itself—is "the ultimate resource" for solving Earth's problems, transcending Earth's limits, and turning scarcity into abundance. But if all the bright ideas generated by a human population of 5.9 billion haven't yet relieved the desperate needfulness of 1.1 billion absolute poor, why should we expect that human ingenuity will do any better for roughly 2 billion poor in the future?

Other writers besides Durning have warned about this deepening class rift. Tom Athanasiou, in *Divided Planet: The Ecology of Rich and Poor,* sees population growth only exacerbating the division, and notes that governments often promote destructive schemes of transmigration and rain-forest colonization as safety valves for the pressures of land hunger and discontent. A young Canadian policy analyst named Thomas F. Homer-Dixon, author of several calm-voiced but frightening articles on the linkage between what he terms "environmental scarcity" and global sociopolitical instability, reports that the amount of cropland available per person is falling in the less-developed countries because of population growth and because millions of hectares "are being lost each year to a combination of problems, including encroachment by cities, erosion, depletion of nutrients, acidification, compacting and salinization and waterlogging from overirrigation." In the cropland pinch and other forms of environmental scarcity, Homer-Dixon foresees potential for "a widening gap" of two sorts—between demands on the state and its ability to deliver, and more basically between rich and poor. In conversation with the journalist Robert D. Kaplan, as quoted in Kaplan's book *The Ends of the Earth,* Homer-Dixon said it more vividly: "Think of a stretch limo in the potholed streets of New York City, where homeless beggars live. Inside the limo are the air-conditioned post-industrial regions of North America, Europe, the emerging Pacific Rim, and a few other isolated places, with their trade summitry and computer information highways. Outside is the rest of mankind, going in a completely different direction."

That direction, necessarily, will be toward ever more desperate exploitation of landscape. When you think of Homer-Dixon's stretch limo on those potholed urban streets, don't assume there will be room inside for

> EVEN NOAH'S ARK ONLY MANAGED TO RESCUE PAIRED ANIMALS, NOT LARGE PARCELS OF HABITAT

15. Planet of Weeds

MAN'S ACCIDENTAL RELOCATION OF CERTAIN SPECIES HAS LONG CREATED PROFOUND DISLOCATIONS IN NATURE

tropical forests. Even Noah's ark only managed to rescue paired animals, not large parcels of habitat. The jeopardy of the ecological fragments that we presently cherish as parks, refuges, and reserves is already severe, due to both internal and external forces: internal, because insularity itself leads to ecological unraveling; and external, because those areas are still under siege by needy and covetous people. Projected forward into a future of 10.8 billion humans, of which perhaps 2 billion are starving at the periphery of those areas, while another 2 billion are living in a fool's paradise maintained by unremitting exploitation of whatever resources remain, that jeopardy increases to the point of impossibility. In addition, any form of climate change in the mid-term future, whether caused by greenhouse gases or by a natural flip-flop of climatic forces, is liable to change habitat conditions within a given protected area beyond the tolerance range for many species. If such creatures can't migrate beyond the park or reserve boundaries in order to chase their habitat needs, they may be "protected" from guns and chainsaws within their little island, but they'll still die.

We shouldn't take comfort in assuming that at least Yellowstone National Park will still harbor grizzly bears in the year 2150, that at least Royal Chitwan in Nepal will still harbor tigers, that at least Serengeti in Tanzania and Gir in India will still harbor lions. Those predator populations, and other species down the cascade, are likely to disappear. "Wildness" will be a word applicable only to urban turmoil. Lions, tigers, and bears will exist in zoos, period. Nature won't come to an end, but it will look very different.

The most obvious differences will be those I've already mentioned: tropical forests and other terrestrial ecosystems will be drastically reduced in area, and the fragmented remnants will stand tiny and isolated. Because of those two factors, plus the cascading secondary effects, plus an additional dire factor I'll mention in a moment, much of Earth's biological diversity will be gone. How much? That's impossible to predict confidently, but the careful guesses of Robert May, Stuart Pimm, and other biologists suggest losses reaching half to two thirds of all species. In the oceans, deepwater fish and shellfish populations will be drastically depleted by overharvesting, if not to the point of extinction then at least enough to cause more cascading consequences. Coral reefs and other shallow-water ecosystems will be badly stressed, if not devastated, by erosion and chemical runoff from the land. The additional dire factor is invasive species, fifth of the five factors contributing to our current experiment in mass extinction.

That factor, even more than habitat destruction and fragmentation, is a symptom of modernity. Maybe you haven't heard much about invasive species, but in coming years you will. The ecologist Daniel Simberloff takes it so seriously that he recently committed himself to founding an institute on invasive biology at the University of Tennessee, and Interior Secretary Bruce Babbitt sounded the alarm last April in a speech to a weed-management symposium in Denver. The spectacle of a cabinet secretary denouncing an alien plant called purple loosestrife struck some observers as droll, but it wasn't as silly as it seemed. Forty years ago, the British ecologist Charles Elton warned prophetically in a little book titled *The Ecology of Invasions by Animals and Plants* that "we are living in a period of the world's history when the mingling of thousands of kinds of organisms from different parts of the world is setting up terrific dislocations in nature." Elton's word "dislocations" was nicely chosen to ring with a double meaning: species are being moved from one location to another, and as a result ecosystems are being thrown into disorder.

The problem dates back to when people began using ingenious new modes of conveyance (the horse, the camel, the canoe) to travel quickly across mountains, deserts, and oceans, bringing with them rats, lice, disease microbes, burrs, dogs, pigs, goats, cats, cows, and other forms of parasitic, commensal, or domesticated creature. One immediate result of those travels was a wave of island-bird extinctions, claiming more than a thousand species, that followed oceangoing canoes across the Pacific and elsewhere. Having evolved in insular ecosystems free of predators, many of those species were flightless, unequipped to defend themselves or their eggs against ravenous mammals. *Raphus cucullatus,* a giant cousin of the pigeon lineage, endemic to Mauritius in the Indian Ocean and better known as the dodo, was only the most easily caricatured representative of this much larger pattern. Dutch sailors killed and ate dodos during the seventeenth century, but probably what guaranteed the extinction of *Raphus cucullatus* is that the European ships put ashore rats, pigs, and *Macaca fascicularis,* an opportunistic species of Asian monkey. Although commonly known as the crab-eating macaque, *M. fascicularis* will eat almost anything. The monkeys are still pestilential on Mauritius, hungry and daring and always ready to grab what they can, including raw eggs. But the dodo hasn't been seen since 1662.

The European age of discovery and conquest was also the great age of biogeography—that is, the study of what creatures live where, a branch of biology practiced by attentive travelers such as Carolus Linnaeus, Alexander von Humboldt, Charles Darwin, and Alfred Russel Wallace. Darwin and Wallace even made biogeography the basis of their discovery that species, rather than being created and plopped onto Earth by divine magic, evolve in particular locales by the process of natural selection. Ironically, the same trend of far-flung human travel that gave biogeographers their data also began to muddle and nullify those data, by transplanting the most ready and roguish species to new places and thereby delivering misery unto death for many other species. Rats and cats went everywhere, causing havoc in what for millions of years had been sheltered, less competitive ecosystems. The Asiatic chestnut blight and the European starling came to America; the American muskrat and the Chinese mitten crab got to Europe. Sometimes these human-mediated transfers were unintentional, sometimes merely shortsighted. Nostalgic sportsmen in New Zealand imported British red deer; European brown trout and Coastal rainbows were planted in disregard of the native cutthroats of Rocky Mountain rivers. Prickly-pear cactus, rabbits, and cane toads were inadvisedly welcomed to Australia. Goats went wild in the Galapagos. The bacterium that causes bubonic plague journeyed from China to California by way of a flea, a rat, and a ship. The Atlantic sea lamprey found its own way up into Lake Erie, but only after the Welland Canal gave it a bypass around Niagara Falls. Unintentional or otherwise, all these transfers had unforeseen consequences, which in many cases included the extinction of less competitive, less opportunistic native species. The rosy wolfsnail, a small creature in-

THE SPECIES THAT SURVIVE WILL BE LIKE WEEDS, REPRODUCING QUICKLY AND SURVIVING ALMOST ANYWHERE

WILDLIFE WILL CONSIST OF PIGEONS, COYOTES, RATS, ROACHES, HOUSE SPARROWS, CROWS, AND FERAL DOGS

troduced onto Oahu for the purpose of controlling a larger and more obviously noxious species of snail, which was itself invasive, proved to be medicine worse than the disease; it became a fearsome predator upon native snails, of which twenty species are now gone. The Nile perch, a big predatory fish introduced into Lake Victoria in 1962 because it promised good eating, seems to have exterminated at least eighty species of smaller cichlid fishes that were native to the lake's Mwanza Gulf.

The problem is vastly amplified by modern shipping and air transport, which are quick and capacious enough to allow many more kinds of organisms to get themselves transplanted into zones of habitat they never could have reached on their own. The brown tree snake, having hitchhiked aboard military planes from the New Guinea region near the end of World War II, has eaten most of the native forest birds of Guam. Hanta virus, first identified in Korea, burbles quietly in the deer mice of Arizona. Ebola will next appear who knows where. Apart from the frightening epidemiological possibilities, agricultural damages are the most conspicuous form of impact. One study, by the congressional Office of Technology Assessment, reports that in the United States 4,500 nonnative species have established free-living populations, of which about 15 percent cause severe harm; looking at just 79 of those species, the OTA documented $97 billion in damages. The lost value in Hawaiian snail species or cichlid diversity is harder to measure. But another report, from the U.N. Environmental Program, declares that almost 20 percent of the world's endangered vertebrates suffer from pressures (competition, predation, habitat transformation) created by exotic interlopers. Michael Soulé, a biologist much respected for his work on landscape conversion and extinction, has said that invasive species may soon surpass habitat loss and fragmentation as the major cause of "ecological disintegration." Having exterminated Guam's avifauna, the brown tree snake has lately been spotted in Hawaii.

Is there a larger pattern to these invasions? What do fire ants, zebra mussels, Asian gypsy mothers, tamarisk trees, maleleuca trees, kudzu, Mediterranean fruit flies, boll weevils, and water hyacinths have in common with crab-eating macaques or Nile perch? Answers: They're *weedy* species, in the sense that animals as well as plants can be weedy. What that implies is a constellation of characteristics: They reproduce quickly, disperse widely when given a chance, tolerate a fairly broad range of habitat conditions, take hold in strange places, succeed especially in disturbed ecosystems, and resist eradication once they're established. They are scrappers, generalists, opportunists. They tend to thrive in human-dominated terrain because in crucial ways they resemble *Homo sapiens:* aggressive, versatile, prolific, and ready to travel. The city pigeon, a cosmopolitan creature derived from wild ancestry as a Eurasian rock dove (*Columba livia*) by way of centuries of pigeon fanciers whose coop-bred birds occasionally went AWOL, is a weed. So are those species that, benefiting from human impacts upon landscape, have increased grossly in abundance or expanded their geographical scope without having to cross an ocean by plane or by boat—for instance, the coyote in New York, the raccoon in Montana, the while-tailed deer in northern Wisconsin or western Connecticut. The brown-headed cowbird, also weedy, has enlarged its range from the Eastern United States into the agricultural Midwest at the expense of migratory songbirds. In gardening usage the word "weed" may be utterly subjective, indicating any plant you don't happen to like, but in ecological usage it has these firmer meanings. Biologists frequently talk of weedy species, meaning animals as well as plants.

Paleontologists, too, embrace the idea and even the term. Jablonski himself, in a 1991 paper published in *Science,* extrapolated from past mass extinctions to our current one and suggested that human activities are likely to take their heaviest toll on narrowly endemic species, while causing fewer extinctions among those species that are broadly adapted and broadly distributed. "In the face of ongoing habitat alteration and fragmentation," he wrote, "this implies a biota increasingly enriched in widespread, weedy species—rats, ragweed, and cockroaches—relative to the large number of species that are more vulnerable and potentially more useful to humans as food, medicines, and genetic resources." Now, as we sit in his office, he repeats: "It's just a question of how much the world becomes enriched in these weedy species." Both in print and in talk he uses "enriched" somewhat caustically, knowing that the actual direction of the trend is toward impoverishment.

Regarding impoverishment, let's note another dark, interesting irony: that the two converse trends I've described—partitioning the world's landscape by habitat fragmentation, and unifying the world's landscape by global transport for weedy species—produce not converse results but one redoubled result, the further loss of biological diversity. Immersing myself in the literature of extinctions, and making dilettantish excursions across India, Madagascar, New Guinea, Indonesia, Brazil, Guam, Australia, New Zealand, Wyoming, the hills of Burbank, and other semi-wild places over the past decade, I've seen those redoubling trends everywhere, portending a near-term future in which Earth's landscape is threadbare, leached of diversity, heavy with humans, and "enriched" in weedy species. That's an ugly vision, but I find it vivid. Wildlife will consist of the pigeons and the coyotes and the white-tails, the black rats (*Rattus rattus*) and the brown rats (*Rattus norvegicus*) and a few other species of worldly rodent, the crab-eating macaques and the cockroaches (though, as with the rats, not *every* species—some are narrowly endemic, like the giant Madagascar hissing cockroach) and the mongooses, the house sparrows and the house geckos and the houseflies and the barn cats and the skinny brown feral dogs and a short list of additional species that play by our rules. Forests will be tiny insular patches existing on bare sufferance, much of their biological diversity (the big predators, the migratory birds, the shy creatures that can't tolerate edges, and many other species linked inextricably with those) long since decayed away. They'll essentially be tall woody gardens, not forests in the richer sense. Elsewhere the landscape will have its strips and swatches of green, but except on much-poisoned lawns and golf courses the foliage will be infested with cheat-grass and European buckthorn and spotted knapweed and Russian thistle and leafy spurge and salt meadow cordgrass and Bruce Babbitt's purple loosestrife. Having recently passed the great age of biogeography, we will have entered the age *after* biogeography, in that virtually everything will live virtually everywhere, though the list of species that constitute "everything" will be small. I see this world implicitly foretold in the U.N. population projections, the FAO reports on deforestation, the northward advance into Texas of Africanized honeybees, the rhesus monkeys that haunt the parapets of public buildings

Homo sapiens— REMARKABLY WIDESPREAD, PROLIFIC, AND ADAPTABLE—IS THE CONSUMMATE WEED

15. Planet of Weeds

in New Delhi, and every fat gray squirrel on a bird feeder in England. Earth will be a different sort of place—soon, in just five or six human generations. My label for that place, that time, that apparently unavoidable prospect, is the Planet of Weeds. Its main consoling felicity, as far as I can imagine, is that there will be no shortage of crows.

Now we come to the question of human survival, a matter of some interest to many. We come to a certain fretful leap of logic that otherwise thoughtful observers seem willing, even eager, to make: that the ultimate consequence will be the extinction of us. By seizing such a huge share of Earth's landscape, by imposing so wantonly on its providence and presuming so recklessly on its forgiveness, by killing off so many species, they say, we will doom our own species to extinction. My quibbles with the idea are that it seems ecologically improbable and too optimistic. But it bears examining, because it's frequently offered as the ultimate argument against proceeding as we are.

Jablonski also has his doubts. Do you see *Homo sapiens* as a likely survivor, I ask him, or as a casualty? "Oh, we've got to be one of the most bomb-proof species on the planet," he says "We're geographically widespread, we have a pretty remarkable reproductive rate, we're incredibly good at co-opting and monopolizing resources. I think it would take really serious, concerted effort to wipe out the human species." The point he's making is one that has probably already dawned on you: *Homo sapiens* itself is the consummate weed. Why shouldn't we survive, then, on the Planet of Weeds? But there's a wide range of possible circumstances, Jablonski reminds me, between the extinction of our species and the continued growth of human population, consumptions, and comfort. "I think we'll be one of the survivors," he says, "sort of picking through the rubble." Besides losing all the pharmaceutical and generic resources that lay hidden within those extinguished species, and all the spiritual and aesthetic values they offered, he foresees unpredictable levels of loss in many physical and biochemical functions that ordinarily come as benefits from diverse, robust ecosystems—functions such as cleaning and recirculating air and water, mitigating droughts and floods, decomposing wastes, controlling erosion, creating new soil, pollinating crops, capturing and transporting nutrients, damping short-term temperature extremes and longer-term fluctuations of climate, restraining outbreaks of pestiferous species, and shielding Earth's surface from the full brunt of ultraviolet radiation. Strip away the ecosystems that perform those services, Jablonski says, and you can expect grievous detriment to the reality we inhabit. "A lot of things are going to happen that will make this a crummier place to live—a more stressful place to live, a more difficult place to live, a less resilient place to live—before the human species is at any risk at all." And maybe some of the new difficulties, he adds, will serve as incentive for major changes in the trajectory along which we pursue our aggregate self-interests. Maybe we'll pull back before our current episode matches the Triassic extinction or the K-T event. Maybe it will turn out to be no worse than the Eocene extinction, with a 35 percent loss of species.

"Are you hopeful?" I ask.

Given that hope is a duty from which paleontologists are exempt, I'm surprised when he answers, "Yes, I am."

I'm not. My own guess about the mid-term future, excused by no exemption, is that our Planet of Weeds will indeed be a crummier place, a lonelier and uglier place, and a particularly wretched place for the 2 billion people comprising Alan Durning's absolute poor. What will increase most dramatically as time proceeds, I suspect won't be generalized misery or futuristic modes of consumption but the gulf between two global classes experiencing those extremes. Progressive failure of ecosystem functions? Yes, but human resourcefulness of the sort Julian Simon so admired will probably find stopgap technological remedies, to be available for a price. So the world's privileged class—that's your class and my class—will probably still manage to maintain themselves inside Homer-Dixon's stretch limo, drinking bottled water and breathing bottled air and eating reasonably healthy food that has become incredibly precious, while the potholes on the road outside grow ever deeper. Eventually the limo will look more like a lunar rover. Ragtag mobs of desperate souls will cling to its bumpers, like groupies on Elvis's final Cadillac. The absolute poor will suffer their lack of ecological privilege in the form of lowered life expectancy, bad health, absence of education, corrosive want, and anger. Maybe in time they'll find ways to gather themselves in localized revolt against the affluent class. Not likely, though, as long as affluence buys guns. In any case, well be-

WHAT WILL HAPPEN AFTER THIS MASS EXTINCTION, AFTER WE DESTROY TWO THIRDS OF ALL LIVING SPECIES?

fore that they will have burned the last stick of Bornean dipterocarp for firewood and roasted the last lemur, the last grizzly bear, the last elephant left unprotected outside a zoo.

Jablonski has a hundred things to do before leaving for Alaska, so after two hours I clear out. The heat on the sidewalk is fierce, though not nearly as fierce as this summer's heat in New Delhi or Dallas, where people are dying. Since my flight doesn't leave until early evening, I cab downtown and take refuge in a nouveau-Cajun restaurant near the river. Over a beer and jambalaya, I glance again at Jablonski's 1991 *Science* paper, titled "Extinctions: A Paleontological Perspective." I also play back the tape of our conversation, pressing my ear against the little recorder to hear it over the lunch-crowd noise.

Among the last questions I asked Jablonski was, What will happen *after* this mass extinction, assuming it proceeds to a worst-case scenario? If we destroy half or two thirds of all living species, how long will it take for evolution to fill the planet back up? "I don't know the answer to that," he said. "I'd rather not bottom out and see what happens next." In the journal paper he had hazarded that, based on fossil evidence in rock laid down atop the K-T event and others, the time required for full recovery might be 5 or 10 million years. From a paleontological perspective, that's fast. "Biotic recoveries after mass extinctions are geologically rapid but immensely prolonged on human time scales," he wrote. There was also the proviso, cited from another expert, that recovery might not begin until *after* the extinction-causing circumstances have disappeared. But in this case, of course, the circumstances won't likely disappear until *we* do.

Still, evolution never rests. It's happening right now, in weed patches all over the planet. I'm not presuming to alert you to the end of the world, the end of evolution, or the end of nature. What I've tried to describe here is not an absolute end but a very deep dip, a repeat point within a long, violent cycle. Species die, species arise. The relative pace of those two processes is what matters. Even rats and cockroaches are capable—given the requisite conditions, namely, habitat diversity and time—of speciation. And speciation brings new diversity. So we might reasonably imagine an Earth upon which, 10 million years after the extinction (or, alternatively, the drastic transformation) of *Homo sapiens,* wondrous forests are again filled with wondrous beasts. That's the good news.

David Quammen is the author of eight books, including The Song of the Dodo *and, most recently,* Wild Thoughts from Wild Places. *His last article for* Harper's *Magazine, "Brazil's Jungle Blackboard," appeared in the March 1998 issue.*

Invasive Species: Pathogens of Globalization

by Christopher Bright

World trade has become the primary driver of one of the most dangerous and least visible forms of environmental decline: Thousands of foreign, invasive species are hitch-hiking through the global trading network aboard ships, planes, and railroad cars, while hundreds of others are traveling as commodities. The impact of these bioinvasions can now be seen on every landmass, in nearly all coastal waters (which comprise the most biologically productive parts of the oceans), and probably in most major rivers and lakes. This "biological pollution" is degrading ecosystems, threatening public health, and costing billions of dollars annually. Confronting the problem may now be as critical an environmental challenge as reducing global carbon emissions.

Despite such dangers, policies aimed at stopping the spread of invasive "exotic" species have so far been largely ineffective. Not only do they run up against far more powerful policies and interests that in one way or another encourage invasion, but the national and international mechanisms needed to control the spread of non-native species are still relatively undeveloped. Unlike chemical pollution, for instance, bioinvasion is not yet a working category of environmental decline within the legal culture of most countries and international institutions.

In part, this conceptual blindness can be explained by the fact that even badly invaded landscapes can still look healthy. It is also a consequence of the ancient and widespread practice of introducing exotic species for some tangible benefit: A bigger fish makes for better fishing, a faster-growing tree means more wood. It can be difficult to think of these activities as a form of ecological corrosion—even if the fish or the tree ends up demolishing the original natural community.

The increasing integration of the world's economies is rapidly making a bad situation even worse. The continual expansion of world trade—in ways that are not shaped by any real understanding of their environmental effects—is causing a degree of ecological mixing that appears to have no evolutionary precedent. Under more or less natural conditions, the arrival of an entirely new organism was a rare event in most times and places. Today it can happen any time a ship comes into port or an airplane lands. The real problem, in other words, does not lie with the exotic species themselves, but with the economic system that is continually showering them over the Earth's surface. Bioinvasion has become a kind of globalization disease.

THEY CAME, THEY BRED, THEY CONQUERED

Bioinvasion occurs when a species finds its way into an ecosystem where it did not evolve. Most of the time when this happens, conditions are not suitable for the new arrival, and it enjoys only a brief career. But in a small percentage of cases, the exotic finds everything it needs—and nothing capable of controlling it. At the very least, the invading organism is liable to suppress some native species by consuming resources that they would have used instead. At worst, the invader may rewrite some basic ecosystem "rules"—checks and balances that have developed between native species, usually over many millennia.

Although it is not always easy to discern the full extent of havoc that invasive species can wreak upon an ecosystem, the resulting financial damage is becoming increasingly difficult to ignore. Worldwide, the losses to agriculture might be anywhere from $55 billion to nearly $248 billion annually. Researchers at Cornell University recently concluded that bioinvasion might be costing the United States alone as much as $123 billion per year. In South and Central America, the growth of specialty export crops—upscale vegetables and fruits—has spurred the spread of whiteflies, which are capable of transmitting at least 60 plant viruses. The spread of these viruses has forced the abandonment of more than 1 million hectares of cropland in South America. In the wetlands of northern Nigeria, an exotic cattail is strangling rice paddies, ruining fish habitats, and slowly choking off the Hadejia-Nguru river system. In southern India, a tropical American shrub, the bush morning glory, is causing similar chaos throughout the basin of the Cauvery, one of the region's biggest rivers. In the late 1980s, the accidental release into the Black Sea of *Mnemiopsis leidyi*—a comb jelly native to the east coast of the Americas—provoked the collapse of the already highly stressed Black Sea fisheries, with estimated financial losses as high as $350 million.

Controlling such exotics in the field is difficult enough, but the bigger problem is preventing the machinery of the world trading system from releasing them in the first place. That task is becoming steadily more formidable as the trading system continues to grow. Since 1950, world trade has expanded sixfold in terms of value. More important in terms of potential invasions is the vast increase in the volume of goods traded. Look, for instance, at the ship, the primary mechanism of trade—80 percent of the world's goods travel by ship for at least part of their journey from manufacturer to consumer. From 1970 to 1996, the volume of seaborne trade nearly doubled.

Ships, of course, have always carried species from place to place. In the days of sail, shipworms bored into the wooden hulls, while barnacles and seaweeds attached themselves to the sides. A small menagerie of other creatures usually took up residence within these "fouling communities." Today, special paints and rapid crossing times have greatly reduced hull fouling, but each of the 28,700 ships in the world's major merchant fleets represents a honeycomb of potential habitats for all sorts of life, both terrestrial and aquatic.

Controlling invasive species is difficult enough, but the bigger problem is preventing the machinery of the world trading system from releasing them in the first place.

The most important of these habitats lies deep within a modern ship's plumbing, in the ballast tanks. The ballast tanks of a really big ship—say, a supertanker—may contain more than 200,000 cubic meters of water—equivalent to 2,000 Olympic-sized swimming pools. When those tanks are filled, any little creatures in the nearby water or sediment may suddenly become inadvertent passengers. A few days or weeks later, when the tanks are discharged at journey's end, they may become residents of a coastal community on the other side of the world. Every year, these artificial ballast currents move some 10 billion cubic meters of water from port to port. Every day, some 3,000 to 10,000 different species are thought to be riding the ballast currents. The result is a creeping homogenization of estuary and bay life. The same creatures come to dominate one coastline after another, eroding the biological diversity of the planet's coastal zones—and jeopardizing their ecological stability.

Some pathways of invasion extend far beyond ships. Another prime mechanism of trade is the container: the metal box that has revolutionized the transportation of just about every good not shipped in bulk. The container's effect on invasion ecology has been just as profound. For centuries, shipborne exotics were largely confined to port areas—but no longer. Containers move from ship to harbor crane to the flatbed of a truck or railroad car and then on to wherever a road or railroad leads. As a result, all sorts of stowaways that creep aboard containers often wind up far inland. Take the Asian tiger mosquito, for example, which can carry dengue fever, yellow fever, and encephalitis. The huge global trade in containers of used tires—which are, under the right conditions, an ideal mosquito habitat—has dispersed this species from Asia and the Indo-Pacific into Australia, Brazil, the eastern United States, Mozambique, New Zealand, Nigeria, and southern Europe. Even packing material within containers can be a conduit for exotic species. Untreated wood pallets, for example, are to forest pests what tires are to mosquitoes. One creature currently moving along this pathway is the Asian longhorn beetle, a wood-boring insect from China with a lethal appetite for deciduous trees. It has turned up at more than 30 locations around the United States and has also been detected in Great Britain. The only known way to eradicate it is to cut every tree suspected of harboring it, chip all the wood, and burn all the chips.

As other conduits for global trade expand, so does the potential for new invasions. Air cargo service, for example, is building a global network of virtual canals that have great potential for transporting tiny, short-lived creatures such as microbes and insects. In 1989, only three airports received more than 1 million tons of cargo; by 1996, there were 13 such airports. Virtually everywhere you look, the newly constructed infrastructure of the global economy is forming the groundwork for an ever-greater volume of biological pollution.

The Global Supermarket

Bioinvasion cannot simply be attributed to trade in general, since not all trade is "biologically dirty." The natural resource industries—especially agriculture, aquaculture, and forestry—are causing a disproportionate share of the problem. Certain trends within each of these industries are liable to exacerbate the invasion pressure. The migration of crop pests can be attributed, in part, to a global agricultural system that has become increasingly uniform and integrated. (In China, for example, there were about 10,000 varieties of wheat being grown at mid-century; by 1970 there were only about 1,000.) Any new pest—or any new form of an old pest—that emerges in one field may eventually wind up in another.

The key reason that South America has suffered so badly from white-flies, for instance, is because a pesticide-resistant biotype of that fly emerged in California in the 1980s and rapidly became one of the world's most virulent crop pests. The fly's career illustrates a common dynamic: A pest can enter the system, disperse throughout it, and then develop new strains that reinvade other parts of the system. The displacement of traditional developing-world crop varieties by commercial, homogenous varieties that require more pesticide, and the increasing development of pesticide resistance among all the major pest categories—insects, weeds, and fungi—are likely to boost this trend.

Similar problems pertain to aquaculture—the farming and exporting of fish, shellfish, and shrimp. Partly because of the progressive depletion of the world's most productive fishing grounds, aquaculture is a booming business. Farmed fish production exceeded 23 million tons by 1996, more than triple the volume just 12 years before. Developing countries in particular see aquaculture as a way of increasing protein supply.

But many aquaculture "crops" have proved very invasive. In much of the developing world, it is still common to release exotic fish directly into natural waterways. It is hardly surprising, then, that some of the most popular aquaculture fish have become true cosmopolitans. The Mozambique tilapia, for example, is now established in virtually every tropical and subtropical country. Many of these introductions—not just tilapia, but bass, carp, trout, and other types of fish—are implicated in the decline of native species. The constant flow of new introductions catalogued with such enthusiasm in the industry's publications are a virtual guarantee that tropical freshwater ecosystems are unraveling beneath the surface.

Aquaculture is also a spectacularly efficient conduit of disease. Perhaps the most virulent set of wildlife epidemics circling the Earth today involves shrimp production in the developing world. Unlike fish, shrimp are not a subsistence crop: They are an extremely lucrative export business that has led to the bulldozing of many tropical coasts to make way for shrimp ponds. One of the biggest current developments is an Indonesian operation that may eventually cover 200,000 hectares. A horde of shrimp pathogens—everything from viruses to protozoa—is chasing these operations, knocking out ponds, an occasionally ruining entire national shrimp industries: in Taiwan in 1987, in China in 1993, and in India in 1994. Shrimp farming has become, in effect, a form of "managed invasion." Since shrimp are important components of both marine and freshwater ecosystems worldwide, it is anybody's guess at this point what impact shrimpborne pathogens will ultimately have.

Managed invasion is an increasingly common procedure in another big biopolluting industry: forestry. Industrial roundwood production (basically, the cutting of logs for uses other than fuel) currently hovers at around 1.5 billion cubic meters annually, which is more than twice the level of the 1950s. An increasing amount of wood and wood pulp is coming out of tree plantations (not inherently a bad idea, given the rate at which the world is losing natural forests). In North America and Europe, plantation forestry generally uses native species, so the gradation from natural forest to plantation is not usually as stark as it is in developing countries, where exotics are the rule in industrial-plantation development.

For the most part, these developing-country plantations bear about as much resemblance to natural forests as corn fields do to undisturbed prairies. And like corn fields, they are maintained with heavy doses of pesticides and subjected to a level of disturbance—in particular, the use of heavy equipment to harvest the trees—that tends to degrade soil. Some plantation trees have launched careers as king-sized weeds. At least 19 species of exotic pine, for example, have invaded various regions in the Southern Hemisphere, where they have displaced native vegetation and, in some areas, apparently lowered the water tables by "drinking" more water than the native vegetation would consume. Even where the trees have not proved invasive, the exotic plantations themselves are displacing natural forest and traditional forest peoples. This type of tree plantation is almost entirely designed to feed wood to the industrialized world, where 77 percent of industrial roundwood is consumed. As with shrimp production, local ecological health is being sacrificed for foreign currency.

There is another, more poignant motive for the introduction of large numbers of exotic trees into the developing world. In many countries severely affected by forest loss, reforestation is recognized as an important social imperative. But the goal is often nothing more than increasing tree cover. Little distinction is made between plantation and forest or between foreign and native species. Surayya Khatoon, a botanist at the University of Karachi, observes that "awareness of the dangers associated with invasive species is almost nonexistent in Pakistan, where alien species are being planted on a large scale in so-called afforestation drives."

The industrial sources of biological pollution are very diverse, but they reflect a common mindset. Whether it is a tree plantation, a shrimp farm, or even a bit of landscaping in the back yard, the Earth has become a sort of "species supermarket"; if a species looks good for whatever it is that you have in mind, pull it off the shelf and take it home. The problem is that many of the traits you want the most—adaptability, rapid growth, and easy reproduction—also tend to make the organism a good candidate for invasion.

Launching a Counter-Attack

Since the processes of invasion are deeply embedded in the globalizing economy, any serious effort to root them out will run the risk of exhausting itself. Most industries and policymakers are striving to open borders, not erect new barriers to trade. Moreover, because bioinvasion is not yet an established policy category, jurisdiction over it is generally badly fragmented—or even absent—on both the national and international levels. Most countries have some relevant legislation—laws intended to discourage the movement of crop pests, for example—but very few have any overall legislative authority for dealing with the problem. (New Zealand is the noteworthy exception: Its Biosecurity Act of 1993 and its Hazardous Substances and New Organisms Act of 1996 do establish such an authority.) Although it is true that there are many treaties that bear on the problem in one way or another—23 at least count—there is not such thing as a bioinvasion treaty.

Even agreements that focus specifically on ecological problems have generally given bioinvasion short shrift. Agenda 21, for example—the blueprint for sustainable development that emerged from the 1992 Earth Summit in Rio de Janeiro—reflects little awareness of the dangers of exotic forestry and aquaculture. Among international agencies, only certain types of invasion seem to get much attention. There are treaties—such as the 1951 International Plant Protection Convention—that limit the movement of agricultural pests, but there is currently no clear international mechanism for dealing with ballast water releases. Obviously, in such a context, you need to pick your fights carefully.

They have to be important, winnable, and capable of yielding major opportunities elsewhere. The following three-point agenda offers some hope of slowing invasion over the near term.

Even international agreements that focus specifically on ecological problems have generally given bioinvasion short shrift.

The first item: Plug the ballast water pathway. As a technical problem, this objective is probably just on the horizon of feasibility, making it an excellent policy target. Strong national and international action could push technologies ahead rapidly. At present, the most effective technique is ballast water exchange, in which the tanks of a ship are pumped out and refilled in the open sea. (Coastal organisms, pumped into the tanks at the ship's last port of call, usually will not survive in the open ocean; organisms that enter the tanks in mid-ocean probably will not survive in the next port of call.) But it can take several days to exchange the water in all of a ship's ballast tanks, so the procedure may not be feasible for every leg of a journey, and the tanks never empty completely. In bad weather, the process can be too dangerous to perform at all. Consequently, other options will be necessary—filters or even toxins (that may not sound very appealing, but some common water treatment compounds may be environmentally sound). It might even be possible to build port-side ballast water treatment plants. Such a mixture of technologies already exists as the standard means of controlling chemical pollution.

This objective is drifting into the realm of legal possibility as well. As of July 1 this year, all ships entering U.S. waters must keep a record of their ballast water management. The United States has also issued voluntary guidelines on where those ships can release ballast water. These measures are a loose extension of the regulations that the United States and Canada have imposed on ship traffic in the Great Lakes, where foreign ballast water release is now explicitly forbidden. In California, the State Water Resources Control Board has declared San Francisco Bay "impaired" because it is so badly invaded—a move that may allow authorities to use regulations written for chemical pollution as a way of controlling ballast water. Australia now levies a small tax on all incoming ships to support ballast water research.

Internationally, the problem has acquired a high profile at the UN International Maritime Organization (IMO), which is studying the possibility of developing a ballast management protocol that would have the force of international law. No decision has been made on the legal mechanism for such an agreement, although the most likely possibility is an annex to MARPOL, the International Convention for the Prevention of Pollution from Ships.

Within the shipping industry, the responses to such proposals have been mixed. Although industry officials concede the problem in the abstract, the prospect of specific regulations has tended to provoke unfavorable comment. After an IMO meeting last year on ballast water management, a spokesperson for the International Chamber of Shipping argued that rigorous ballast exchange would cost the industry millions of dollars a year—and that internationally binding regulations should be avoided in favor of local regulation, wherever particular jurisdictions decide to address the problem. Earlier this year in California, a proposed bill that would have essentially prohibited foreign ballast water release in the state's ports provoked outcries from local port representatives, who argued that such regulations might encourage ship traffic to bypass California ports in favor of the Pacific Northwest or Mexico. Of course, any management strategy is bound to cost something, but the important question is: What impact will this additional cost have? It may not have much impact at all. In Canada, for example, the Vancouver Port Authority reported that its ballast water program has had no detectable effect on port revenues.

The second item on the agenda: Fix the World Trade Organization (WTO) Agreement on the Application of Sanitary and Phytosanitary Measures. This agreement, known as the SPS, was part of the diplomatic package that created the WTO in 1994. The SPS is supposed to promote a common set of procedures for evaluating risks of contamination in internationally traded commodities. The contaminants can be chemical (pesticide residues in food) or they can be living things (Asian longhorn beetles in raw wood).

One of the procedures required by the SPS is a risk assessment, which is supposed to be done before any trade-constricting barriers are imposed to prevent a contaminated good from entering a country. If you want to understand the fundamental flaw in this approach as it applies to bioinvasion, all you have to do is recall the famous observation by the eminent biologist E.O. Wilson: "We dwell on a largely unexplored planet." When it comes to the largest categories of living things—insects, fungi, bacteria, and so on—we have managed to name only a tiny fraction of them, let alone figure out what damage they can cause. Consider, for example, the rough, aggregate risk assessments done by the United States Department of Agriculture (USDA) for wood imported into the United States from Chile, Mexico, and New Zealand. The USDA found dozens of "moderate" and "high" risk pests and pathogens that have the potential for doing economic damage on the order of hundreds of millions of dollars at least—and ecological damage that is incalculable. But even with wide-open thoroughfares of invasion such as these, the SPS requirement in its current form is likely to make preemptive action vulnerable to trade complaints before the WTO.

Another SPS requirement intended to insure a consistent application of standards is that a country must not set up barriers against an organism that is already living within its borders unless it has an "official control program" for that species. This approach is unrealistic for both biological and financial reasons. Thousands of exotic species are likely to have invaded most of the world's countries and not even the wealthiest country could possibly afford to fight them all. Yet it certainly is possible to exacerbate a problem by introducing additional infestations of a pest, or by boosting the size of existing infestations, or even by increasing the genetic vigor of a pest population by adding more "breeding stock." The SPS does not like "inconsistencies"—if you are not controlling a pest, you have no right to object to more of it; if you try to block one pathway of invasion, you had better be trying to block all the equivalent pathways. Such an approach may be theoretically neat, but in the practical matter of dealing with exotics, it is a prescription for paralysis.

In the near term, however, any effort to repair the SPS is likely to be difficult. The support of the United States, a key member of the WTO, will be critical for such reforms. And although the United States has demonstrated a heightened awareness of the problem—as evidenced by President Bill Clinton's executive order to create an Invasive Species Council—it is not clear whether that commitment will be reflected in the administration's trade policy. During recent testimony before Congress, the U.S. Trade Representative's special trade negotiator for agricultural issues warned that the United States was becoming impatient with the "increasing use of SPS barriers as the 'trade barrier of choice.'" In the developing world, it is reasonable to assume that any country with a strong export sector in a natural resource industry would not welcome tougher regulations. Some developed countries, however, may be sympathetic to change. The European Union (EU) has sought very strict standards in its disputes with the United States over bans on beef from cattle fed with growth hormones and on genetically altered foods. It is possible that the EU might be willing to entertain a stricter SPS. The same might be true of Japan, which has attempted to secure stricter testing of U.S. fruit imports.

The third item: Build a global invasion database. Currently, the study of bioinvasion is an obscure and rather fractured enterprise. It can be difficult to locate critical information or relevant expertise. The full magnitude of the issue is still not registering on the public radar screen. A global database would consolidate existing information, presumably into some sort of central research institution with a major presence on the World Wide Web. One could "go" to such a place—either physically or through cyberspace—to learn about everything from the National Ballast Water Information Clearinghouse that the U.S. Coast Guard is setting up, to the database on invasive woody plants in the tropics that is being assembled at the University of Wales. The database would also stimulate the production of new media to encourage additional research and synthesis. It is a telling indication of how fragmented this field is that, after more than 40 years of formal study, it is just now getting its first comprehensive journal: *Biological Invasions*.

Better information should have a number of practical effects. The best way to control an invasion—when it cannot be prevented outright—is to go after the exotic as soon as it is detected. An emergency response capability will only work if officials know what to look for and what to do when they find it. But beyond such obvious applications, the database could help bring the big picture into focus. In the struggle with exotics, you can see the free-trade ideal colliding with some hard ecological realities. Put simply: It may never be safe to ship certain goods to certain places—raw wood from Siberia, for instance, to North America. The notion of real, permanent limits to economic activity will for many politicians (and probably some economists) come as a strange and unpalatable idea. But the global economy is badly in need of a large dose of ecological realism. Ecosystems are very diverse and very different from each other. They need to stay that way if they are going to continue to function.

Want to Know More?

Although the scientific literature on bioinvasion is enormous and growing rapidly, most of it is too technical to attract a readership outside the field. For a nontechnical, broad overview of the problem, readers should consult Robert Devine's *Alien Invasion: America's Battle with Non-Native Animals and Plants* (Washington: National Geographic Society, 1998) or Christopher Bright's *Life Out of Bounds: Bioinvasion in a Borderless World* (New York: W.W. Norton & Company, 1998).

If you have a long-term interest in bioinvasion, you will want to get acquainted with the book that founded the field: Charles Elton's *The Ecology of Invasions by Animals and Plants* (London: Methuen, 1958). A historical overview of bioinvasions can be found in Alfred Crosby's book *Ecological Imperialism: The Biological Expansion of Europe, 900–1900* (Cambridge: Cambridge University Press, 1986).

Many studies focus on invasion of particular regions. The focus can be very broad, as in P.S. Ramakrishnan, ed., *Ecology of Biological Invasions in the Tropics,* proceedings of an international workshop held at Nainital, India, (New Delhi: International Scientific Publications, 1989). Generally, however, the coverage is much narrower, as in Daniel Simberloff, Don Schmitz, and Tom Brown, eds., *Strangers in Paradise: Impact and Management of Nonindigenous Species in Florida* (Washington: Island Press, 1997). The other standard research tack has been to look at a particular type of invader. The most accessible results of this exercise are encyclopedic surveys such as Christopher Lever's *Naturalized Mammals of the World* (London: Longman, 1985) and his companion volumes on naturalized birds and fish. In the plant kingdom, the genre is represented by Leroy Holm, et al., *World Weeds: Natural Histories and Distribution* (New York: John Wiley and Sons, 1997).

There are many worthwhile documents available for anyone who is interested not just in the ecology of invasion, but also in its economic, social, and epidemiological implications. Just about every aspect of the problem is discussed in Odd Terje Sandlund, Peter Johan Schei, and Aslaug Viken, eds., *Proceedings of the Norway/UN Conference on Alien Species* (Trondheim: Directorate for Nature Management and Norwegian Institute for Nature Research, 1996). A groundbreaking study of invasion in the United States, with particular emphasis on economic effects, is *Harmful Nonindigenous Species in the United States* (Washington: Office of Technology Assessment, September 1993). An assessment of the ballast water problem is available from the National Research Council's Commission on Ships' Ballast Operations' *Stemming the Tide: Controlling Introductions of Nonindigenous Species by Ships' Ballast Water* (Washington: National Academy Press, 1996). Readers who are interested in exotic tree plantations as a form of "managed invasion" might look through Ricardo Carrere and Larry Lohmann's *Pulping the South: Industrial Tree Plantations and the World Paper Economy* (London: Zed Books, 1996) and the World Rainforest Movement's *Tree Plantations: Impacts and Struggles* (Montevideo: WRM, 1999). Unfortunately, there are no analogous studies of shrimp farms.

For links to relevant Web sites, as well as a comprehensive index of related FOREIGN POLICY articles, access **www.foreignpolicy.com**.

CHRISTOPHER BRIGHT *is a research associate at the Worldwatch Institute in Washington, DC, and author of* Life Out of Bounds: Bioinvasion in a Borderless World *(New York: W.W. Norton & Company, 1998).*

Mass Extinction

By David Hosansky

THE ISSUES

Zoo biologist Edward J. Maruska can remember exploring the rain forests of Costa Rica in the late 1970s, when thousands of shimmering golden toads gathered in ponds of the mist-shrouded Monteverde Cloud Forest Reserve to breed.

But all the toads and vanished by the 1980s. Scientists believe that the spectacular toads, easily recognized by the males' bright orange color, fell victim to disease, changing climate patterns or pollution.

"They were so unique," recalls Maruska, executive director of the Cincinnati Zoo and Botanical Garden. "Then they were gone."

Last sighted by scientists in 1989, the golden toad is among thousands of species that have become extinct in recent years. Humans are wiping out much of the Earth's plant and animal live by paving over open space for homes and factories; clearing forests for cultivation and grazing; polluting the air and water; and introducing non-native species into fragile ecological areas.

Even primates, which belong to the taxonomic group that includes human beings, are not immune. Scientists recently concluded that a West African monkey known as Miss Waldron's red colobus has been wiped out because of deforestation and hunting. The red-cheeked monkey lived in the rain forest canopy of Ghana and the Ivory Coast. It was the first time in several centuries that a primate had become extinct.[1]

In fact, civilization's unrelenting march across unspoiled lands has had such a profound effect on nature that scientists warn we have entered an age of mass extinction the likes of which have not been seen since the demise of dinosaurs some 65 million years ago.

The Earth, scientists say, has experienced wholesale loss of life on such a colossal scale only five times before. Some contend it will bring irrevocable changes for the planet's dominant species—humans—altering everything from food supplies to medical breakthroughs to the weather.

"There's scientific debate about the rate and the extent of species loss, but I don't think there's much remaining debate that we're in a period of mass extinction," says Eleanor Sterling, director of the Center for Biodiversity and Conservation at the American Museum of Natural History in New York City. "It's absolutely one of the most critical issues that's facing us today."

Biologists cannot quantify the rate of extinction because they do not know the total number of species that exist on Earth—let alone the numbers of mostly unknown animals, plants, fungi and other organisms that are vanishing. But leading scientists believe that based on the pace of destruction of the richest habitats, such as tropical rain forests and coral reefs, the world is losing species at a rate of 100 to 10,000 times the normal, or "background," rate. They warn that 50 percent or more of all species will be gone by the end of the current century.[2]

In the United States alone, the U.S. Fish and Wildlife Service lists 1,233 plants and animals as threatened or endangered. That probably greatly understates the full number of vanishing species because the government lacks the resources to search for all types of endangered organisms.[3] Around the globe, human activities are threatening countless species of frogs, tropical beetles, freshwater fish, birds, flowers and trees as well as familiar mammals such as tigers, gorillas and giant pandas.

"We anticipate we've lost a whole host of species, sometimes before we even documented them," says David Olsen, a conservation biologist with the World Wildlife Fund (WWF). "We know enough about patterns of biodiversity around the world and the loss of natural habitats to say that we are in the midst of a very serious event."

Although not all biologists agree on the extent of the loss, they generally regard mass extinction as one of the gravest issues facing humanity. According to a 1998 poll by the American Museum of Natural History, most scientists in the United States believe that the world is in the midst of a mass extinction. Moreover, they rate it a greater threat to society than more publicized problems such as pollution, global warming and the thinning of the ozone layer.

Scientists believe the loss of so many species of animals, plants and microorganisms could have profound and unpredictable effects on the United States and every other nation. Many plants and animals provide food, fibers and building materials, as well as new medicines. Others regulate the flow of water, influence weather patterns, fertilize crops, prevent topsoil erosion and reduce the amount of carbon dioxide in the atmosphere. Even tiny creatures such as insects, regarded by many as pests, are probably essential for the survival of Homo sapiens by filling critical niches in the global ecosystem.

"So important are insects and other land-dwelling arthropods that if all were to disappear, humanity probably could not last more than a few months," Harvard biologist Edward O. Wilson wrote in his influential 1992 book, *The Diversity of Life*.[4]

World leaders have responded to the threat of biodiversity loss with a series of international agreements. The 1973 Convention on International Trade in Endangered Species of Wild Fauna and Flora (CITES) has helped

Vast Regions of Earth Are 'In Danger'

In vast portions of the world, including Australia, India, China and much of South America, 7 percent or more of the mammal and bird species are in danger, the World Wildlife Fund says. * According to the WWF's Living Planet Index, the Earth has lost 30 percent of its plant and animal species since 1970, based on the loss of forest cover and the decline in abundance of marine and freshwater species.

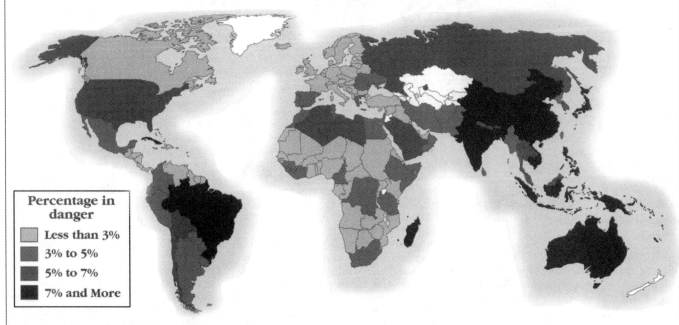

Mammal and bird species in danger

Percentage in danger
- Less than 3%
- 3% to 5%
- 5% to 7%
- 7% and More

*Many places where losses are believed to be severe, such as Ivory Coast and other areas in Africa, don't appear on the map because of counting anomalies.

Source: "Living Planet Report," World Wildlife Fund, September 2000.

preserve well-known species such as elephants and sea turtles. In the United States, the Endangered Species Act (ESA), along with other government measures such as pesticide regulations and passage of the Clean Water Act, are credited with fostering the recovery of several species on the brink of extinction, including the bald eagle.

A series of high-profile conferences, including the 1992 Earth Summit in Rio de Janeiro, Brazil, has helped focus attention on the worldwide loss of biodiversity. In recent years, powerful organizations such as the World Bank have begun to provide grants to developing countries for projects that help restore the environment, and industrialized nations have provided incentives, such as debt forgiveness, to developing countries to preserve biologically important habitat. Some environmental groups promote ecotourism as a way to help local communities profit by limiting harmful development.

However, ecosystems are threatened by such a multitude of factors that even optimistic conservationists warn that numerous species are doomed. "We're at the beginning of the [extinction] curve, and the question is whether we'll get it together soon enough to avoid a lot of the loss," says Thomas Lovejoy, chief biodiversity adviser at the World Bank.

The No. 1 cause of extinction, both in the United States and worldwide, is habitat destruction. As people clear forests, drain wetlands and dam rivers, they destroy the homes of countless organisms. The United States, for example, has lost almost half its wetlands since the 18th century, and tropical countries have lost more than half of their rain forests.[5]

The introduction of invasive species such as fire ants and kudzu, which proliferate rapidly and overcome native species, also has increased the pace of extinctions. Island ecosystems are particularly sensitive. In Guam and Hawaii, for example, newly introduced rats, cats and snakes have wiped out many species of native birds.

Pollution, overfishing and hunting also threaten wildlife habitat. And more potential threats loom. Environmentalists now fear global climatic changes could decimate habitats in natural parks before they can migrate to more suitable surroundings.

"We have quite static ways of protecting biodiversity at the moment, drawing boundaries about a population and saying, 'This population is now conserved,'" says the American Museum of Natural History's Sterling. "As the temperature changes, you're going to have an empty preserve and a species that has migrated out."

But while scientists have expressed growing alarm about the rapid disappearance of more and more species, the Museum of Natural History poll indi-

17. Mass Extinction

THE TEN MOST WANTED

The World Wildlife Fund has identified 10 species of plants and animals that are particularly threatened by illegal and unsustainable trade. The list was presented at the Convention on International Trade in Endangered Species of Wild Flora and Fauna (CITIES) annual conference in Nairobi, Kenya, last April.

Tiger
- Habitat: Russian Far East, Southeast Asia
- Threat: Poaching for skin, bones and body parts
- Population: under 6,000

Whale Shark
- Habitat: Highly migratory, coastal
- Threat: Overfishing for fins, meat
- Population: Unknown, decline in sightings

Javan Pangolin
- Habitat: Indonesia
- Threat: Skin, meat and scales, used in traditional Asian medicine
- Population: Unknown

Asian Box Turtles
- Habitat: South east Asia
- Threat: Hunted for shell, which is used in traditional medicine
- Population: Nearing extinction

Horned Parakeet
- Habitat: New Caledonia
- Threat: Habitat loss, illegal capture for the pet trade
- Population: under 1,700

Tibetan Antelope
- Habitat: East Asia
- Threat: Wool used to make highly priced Shahtoosh shawls. Up to five killed to make one shawl
- Population: under 75,000

Hawksbill Sea Turtle
- Habitat: Caribbean
- Threat: Global market for tortoise shell, exploitation for eggs and meat
- Population: Unknown

Asian Ginseng
- Habitat: Russia, China, Korean Peninsula
- Threat: High prices paid for wild roots ($3,000 per ounce)
- Population: Isolated areas

Sumatran Rhinoceros
- Habitat: Sumatra, Borneo, Peninsular Malaysia
- Threat: Hunted for use in traditional medicine
- Population: under 300

Giant Panda
- Habitat: China
- Threat: Habitat loss and poaching. Demand from overseas zoos
- Population: 1,000

Source: World Wildlife Fund

Reuters Graphics

cates the issue of mass extinction has not yet resonated with the general public. Unlike more visible environmental problems, such as air pollution and contaminated drinking water, the loss of obscure beetles and salamanders does not affect the daily lives of most people.

Furthermore, the business community is reluctant to support government regulations that restrict development to protect plants and creatures that seem to have little significance. That was dramatically illustrated in the late 1980s and early '90s, when loggers heatedly protested plans to set aside forests that were habitat to the rare northern spotted owl.

"Let's prioritize what the real costs are, because our resources are limited," says William L. Kovacs, vice president of environmental and regulatory affairs for the U.S. Chamber of Commerce. "We can spend tens of billions of dollars in trying to protect something that has very little benefit to man."

Complicating the issue, scientists disagree about both the extent and the implications of mass extinction. Biologists, who have identified about 1.75 million species worldwide, estimate the actual total ranges anywhere from 3.6 million to more than 100 million.

Harvard's Wilson believes that more than half of these will be gone by the end of the century unless strong conservation measures are undertaken. He and other biologists have arrived at such estimates by calculating the number of species in biologically diverse habitats—especially coral reefs and tropical rain forests—and the rate at which those habitats are being destroyed by human actions.

"Biologists who explore biodiversity see it vanishing before their eyes," he wrote in a special Earth Day edition of *Time* magazine earlier this year.[6]

But some contend that concerns about extinction are overstated. Michael Gilpin, a University of San Diego biology professor, predicts that future extinctions will be largely confined to obscure organisms in developing countries. "It's not that we're going to lose zebras and wildebeests," he says. "Beetles we're going to lose, but we're never going to know what they are."

Others go as far as to argue that human disruptions of nature actually may be good in the long run. They contend that past extinctions and other disturbances that alter ecosystems created openings for new organisms, often increasing overall diversity. "Without extinction, without a loss of current variety, future variation diminishes," two professors argued in a controversial 1992 article.[7]

However, few in the scientific community take such a sanguine view. Even a skeptic such as Gilpin warns the loss of species deprives the world of genetic diversity, making it harder for scientists to develop new medicines and more disease-resistant strains of crops. "We're burning our genetic library," he says.

Others believe that a mass extinction will result in a natural world dominated by disease-carrying animals that have learned to adapt to human society. What sorts of animals are likely to proliferate if the pace of extinction continues? Sterling predicts "rats and cockroaches," among others.

As policy-makers confront the specter of mass extinction, here are key questions being asked:

Should we preserve endangered plants and animals?

As an aspiring botanist growing up in the San Francisco area, Peter H. Raven enjoyed tromping about and identifying numerous species of local plants. Now he finds many formerly wild areas paved over, and the plants either gone or dying out.

Raven, who is director of the Missouri Botanical Garden and a leading advocate for protecting biodiversity, warns that the same pattern is being repeated around the globe, threatening to impoverish human society on a vast scale because scientists constantly turn to nature to develop new foods, medicines and other products.

"As we lose biodiversity, we lose many opportunities for a rich and sustainable and healthy life," Raven says. "It's biodiversity that makes this a living planet. Most of the species that are likely to be lost in the coming century have never been seen by anybody. We will never be able to exploit their potential."

Raven and other conservationists cite three major reasons why it is critical for world leaders to do everything possible to preserve fast-disappearing species:

• Plants and animals provide humans with essentials such as food, clothing, shelter and medicine. Food supplies could dwindle if not for wild plants that can be crossbred with cultivated varieties to provide hardier crops. For example, a wild species of Mexican maize, nearly extinct when found in the 1970s, has been used to develop disease-resistant corn.

Similarly, endangered plants have helped to spur pharmaceutical breakthroughs. The potent anti-cancer drug Taxol, for example, is derived from the bark of the Pacific yew tree. A study by the National Institutes of Health and other agencies concluded that about 40 percent of the most commonly prescribed drugs are developed from natural sources.[8]

• Living organisms stabilize the environment. Scientists are finding that forests, wetlands and other ecosystems play a major role in regulating water drainage, preventing landslides and soil erosion, and even influencing rain patterns and other types of weather. When the environment is degraded, catastrophe may occur. Environmentalists point to Haiti, now virtually deforested and gradually turning into a desert.

• Society has ethical and aesthetic obligations to preserve the environment. In essence, conservationists—and some religious leaders—argue, every organism has an intrinsic beauty and a place in the world that should not be disturbed by human actions. In fact, some go so far as to say that human beings can never feel at home in a world scarred by environmental degradation.

"Our brains are formed around biodiversity," says Raven of the Missouri Botanical Garden. "A lot of our art is based on it. We are related to it."

"If we don't pay more attention to these issues, then the consequences down the road for ourselves and our future generations will be very significant," says Mark Schaefer, president of the Association for Biodiversity Information, a conservation research organization.

Perhaps surprisingly, some skeptics argue that preservation efforts, by disrupting natural processes, actually can damage the environment. They point out that extinction, after all, is a vital part of evolution: Certain species vanish, and others take their place. "I would make the argument that species live and die and evolve," says the Chamber of Commerce's Kovacs. "Whether you like it or not, you and I are part of evolution. At some point of time, our ancestors died out and changed."

Some also question whether the government should continue to stress the protection of species in an age when technological breakthroughs are allowing scientists to develop genetically modified crops that are resistant to disease and adverse weather. Scientists are even beginning to try to recover species through cloning.

Business leaders also are concerned about the cost of protecting endangered species. They point out that the construction of roads, hospitals, housing developments and other structures have been blocked for relatively trivial environmental concerns, sometimes at the cost of human health.

In 1998, for example, senators battled over building a single-lane gravel road through Alaska's Izembek National Wildlife Refuge, disrupting a pristine habitat but giving residents of Cold Bay, an isolated fishing village, year-round access to a healthcare facility. In the end, lawmakers decided to leave the preserve intact while agreeing to spend $37.5 million on an all-weather airport building and other facilities for the community.[9]

Kovacs says the government should pay more attention to the needs of society, even if environmentalists protest. "Human beings are species too," he says, "and they have some rights on the planet."

Are environmental laws helping rare species?

A boater on the Potomac River just outside Washington, D.C., can expect to see something that would have been remarkable just three decades ago: pairs of nesting bald eagles. The majestic birds, which nearly vanished from the continental United States in the 1960s, have staged a strong comeback thanks to the 1970 federal ban on the highly toxic pesticide DDT and the 1973 Endangered Species Act.

"The return of the bald eagle is a fitting cap to a century of environmental stewardship," President Clinton declared at a White House ceremony celebrating the comeback of the bird that has symbolized the nation for 200 years.[10]

Although the bald eagle remains comparatively rare because of the destruction of its habitat, it nevertheless symbolizes the success of environmental laws. Other successes include the grizzly bear, the gray whale and the gray wolf.

The Endangered Species Act is part of a network of environmental laws that preserve habitats and rare species. The 1972 Marine Mammal Protection Act restricts the hunting of seals, polar bears and other marine mammals; the 1972 Clean Water Act helps rehabilitate waterways and restore many species of fish, and the 1964 Wilderness Act protects remote areas from road building and other development that can fragment habitat and isolate species.

Congress in 1990 took a step toward stopping the introduction of non-native species, which can wipe out native species by preying upon them or outcompeting with them for food, by passing the Non-Indigenous Aquatic Nuisance Act. The law established a task force that coordinates federal efforts to keep out non-native aquatic species, such as the zebra mussel, which is notorious for proliferating in the pipes of drinking water systems, hydroelectric plants and industrial facilities, constricting water flow and affecting heating and cooling systems.

The Endangered Species Act, however, is the only U.S. law targeted specifically at helping rare animals and plants. Policy-makers are divided on whether it has been successful.

Since its passage, more than 1,200 species in the United States have been listed as either endangered or threatened. Of that number, just 11 species have recovered sufficiently to be taken off the list; nine were removed because of improved data, such as the discovery of additional populations; and seven have become extinct.[11]

Based on these results, the law's critics brand it a failure. "It hasn't really been effective in achieving its goals," says Duane Desiderio, assistant staff vice president of the National Association of Home Builders. "It's become little more than an act with a list."

He contends that environmentalists are more intent on stopping development by "adding more and more species to the list" than they are on fostering the recovery of rare organisms.

Environmentalists, however, say that many more animals and plants would be extinct today if it were not for the law. For example, the number of black-footed ferrets—a Western weasel that feeds on prairie dogs—had dwindled to 18 in the mid-1980s before the government stepped in and helped nurture the population back to several thousand.

"If you look at it in terms of preventing extinctions and getting species to a point where at least there is a chance to bring them back, then on that level it's been effective," says Christopher Williams, senior program officer for wildlife conservation policy at the WWF.

The Endangered Species Act is controversial because it imposes strict constraints on property owners who want to develop land that is home to rare plants and animals. As a result, it can have the perverse effect of encouraging property owners to destroy habitat on their land to make sure that it never becomes home to an endangered species.

In Austin, Texas, for example, some landowners in the 1990s bulldozed juniper trees on their property as a pre-emptive strike to prevent their land from being listed by the federal government as critical habitat for two endangered songbirds, the golden-cheeked warbler and black-capped vireo.[12] The rare birds nest only in junipers.

"The incentives are all wrong. The more wildlife habitat that a landowner leaves on his land, the more likely it is that he's going to be prevented from using his own land," says R.J. Smith, senior environmental scholar at the Competitive Enterprise Institute, a free-market-oriented think tank. "It punishes landowners for being good stewards."

Critics also contend that environmentalists are more interested in using the ESA to stop development than to protect rare species. They trace the tactic back to the late 1970s, when residents along the Little Tennessee River who were trying to halt construction of a $31 million dam discovered that the river was home to a tiny endangered fish, the snail darter trout. By invoking the Endangered Species Act, the Tennessee residents won court rulings that temporarily blocked the project. Congress eventually overruled the courts and let construction proceed.

The law originally was conceived to help majestic animals such as the bald eagle. But, to the exasperation of developers, it often is imposed instead to protect rodents and insects such as the Delhi fly and the Indiana bat—even though few people appear concerned about such species.[13]

"Should you stop an entire road or an entire hospital that serves a region just because you've found some flies that scientists say they can't find any use for?" asks Kovach of the Chamber of Commerce. "The program has gotten ridiculous. The environmentalists have really lost control of their common sense."

But environmentalists say it is essential to protect all species, even the most obscure, to preserve an ecosystem. "Each plant or animal, of course, has its own unique set of genes. Once that organism is lost, so is that unique genome," says Schaffer of the Association of Biodiversity Information. "We simply don't know where the next critical bacterium or fungi or plant may be found that contains some unique chemical substance encoded in its gene."

U.S. laws aside, policy-makers debate the effectiveness of international efforts to protect endangered species, including the Convention on International Trade in Endangered Species of Wild Fauna and Flora, which restricts or even bans trading in rare animals such as tigers and rhinoceroses. Like the Endangered Species Act, CITES has produced mixed results. Its ban on ivory trading, for example, inadvertently may have encouraged ivory poachers to instead hunt rhinoceroses for their horns.

Still, says WWF Vice President Richard N. Mott, "It's been instrumental in averting the decline of marine turtles, elephants" and other species.

Will humans survive the current wave of extinctions?

Ever since biologists began studying the extinction of species, they have pondered how long humans will endure. Since animal species typically survive for a few million years, and Homo sapiens evolved only about 100,000 years ago, the odds would appear to be good that humans will be around for quite a while.

However, if extinctions occur at up to 1,000 times the normal rate in the 21st century, could human beings disappear as well? The reassuring answer from most biologists: Not likely. To be sure, humans have the capability of destroying themselves through nuclear or biological warfare, and some scientists even speculate that machines or genetically engineered versions of humans could take over the world.

But, apart from such extraordinary scenarios, the laws of nature suggest that humans are well-positioned to

When Mass Extinctions Ruled the Earth

Extinction is a natural event that usually occurs at a low level, periodically claiming a relative handful of species. A mass extinction, by contrast, wipes out large numbers of plants and animals in a comparatively short time period in geologic terms—perhaps "only" a few hundred thousand years.

Scientists believe that there have been at least five mass extinctions since complex life evolved on Earth a half-billion years ago. In each case, at least one-quarter of all species are thought to have perished, and the toll often was more than one-half. It may have taken millions of years for the diversity of life to recover each time.

Scientists are not sure what caused most of the mass die-offs. They have speculated about giant asteroids striking the planet or massive volcanic eruptions that could have caused devastating climate shifts and changes in the earth's atmosphere. Some have even raised the possibility of deadly cosmic radiation.[1]

The five previous mass extinctions occurred during the following geological eras:

■ **Ordovician (440 million years ago)**—Glaciation may have wiped out an estimated 25 percent of Earth's plant and animal families, some of which may have had thousands of species. Freezing temperatures lowered sea levels and eliminated much marine habitat. Groups of marine organisms, such as trilobites, suffered especially heavy losses.

■ **Devonian (370 million years ago)**—A massive loss of biodiversity may have occurred within a period of 500,000 years, or there may have been several smaller extinction episodes over a period as long as 15 million years. As many as 70 percent of all species disappeared, with marine organisms more affected than those living in freshwater. Scientists speculate that climate change may have played a role, as well as a drop in oxygen levels in shallow waters.

■ **Permian (250 million years ago)**—The most catastrophic of all mass extinctions may have wiped out 96 percent of all marine species and more than three-fourths of the vertebrate families on land. Scientists speculate that the cause may have been volcanic activity, a change in ocean salinity or climate shifts.

■ **Triassic (210 million years ago)**—Although far less devastating than the Permian extinction, the Triassic is important because it opened a niche for dinosaurs and early mammals. The cause of this extinction, which wiped out many sponges, insects and vertebrate groups, is poorly understood. But some scientists speculate it was caused by dramatic climate change.

■ **Cretaceous (65 million years ago)**—An estimated 85 percent of Earth's species, including all dinosaurs and flying lizards (pterosaurs) were wiped out during this much studied period. However, mammals, birds and many reptiles, including crocodiles, survived virtually unscathed. Since 1980, scientists have found evidence that a large asteroid crashing into the Earth caused the Cretaceous extinction, although some speculate that volcanic eruptions or other causes played a role.

[1] Background taken from *National Geographic*, February 1999, pp. 48–49 and British Broadcasting Corporation, www.bbc.co.uk/education/darwin/esfiles/massintro.htm.

---

survive a mass extinction. "We are by far the most widely distributed . . . species on the planet and, with our technology, I believe the most unassailable," University of Washington geologist and paleontologist Peter D. Ward wrote in a 1997 book on extinction, *The Call of Distant Mammoths*. "We are the least endangered species on the planet."[14]

Plant and animal species most vulnerable to extinction usually are few in number, live in a limited area and lack the ability to adjust to change. In contrast, there are more than 6 billion people who live throughout the world and eat all types of food. It is hard to picture a natural scenario—even repeated volcanic eruptions, massive flooding or global climate change—in which the Earth's environment is so altered that people can no longer survive, paleontologists say.

However, some experts warn that the loss of biodiversity could undermine the well-being of society because it may become harder to grow crops and develop new medicines and other products.

"I can't see a scenario where the destruction of biodiversity is going to drive human beings to extinction," says Raven of the Missouri Botanical Garden, "but I see a world that is dull, gray, homogenized and bleak, with many fewer possibilities for developing new products and with many fewer interesting things to do. As we lose biodiversity, we lose many opportunities for rich and sustainable and healthy lives."

Raven and other scientists also believe that the Earth may not be able to sustain its population, which is expected to reach anywhere from 7.3 billion to 10.7 billion by 2050, according to the United Nations projections.[15] In particular, they say that people in wealthy countries like the United States have to consume less or risk depleting the world's resources. "If everybody in the world were consuming at the level of the United States, we would need about three planets like Earth to support them," he says.

Others, however, reject such bleak scenarios. They believe that, thanks to the advances in genetics, scientists will be able to develop better crops and more effective medicines despite the loss of wild species. Indeed, they speculate, genetically modified organisms may even help speed up habitat restoration.

The debate may be moot. Scientists are not certain of what prompted previous mass extinctions, but they know that the causes must have been cataclysmic. If a massive asteroid were to strike the Earth—which may have wiped out the dinosaurs 65 million years ago—humans could perish along with most other plants and animals.

University of Chicago statistical paleontologist David M. Raup estimated the odds of a significant asteroid or comet striking Earth during a person's 75-year lifespan at 1 in 4,000. Although that may indicate that humans are pretty safe, Raup writes that experiencing a "civilization-destroying impact" would appear to be a far greater possibility than dying in an airplane crash.

"We don't know," Raup concludes, "whether we chose a safe planet."[16]

Notes

1. Andrew C. Revkin, "A West African Monkey Is Extinct, Scientists Say," *The New York Times,* Sept. 12, 2000, p. A20.
2. For more detail, see Edward O. Wilson, "Vanishing Before Our Eyes," *Time,* special Spring 2000 Earth Day edition, p. 29; M. Lynne Corn, "Endangered Species: Continuing Controversy," Congressional Research Service Report, May 8, 2000.
3. See U.S. Fish and Wildlife Service, "General Statistics on Endangered Species," http://endangered.fws.gov/stats/genstats.html.
4. Edward O. Wilson, *The Diversity of Life* (1992). p. 133.
5. For background, see Jeffrey A. Zinn and Claudia Copeland, "Wetland Issues," Congressional Research Service Report, May 1, 2000, viewed at http://www.cnie.org/nle/wet-5.html#_1_3; David Hosansky, "Saving the Rain Forests," *The CQ Researcher,* June 11, 1999, pp. 497–520.
6. Wilson, "Vanishing Before Our Eyes" *op. cit.,* p. 30.
7. Julian L. Simon and Aaron Wildavsky, cited in Brenda Stalcup, ed., *Endangered Species Opposing Viewpoints* (1996), p. 24. Simon, now deceased, was a University of Maryland business administration professor; Wildavsky is a University of California, Berkeley, political science and public policy professor.
8. Hosansky, *op. cit.,* p. 505.
9. "Congress Compiles a Modest Record In a Session Sidetracked by Scandal," *CQ Weekly,* Nov. 14, 1998, p. 3987.
10. Edwin Chen, "Eagle Lauded With an Eye Toward Future Environment," *Los Angeles Times,* July 3, 1999, p. A17.
11. "Endangered Species: Continuing Controversy," Congressional Research Service Report, May 8, 1999, www.cnie.org/nle/biodv-1.html#_1_5.
12. Geneva Overholser, "A New Tool to Help the Environment," *Austin American-Statesman,* Feb. 4, 2000, p. A15.
13. H. Josef Herbert, "On 25th Anniversary, Endangered Species Act Elicits Admiration, Ire," *Los Angeles Times,* March 14, 1999, p. B4.
14. Peter D. Ward, *The Call of Distant Mammoths* (1997), p. 95.
15. Jeffrey Kluger, "The Big Crunch," *Time,* special Spring 2000 Earth Day edition, p. 46.
16. David M. Raup, *Extinction: Bad Genes or Bad Luck?* (1991), p. 199.

David Hosansky is a freelance writer in Denver who specializes in environmental issues. He previously was a senior writer at *CQ Weekly* and a reporter at the *Florida Times Union* in Jacksonville, where he was twice nominated for a Pulitzer Prize.

Watching vs. Taking

We are seeing a shift in human relationships with wildlife, as millions turn from taking other species for furs, food, or sport to just watching. In a way, it's a new kind of hunt.

by Howard Youth
Illustrations by Mark Geyer

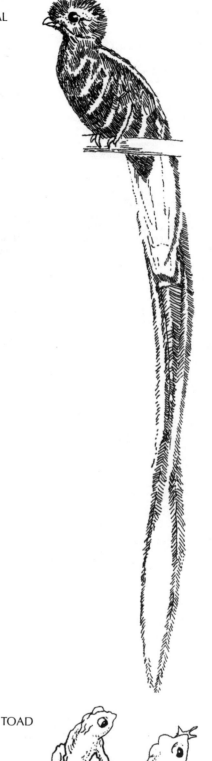

RESPLENDENT QUETZAL

"Get over here ... It's back," whispers a stone-faced Philadelphia man into his walkie-talkie. He's crouched at a campsite in the Bentsen-Rio Grande Valley State Park in southernmost Texas. A short distance away, two forty-something men launch into a fervent but silent race-walk. The site, festooned with bird feeders, backs up to a dense curtain of granjeno, catclaw, and other native brush that harbors some of North America's rarest birds. Eight binocular-toting bird watchers (or "birders," as they call themselves) already ring the site, peering through their optics at a brown, sparrow-sized bird—a female blue bunting. Unaccompanied by a colorful male companion, this Mexican bird is the only member of her species known to be visiting the United States.

"Where's the blue?" asks a woman, a native Texan who is camping nearby.

"You won't find any. That's the female," says a Maryland birder who, tipped off by the local rare bird alert, pulled up in his rental car just a few minutes before.

"Yes!" whispers one of the just-arrived racewalkers, his teeth and fist clenched.

"That's a big tick for you," comments his friend—the "tick" referring to a check on his lifetime list of birds seen.

"Look at the warm brown tone all over the bird," says another watcher, perhaps talking herself out of any disappointment at not seeing the bright blue of the missing male. **"She's a real beauty in her own right."**

The blue bunting encounter, which I witnessed during a December 1999 trip to South Texas, could have taken

GOLDEN TOAD

18. Watching vs. Taking

BLUE BUNTING Only one has been seen in the United States during the past year.

place in any of a thousand locations across North America. The species of bird might vary—it might be a piping plover or an elegant trogon or a northern hawk owl—but the intensity of the fascination would be much the same. Birding has become one of the continent's fastest growing outdoor pastimes, and it's leading a whole parade of newly popular wildlife-watching avocations: there are also people (and organizations) devoted to sighting butterflies, wildflowers, wolves, mountain lions, and whales. Nor is this growing fascination with wildlife confined to North America; it appears to be a global phenomenon, with large economic and ecological implications.

It has come on fast. Today, some species are worth as much—if not more—in their natural habitat, alive and free, than they bring as game. In parts of Africa where tourists pay well to see lions, for example, a lion that remains alive has been estimated to be worth $575,000. In many cases, where wildlife-watching tourism has grown, poaching has declined. In Belize, as manatee-watching tours have brought a growing income, illegal hunting of the big mammals has become less of a problem. And in South Texas, there is now hope that the remaining habitats of the blue bunting, the small wild cats called ocelots, and other rare species won't fall to bulldozers.

The growing interest in wildlife watching draws attention to a growing perceptual chasm, or difference of fundamental values and sensibilities, between those who view wildlife as a resource to be exploited, and those who share a resurgent awareness of wild animals as our co-inhabitants on a fragile planet. The interest in watching may also reflect a new kind of frontier for human curiosity. Over the past millennium, the goals of exploration were geographical—and strange beasts or trees encountered along the way were seen mainly as curiosities. Now, the goals of exploration are increasingly those of understanding the nature of the world we have conquered. The geographical mysteries have given way to new ecological and biological ones. Close watching of wildlife has led to the realization that many animals have surprising levels of intelligence and social organization, and that many plants play powerful and complex ecological roles as well. In an era of dangerously unsustainable human domination, it is of growing interest that many of the animals we have considered our inferiors have in fact thrived for millions of years longer than we have, and have proven to be masters of adaptability and survival. Whatever our reasons for watching, we are less inclined now to take wildlife for granted.

Down on the Rio Grande

Once best known for citrus, cotton, cabbage, and as a haven for winter-fleeing "snowbirds" (people in Winnebagos or Airstreams) from the north, towns in the southern tip of Texas now herald themselves as wildlife-watching paradises. Birders, butterfly watchers, and other nature lovers arrive to take in subtropical sights that can be found nowhere else north of the U.S.-Mexican border. The new tourism may have arrived just in time. By 1999, after decades of conversion to cattle pastures, farm fields, and housing subdivisions, more than 95 percent of the natural environment of the region had disappeared. Today, Bentsen-Rio Grande Valley State Park and other protected areas account for most of the remaining natural habitat of the Lower Rio Grande Valley.

Although birders have been coming to southern Texas for more than 30 years, their growing numbers only recently caught the attention of many local businesses. Now, visitors to the Best Western Inn in Harlingen are greeted by paintings *not* of cowboys, bucking broncos, and ten gallon hats, but of the red-crowned parrot and ringed king-

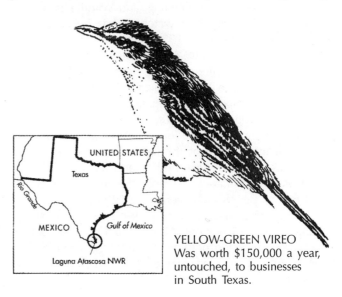

YELLOW-GREEN VIREO Was worth $150,000 a year, untouched, to businesses in South Texas.

fisher. A Harlingen Area Chamber of Commerce brochure, titled "Hooters, Hawks, and Hummingbirds," contains alluring photos of the green jay, chachalaca, kiskadee flycatcher, buff-bellied hummingbird, groove-billed ani, and other local birds. Three South Texas cities—Harlingen, McAllen, and Raymondville—now hold annual birding festivals that lure crowds of out-of-state bird lovers for field trips, seminars, and specialty sales, where they snap up the latest binoculars, books, and spotting scopes. These were among the first such events on the continent; now more than 200 annual nature-oriented festivals are scheduled throughout the United States and Canada. "A lot of things coalesced about five years ago," says Frank Judd, a 30-year Texas resident and biology professor at the University of Texas-Pan American in Edinburg. "Ecotourism is being championed in the local press and other media, and this has made local people aware that they can derive income from it. And it doesn't hurt anything—it just brings in money."

Just how much money? One measure is the worth of the only known U.S. nesting pair of yellow-green vireos, which resided in South Texas's Laguna Atascosa National Wildlife Refuge for several years in the early 1990s. These chickadee-sized songbirds generated an estimated $150,000 per year for local businesses near the refuge. About a three-hour drive to the north, 200 wintering whooping cranes attract an annual $1.2 million in tourist dollars for the small town of Rockport. The cranes arrive on the central Texas coast each fall, after breeding in Canada's Wood Buffalo National Park. Most of the cranes spend the winter at nearby Aransas National Wildlife Refuge. Visitors staying in Rockport can buy tickets for boat rides that ferry them past spots where the birds can be seen searching for crabs, fish, and frogs in the shallow water. Down on the Rio Grande, the Santa Ana National Wildlife Refuge draws 100,000 birders each year, who annually contribute about $14 million to the local economy.

While the full economic impact of wildlife-related activities can't always be so easily quantified, a study by the U.S. Departments of the Interior and Commerce, the "National Survey of Fishing, Hunting, and Wildlife-associated Recreation," suggests that watching has become more than a fringe industry. In 1996, according to the report, 77 million adults—about 40 percent of the U.S. adult population—participated in some form of wildlife-related recreation. Their activities generated $100 billion from sales of equipment, transportation, permits, lodging, food, and other expenses relating to their outdoor interests. Of course, these figures include those for whom wildlife-related recreation means hunting or fishing, and a lot of that $100 billion was spent on guns, bullets, and lures. But if those who *watch* wildlife are broken out, the number still comes to 63 million people, who generated $29 billion. Between 1991 and 1996, wildlife watching trip and equipment expenditures rose 21 percent. Almost equal numbers of men and women reported participating in these activities.

Wildlife watching has gained popularity in other regions of the world as well. In 1994, an Australian gov-

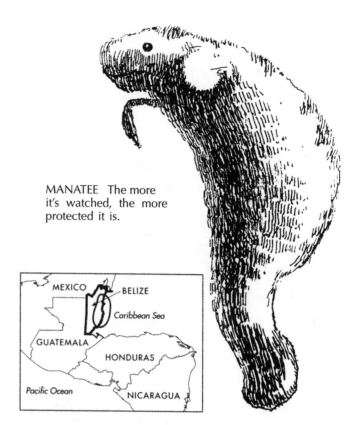

MANATEE The more it's watched, the more protected it is.

ernment report indicated that 53 percent of adult Australians planned to take nature-based trips within the following year. And according to a survey conducted by the Royal Society for the Protection of Birds (RSPB), more than 1 million people in Britain, out of a total population of some 59 million, are regular birders. "Birdwatching in this country is the third most popular leisure pursuit, only beaten by angling and golf," says Graham Madge, an RSPB spokesman. Madge adds that an interest in birds often evolves into broader interests in butterflies, small mammals, and wildflowers that share bird habitats.

Why Watch?

Undoubtedly, at least some of the millions of people who watch birds or whales do so for reasons that have little to do with raising ecological awareness. Keeping personal lists of species sighted can be like keeping score in a game, and weekend outings may be more satisfying for the social experience than for any bonding with nature they induce. Nonetheless, it would be hard to imagine any activity having more potential to heighten people's interest in—and their willingness to seriously protect—the other life of the planet than that of closely observing some of that life. It could be of considerable interest to know what really motivates most wildlife watchers, and to what degree that motivation can be harnessed to larger ends.

Certainly, one factor in this movement has been the sharply increased media coverage of environmental issues over the past three decades—particularly the escalating

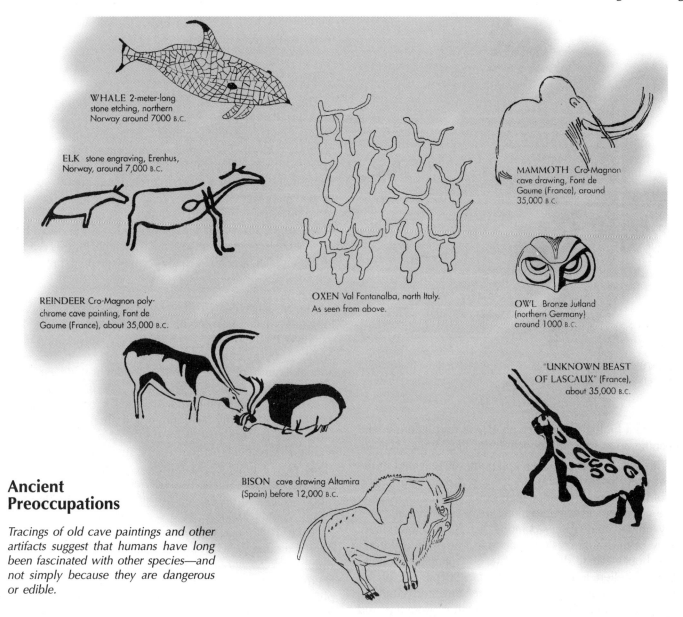

Ancient Preoccupations

Tracings of old cave paintings and other artifacts suggest that humans have long been fascinated with other species—and not simply because they are dangerous or edible.

documentation of global biodiversity loss. Articles in WORLD WATCH over the past few years, for example, have documented sharp declines in thousands of species of birds, fish, reptiles, amphibians, marine mammals, and primates. As a species becomes rarer, it acquires a greater curiosity value. When only a few hundred members of a species remain, those last members may ironically attract thousands of humans who paid little attention when the species was common: witness the crowds that gather to observe captive pandas, gorillas, or California condors.

Threatened species lists—and the rescue movements they engender—have also produced a plethora of TV documentaries and coffee-table books celebrating the wonder those embattled creatures inspire. The movements have been aided by advances in camera technology, allowing photographers and filmmakers to record aspects of the private lives of animals that had never been seen before the last decade or so. And for people who want to see the real thing, the advent of compact field guides has greatly eased the way. Beginning in 1934, Roger Tory Peterson, using arrow-marked and simplified paintings, revolutionized field identification with his North American and European bird field guides. Since then, the market has burgeoned and diversified; you can walk into almost any major bookstore and find the *Field Guide to the Palms of the Americas,* or a *Field Guide to the Orchids of Costa Rica and Panama.*

Beyond the rising concern about threatened species and biodiversity loss, some scientists believe humans may harbor an innate affinity for other species—what the evolutionary biologist E.O. Wilson calls "the urge to affiliate with other forms of life," or "biophilia." Over time, what was a physical connection with its origins in hunting and gathering also became one of culture, religion, and spirit.

4 ❖ BIOSPHERE: ENDANGERED SPECIES: Animals

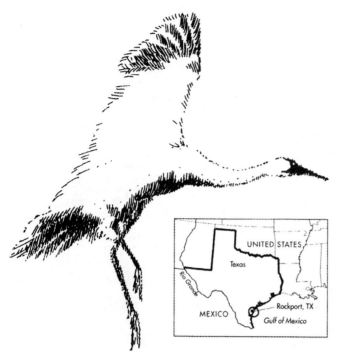

WHOOPING CRANE   One flock brings $1.2 million a year from watchers.

"We are a biological species [that] will find little ultimate meaning apart from the remainder of life," writes Wilson.

This link has always been apparent—as reflected in the great prevalence of animals in the myths, religions, and art of human cultures since prehistoric times. Cro-Magnon cave painters, for example, drew pictures of the creatures they hunted and observed more than 30,000 years ago. Many of the same or similar animals dominate children's story books today. (In 1983, Yale environmental scholar Stephen Kellert found that more than 90 percent of the characters featured in pre-school reading and counting books were animals.) There is the myth of Romulus and Remus, being raised by wolves and then founding Rome. There is the bald eagle becoming a symbol of the United States. Or, there is the turtle recurring as a native American symbol of the world itself. Animals have always been a part of human culture, but what may have happened in the past few decades is that the world's dominant cultures—increasingly overwhelmed by technological and industrial development—have become traumatically separated from what once sustained them.

In any case, we know that across economic, ethnic, and regional lines, people appear to be universally moved by nature—whether in the form of a flying bird, a rolling surf, or the blaze of fall foliage. For many, it takes only a bit of inspiration, perhaps a nature walk led by a school teacher or naturalist, to draw them in. Once hooked, some birding and butterfly-watching enthusiasts research and stalk their quarry almost as though conducting ritualized, non-lethal hunts. Whatever the initial inspiration, most wildlife watchers are driven to learn more about the animals or plants they love.

Citizen Science

This growing interest in watching has brought more than just a surge in hotel reservations and binocular sales. News about impending extinctions, for example, arouses not only curiosity but—often—deep concern. The concern may lead to participation in habitat-saving or species-saving activities, and to a more informed and organized kind of watching—creating a feedback loop that brings still more intensity to the watch. One result has been the rise of a kind of citizen science, in which thousands of watchers collect sightings not just as a pastime, but as a form of data collection for a highly consequential branch of science.

Citizen science now plays a critical role in gathering long-term data on vulnerable wildlife and habitats. For example, the North American Butterfly Association, a nonprofit, 3,500-member conservation organization, holds an annual Fourth of July count, a continent-wide effort in which volunteers identify and tally butterflies living near their homes. The resulting database will provide important insights into butterfly distribution and abundance. Among other findings, the past seven years of observation have enabled researchers to plot major summer concentration areas of the monarch, a widespread migratory butterfly that worries conservationists because much of the population winters in only a few localized and dwindling forests of central Mexico.

Since 1996, the U.S. Geological Survey (USGS) has done the same for toads and frogs with its North American Amphibian Monitoring Program, or NAAMP—an effort spurred by growing global concern over widespread amphibian declines. Volunteers track frog and toad calls during the spring and summer breeding seasons. "It's as close as you're going to get to an idea of what's happening all over the whole patchwork of U.S. and Canadian landscapes," says NAAMP coordinator Linda Weir, who works with states and provinces to set up affiliated monitoring efforts. So far, groups in 29 states have joined, using more than a thousand volunteers to cover about 1,000 designated roadside routes. A similar program recently started in Australia, while regular monitoring programs have been ongoing in Great Britain and a few other European countries. In Britain, for example, the Common Birds Census, conducted since 1970 and carried out primarily by volunteers, has helped create an index on national bird populations. The census has tracked dramatic drops in populations of once-common farmland birds that are thought to be declining due to the destruction of farm hedgerows and increases in pesticide use, and due to harvesting practices that destroy nests and habitat during the birds' breeding seasons. Between 1970 and 1998, for example, the Common Birds Census found an 82 percent decline in grey partridge numbers, a 55 percent decline in song thrush populations, and a 52 percent drop in skylark populations.

One of the oldest citizen science programs is the Christmas Bird Count, which is sponsored by the National Audubon Society and is now in its 100th year. More than 50,000 volunteer birders participated in December 1999 and January 2000, canvassing the wintering grounds of various North American birds. Another well established project is the U.S. Geological Survey's Breeding Bird Survey (BBS), which began in 1966. Volunteers annually cover 3,000 roadside routes, tallying singing and breeding birds across the United States and Canada. Similar surveys are conducted locally in various U.S. states and in Spain, Great Britain, and Australia.

All of this counting of other species could ultimately help drive further shifts toward sustainability in human life itself, by providing inputs to basic policies governing land use, habitat protection, and the like. The Common Birds Census has been instrumental, for example, in persuading the government of the United Kingdom to press for such changes in national agricultural practices as setting harvest times that don't coincide with prime nesting times.

The Backyard Eden

Most of the media's attention to wildlife watching has focused on ecotourism, largely because that's where much of the money is. With tourism booming all over the world, reporters and investors are eager to know what, exactly, people are looking for in their travels. If they'd just as soon choose a ticket for a whooping crane-watching boat as for a theme park ride, that's a significant piece of information. But in fact, the lion's share of nature lovers practice their avocation at home. Of the 63 million wildlife watchers counted in the 1996 National Survey of Fishing, Hunting, and Wildlife-associated Recreation, 44 million reported observing wildlife, mostly birds and mammals, around their own backyards. "Nearly one-third of the adult population of North America dispenses about a billion pounds of birdseed each year, as well as tons of suet and gourmet seed cakes," writes Stephen W. Kress in *Audubon* magazine. Backyard wildlife enthusiasts spend billions of dollars each year on food, feeders, and binoculars. In Britain, the Royal Society for the Preservation of Birds estimates that two out of three people put out food for garden birds during winter months. Many backyard wildlife watchers also landscape their yards with plants that provide food in more natural forms, such as holly berries or cherries.

Citizen science has come to the backyards, as well as to public parks and refuges. Project FeederWatch, a joint effort among the Cornell Laboratory of Ornithology, the National Audubon Society, Bird Studies Canada, and the Canadian Nature Federation, has recruited 14,000 backyard watchers to submit data on their homes' winged visitors. The project tracks long-term distribution patterns such as the expanding range of the recently introduced Eurasian collared dove; short-term irruptions of species outside their normal ranges; and the spread of some easily identified diseases, such as conjunctivitis in finches.

STAG in copper, gold, and silver, late third millennium B.C., Alaca Huyuk Royal Tombs, Turkey

Backyards may also engender a greater sense of responsibility because they are small, and stewardship for a small area may make people more conscious of any threats to the area than they'd be in a large park or wilderness. In Great Britain, for example, many gardens are quite tiny, and birders don't have to look far to see beyond their own

BANDED SWALLOWTAIL BUTTERFLY In Singapore, something to watch besides the stock prices.

4 ❖ BIOSPHERE: ENDANGERED SPECIES: Animals

Detail of internal panel from a silver caldron, middle La Tene culture (Europe), circa first or second century B.C.

property lines. A natural outgrowth of their backyard watching has been the fostering of community projects such as the RSPB's defense of the cirl bunting, a colorful (rust, yellow, and black) European songbird whose last British stronghold is in South Devon. The villages of Bishopsteignton and Stokeinfeignhead, after learning that they were among the cirl bunting's last refuges, each adopted it as their village bird. The upshot is that local farmers have become conscious of its plight and have changed some of their old practices to save it. "There are no real areas of wilderness left in the U.K. and there is not a bit of land that isn't managed for something," says Sue Ellis, a spokesperson for English Nature, an organization that promotes the conservation of England's wildlife and habitats. "Wildlife is on people's doorsteps, so when something happens, people are very, very aware."

Is This a Good Thing, or Not?

While backyard wildlife watchers get less media attention than ecotourists, many of them—in addition to expanding their attentions to community projects—will eventually *become* ecotourists. In 1998, world tourism as a whole generated an estimated $441 billion, according to the Spain-based World Tourism Organization. In her book *Ecotourism and Sustainable Development: Who Owns Paradise?*, Martha Honey calls ecotourism " . . . the most rapidly growing and most dynamic sector of the tourism market." It is big enough to sway critical decisions in biologically rich but cash-poor countries.

A notable example is Costa Rica, which was little known as a tourist destination two or three decades ago. But as word of the country's verdant rainforests, whitewater rivers, and gaudy tropical birds spread, the number of visitors arriving each year from Europe, Japan, and North America jumped from 200,000 to 1 million. Many of them came to see the country's outstanding national parks and reserves, which comprise about a quarter of its total area. Ecotourism has become one of Costa Rica's largest sources of foreign exchange.

In Kenya and Tanzania, too, safari-style ecotourism has become a key revenue source. In 1995, the Kenya Wildlife Service estimated that tourism, 80 percent of which was wildlife watching, was bringing in one-third of the country's foreign exchange. Other countries for which nature-

GOLDEN TOAD People never realized what they were stepping on.

based tourism provides needed foreign exchange include South Africa, Botswana, Belize, Zambia, Ecuador, and Indonesia. In the United States, a group called the Tourism Works for America Council estimated that in 1996, National Park Service lands brought $14.2 billion dollars to local communities and supported almost 300,000 tourism-related jobs.

This isn't always unmitigated good news for local communities and ecosystems, however. The benefits of ecotourism—economic, ecological, and educational—can be offset by all sorts of disbenefits. People eager to see charismatic species can trample less conspicuous ones; building hotels for the visitors cuts large pieces out of the very ecosystems they are coming to see; and the nature trails and jeep roads they use typically lead to increasing fragmentation of what remains. Jet planes and Land Cruisers emit greenhouse gases, the eventual effects of which may weaken ecosystems still further. What ecotourists assume to be harmless observation can turn out to be painful intrusion.

I unwittingly contributed to such an intrusion on Kenya's Masai Mara Reserve, when I joined a safari trip to Kenya in 1995. While traversing the reserve, our van driver suddenly pulled off the road and crashed through the grassland, flushing nesting birds before rumbling within a few feet of a male lion that was lying next to a freshly killed young giraffe. The lion, clearly disturbed by our presence, strained to drag his meal to cover as other safari vehicles rolled through the tail grass towards us. Some of us felt horrible, but I could see that others—their telephoto lenses swiveling—relished being so close.

At the time of my visit, Kenyan tour operators were not forbidden to leave the roads in national reserves. Over the past few years, partly as a result of visitors' outcries, the practice has been banned. But other problems, including water pollution caused by sewage leaching from hotel toilets into nearby wetlands and the widespread gathering of firewood for hotel "safari" campfires and wood-burning furnaces, remain.

As sites grow more popular, other park and reserve managers are beginning to control visitors' movements more carefully. For instance, many parks, including Santa Ana National Wildlife Refuge, now ban vehicular traffic much of the time, with the result that interactions between walkers, bikers, and wildlife are no longer so disruptive. Visitors to Alaska's Denali National Park must leave their cars behind and may only enter the park's wilderness in scheduled buses. The same is true in such parks as Mudumalai Wildlife Sanctuary in India, where visitors ride in on jeeps, buses, or on elephant back to observe wild Asian elephants, tigers, gaur (a kind of undomesticated cattle), and other large, sensitive, and potentially dangerous animals.

Unfortunately, it's often difficult for a group of travelers to know if their tour operator is green or just going *for* the green. "We need to [have] some kind of reviewable ethical standard," says Megan Epler Wood, president of the Vermont-based Ecotourism Society. The Ecotourism Society defines ecotourism as "responsible travel to natural areas that conserves the environment and sustains the well-being of local people." But no global standard or certification process currently exists for tour operations, although Costa Rica and Australia now have strict ecotourism grading standards and efforts are under way to establish them in Kenya.

"We see ecotourism as part of an integrated conservation strategy," says Greta Ryan, manager of ecotourism enterprise development and support at Washington, D.C.-based Conservation International (CI). The nonprofit CI works in 23 countries, with ecotourism now constituting an important component of its work in 17 of them. The key, says Ryan, is to make sure the benefits stay within the community. When this happens, ecotourism tends to reinforce CI's overall program in three ways. By generating local income, it encourages communities to welcome other conservation projects. By alleviating poverty, it reduces the rates of poaching and deforestation. And by making natural assets the centerpieces of the economy, it heightens environmental awareness among both the local people and their visitors.

Often, nature tourists enjoy the sights unaware that their money is being siphoned away from local communities by big-city or out-of-country tour operators who have little ultimate interest in conserving wildlife. "In Kenya, those communities that do not realize a benefit are less likely to consider wildlife positively, and are more likely to want to remove the wildlife from the land," says Neel Inamdar, a director of Eco-Resorts, an ecotourism-oriented travel

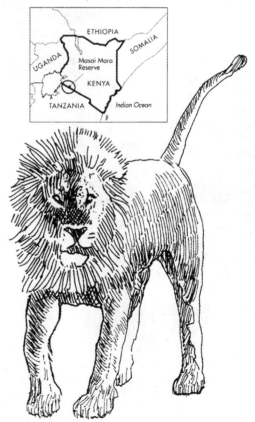

LION Alive and in the wild, it's worth an estimated $575,000.

company, and a board member of the Kenya-based African Center for Conservation. "Basically, a lot of the parks are not viable by themselves—we have to consider the people on the periphery of the parks. After all, wildlife moves from one place to another." Recent initiatives in areas adjacent to African parks include pilot programs in which villagers or farmers run cooperative wildlife reserves on their lands, guiding and hosting paying visitors instead of converting the areas into agricultural land or livestock range.

Some fragile areas, or those within reserves, may simply be unsuited for *any* nature tourism. For instance, there are the rainforest pools once home to golden toads within Costa Rica's Monteverde Cloud Forest Reserve. In the September/October 1990 issue of WORLD WATCH, I quoted Ray Ashton, director of an international consulting firm called Water and Air Research, as saying, "People are tripping over golden toads so they can go see quetzals." Today, the exotic red and green streamer-tailed birds still breed in the reserve, but the golden toad may be extinct. In fact, no one has seen a golden toad since shortly before my article went to press in 1990. While one widely accepted hypothesis suggests that climate change caused the die-off,

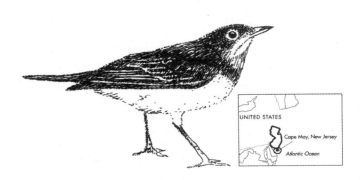

CONNECTICUT WARBLER It helped protect a bean farm from the bulldozers, and the bean farm returned the favor.

another suggests that tourists may have unwittingly contributed to the toad's disappearance by bringing in pathogens on their shoe soles.

On balance, if carefully managed, nature tourism offers large benefits to the environment. Wildlife watchers, a relatively affluent and well educated lot on the whole, are usually willing to pay for their watching—and their economic clout favors protection of the places where they like to do it. A 1995 survey by the Travel Industry Association of America found that 83 percent of U.S. travelers are inclined to support "green" travel companies and are willing to spend, on average, 6.2 percent more for travel services and products provided by environmentally responsible travel suppliers.

In the United States, national parks and wildlife refuges now charge entrance fees, and park fees in other countries are being raised. Private landowners, too, are finding they can charge visitors a fee. South Texas ranches that were off-limits a decade or two ago now court birders as a side business. In Cape May, New Jersey, farmers Les and Diane Rea have managed to supplement the income from their 80-acre lima bean farm, which they feared losing a few years ago because of rising costs and pressure from developers, by maintaining habitat for one of the East Coast's most popular birding areas. In 1999, the Cape May Bird Observatory struck an agreement with the Reas to lease birding rights for the property, paying for the lease with funds collected from permit-buying visitors. Now, each spring and fall they welcome scores of paying visitors. The farm draws a dazzling array of winged migrants, including the elusive Connecticut warbler and about three dozen other warbler species.

"Certainly the precedent existed," says Pete Dunne, vice president for natural history information at the New Jersey Audubon Society, who helped the Reas get the program up and running. "Hunters have been doing this for years. We didn't see any reason not to extend it to birding as a way of showing that birders certainly are willing to pay to support their hobby—and as a way of combatting development pressures. Most farmers want to hold onto

BULL Copper, gold, and silver, late third millennium B.C., Alaca Huyuk Royal Tombs, Turkey

18. Watching vs. Taking

GAUR Yes, it's a wild animal. Compared to a cow, it's huge.

their farms," says Dunne. "In this case, birders are just another cash crop. You don't have to water them, fertilize them, or till them, and on top of that, they will go to your farm stands and buy vegetables."

Birder, writer, ecotourist, and former WORLD WATCH associate editor Howard Youth lives in Rockville, Maryland.

FOR MORE INFORMATION

North American Butterfly Association (NABA) www.naba.org; 4 Delaware Rd. Morristown, NJ 07960, tel: (973) 285-0907

North American Amphibian Monitoring Program (NAAMP) www.mp1-pwrc.usgs.gov/amphibs.html; 12100 Beech Forest Rd., Gabrielson Bldg. 242, Laurel, MD 20708

Royal Society for the Protection of Birds (RSPB) www.rspb.org.uk; The Lodge, Sandy, Bedfordshire SG19 2DL, United Kingdom, tel: 01767-680551

Breeding Bird Survey (BBS) www.mbr.nbs.gov/bbs

Christmas Bird Count (CBC) www.birdsource.org/cbc/index.html; National Audubon Society, 700 Broadway, New York, NY 10003, tel: (212) 353-0347

Project FeederWatch http://birdsource.cornell.edu/pfw; Cornell Lab of Ornithology, P.O. Box 11, Ithica, NY 14851-0011, tel: (607) 254-2473.

Unit 5

Unit Selections

19. **The Tragedy of the Commons: 30 Years Later,** Joanna Burger and Michael Gochfeld
20. **Where Have All the Farmers Gone?** Brian Halweil
21. **When the World's Wells Run Dry,** Sandra Postel
22. **Oceans Are on the Critical List,** Anne Platt McGinn
23. **The Human Impact on Climate,** Thomas R. Karl and Kevin E. Trenberth
24. **Warming Up: The Real Evidence for the Greenhouse Effect,** Gregg Easterbrook

Key Points to Consider

❖ Explain the analogy of the "commons" to describe global resources available to everyone. Is the concept a helpful one in assessing the future of resource management?

❖ Why is the number of farmers in the world decreasing? What kinds of social, economic, and cultural impacts will be produced by decreasing family farmers and increasing amounts of land farmed by corporate farmers?

❖ How is irrigation agriculture contributing to a decrease in the world's groundwater supply? Are there solutions to the problem of increasing food demand that rely on irrigation techniques that will be less demanding of the groundwater resource?

❖ Why and how has overfishing contributed to a decline in the food supply from the oceans? Has oceanic pollution contributed to this decline?

❖ What are some of the uncertainties about the future impact of global temperature increase on human social and economic systems? What reasons are there to develop extensive monitoring systems to identify the causes and effects of global warming?

Links   www.dushkin.com/online/

25. **Global Climate Change**
    http://www.puc.state.oh.us/consumer/gcc/index.html
26. **National Oceanic and Atmospheric Administration**
    http://www.noaa.gov
27. **National Operational Hydrologic Remote Sensing Center**
    http://www.nohrsc.nws.gov
28. **Virtual Seminar in Global Political Economy/Global Cities & Social Movements**
    http://csf.colorado.edu/gpe/gpe95b/resources.html
29. **Websurfers Biweekly Earth Science Review**
    http://shell.rmi.net/~michaelg/index.html

These sites are annotated on pages 4 and 5.

Resources: Land, Water, and Air

The worldwide situations regarding reduction of biodiversity, scarce energy resources, and environmental pollution have received the greatest amount of attention among members of the environmentalist community. But there are a number of other resource issues that demonstrate the interrelated nature of all human activities and the environments in which they occur. One such issue is that of declining agricultural land. In the developing world, excessive rural populations have forced the overuse of lands and sparked a shift into marginal areas, and the total availability of new farmland is decreasing at an alarming rate of 2 percent per year. In the developed world, intensive mechanized agriculture has resulted in such a loss of topsoil that some agricultural experts are predicting a decline in food production. Other natural resources, such as minerals and timber, are declining in quantity and quality as well; in some cases they are no longer usable at present levels of technology. The overuse of groundwater reserves has resulted in potential shortages beside which the energy crisis pales in significance. And the very productivity of Earth's environmental systems—their ability to support human and other life—is being threatened by processes that derive at least in part from energy overuse and inefficiency and from pollution. Many environmentalists believe that both the public and private sectors, including individuals, are continuing to act in a totally irresponsible manner with regard to the natural resources upon which we all depend.

Uppermost in the minds of many who think of the environment in terms of an integrated whole, as evidenced by many of the selections in this unit, is the concept of the threshold or critical limit of human interference with natural systems of land, water, and air. This concept suggests that the environmental systems we occupy have been pushed to the brink of tolerance in terms of stability and that destabilization of environmental systems has consequences that can only be hinted at, rather than predicted. Although the broader issue of system change and instability, along with the lesser issues such as the quantity of agricultural land, the quality of iron ore deposits, the sustained yield of forests, or the availability of fresh water seem to be quite diverse, they are all closely tied to a single concept—that of resource marginality.

Many of these ideas are brought together in the lead article of this unit. Research scholars Joanna Burger and Michael Gochfeld revisit one of the most important of all environmental essays in "The Tragedy of the Commons: 30 Years Later." In the original essay, Garrett Hardin invoked the argument of common resources in support of his argument that as human population increased the global environment would eventually begin to suffer. Burger and Gochfeld note that Hardin's article was seminal in defining the problem of environmental impact in a way that resource managers could deal with and, as a consequence, management of the global commons (fisheries, forests, wildlife, the atmosphere) has received more attention. Whether the commons have received enough attention is for the future to determine.

In "Where Have All the Farmers Gone?" Brian Halweil of the Worldwatch Institute discusses the globalization of industry and trade that is creating a uniform approach to all forms of economic management, including management of agricultural resources. Halweil notes that increasing agribusiness and decreasing numbers of family farmers represent a loss of both biological and cultural diversity.

The second subsection of the unit, *Water*, contains articles concerning water management on two quite different levels: groundwater and the boundless ocean. In each article, the concept of marginality is relevant. In the first selection, "When the World's Wells Runs Dry," Worldwatch researcher Sandra Postel addresses the problem of the world's dwindling groundwater resources as a result of enhanced irrigation technology. Virtually by definition, irrigation agriculture is marginal agriculture and the potential consequences of groundwater withdrawal at rates in excess of recharge are of major concern. The second selection in this subsection also deals with limits to wise use, in this instance, use of the resources of the world's oceans. In "Oceans Are on the Critical List," researcher Anne Platt McGinn contends that the primary threats to the health of the world's oceanic ecosystems are human-induced and include overharvesting of fish, pollution of both coastal and deep-water zones, introduction of alien species and the consequent threat to oceanic biodiversity, and climate change, which also poses threats to biodiversity. Efforts to protect the oceans, McGinn claims, lag far behind what is needed.

In *Air*, the final subsection in this unit, the articles deal with the most critical and controversial of the problems that characterize the global atmosphere: the continuing accumulation of "greenhouse gases," the continuing deterioration of the ozone layer, and the global politics of atmospheric management. In the first article, "The Human Impact on Climate," two of the world's pre-eminent scientists studying climate change, Thomas Karl of the National Climate Data Center and Kevin Trenberth of the National Center for Atmospheric Research, note that while few scientists disagree on the fact of climate change, many disagree on the question of the exact proportion of the change that should be attributed to natural or human-induced causes. With careful monitoring systems in place, we could have accurate answers to the question by the year 2050. That is, according to Karl and Trenberth, entirely too long to wait before taking action. The second article of the subsection also contends that in spite of scientific and nonscientific rhetoric and argument about global warming, the fact that the globe is becoming warmer is indisputable. It is equally indisputable that human activities are playing some role in the process. In "Warming Up: The Real Evidence for the Greenhouse Effect", Gregg Easterbrook of *The New Republic* suggests that arguments over causes are futile. Efforts should be directed toward the development of clean energy systems if for no other reason than the significant competitive advantage that will accrue to countries or corporations that possess such systems.

There are two possible solutions to all these problems posed by the use of increasingly marginal and scarce resources and by the continuing pollution of the global atmosphere. One is to halt the basic cause of the problems—increasing population and consumption. The other is to provide incentives and techniques for the conservation and management of existing resources and for the discovery of alternative resources to eliminate the demand for more marginal resources and the use of heavily polluting ones.

The Tragedy of the Commons:

30 Years Later

by Joanna Burger and Michael Gochfeld

How do we manage resources that seem to belong to everyone? Fish swimming in lakes, game mammals wandering the open plains, and birds migrating overhead belong to everyone and yet are protected by no one. For the sturgeon and bison this lack of protection spelled disaster, for the passenger pigeon, extinction. Today, protecting such common-pool resources has become a challenge, not only on the local scale but on national and global ones as well.

Thirty years ago this December, ecologist Garrett Hardin invoked the analogy of a "commons" in support of his thesis that as human populations increased, there would be increasing pressure on finite resources at both the local and particularly the global levels, with the inevitable result of overexploitation and ruin. He termed this phenomenon the "tragedy of the commons."[1] More specifically, this phrase means that an increase in human population creates an increased strain on limited resources, which jeopardizes sustainability. Hardin argued that common resources could be exploited by anyone who could assert their rights to do so. He painted a bleak picture, emphasizing that the solutions were social rather than technical, and called for privatization or exclusion and for rigorous and even coercive regulation of human population.[2] Recently, he reaffirmed this position.[3]

This article looks at both the positive and negative management of common resources and the legal and ecological progress that has been made since Hardin's original article was written. (See the box, "Hardin's Tragedy of the Commons Thesis.")

Birth of a Discipline

Hardin's original paper was widely cited and stimulated many examples showing that increasing populations did lead to overexploitation, habitat degradation, and species extinctions.[4] Even those ecologists who found Hardin's reliance on coercion distasteful emphasized the consequences of the imbalance between population and resources.[5] Hardin's paper also stimulated many social scientists to alter their perspectives in relation to commons issues, with the result that many examples of both successful and unsuccessful maintenance of common resources have now been published.

The concept of commons is a useful model for understanding environmental management and sustainability. While Hardin believed that ruin was inevitable without coercive population control—an option at odds with our cherished democratic beliefs—recent works by a range of interdisciplinary scientists have identified systems and institutions that do not inevitably lead to overexploitation but that in some cases result in the sustainable use of selected resources, at least on local scales.[6]

While 30 years of research has shown that Hardin's initial thesis emphasizing inevitability and ruin was perhaps too bleak on the local scale, it has been enormously helpful in generating thought-provoking analyses across a wide range of disciplines. His work was widely cited, first by natural scientists and later by social scientists, yet unlike most scientific papers the rate of citation is increasing even 30 years later (see Figure 1). Perhaps its most useful role has been in illustrating the importance of integrating social and political theory with biological data. The traditional theory regarding resource users as unbridled appropriators is being replaced by the recognition that users can communicate and cooperate when it is in their interest to do so and when the resources at their disposal and the sociopolitical context permits it.[7]

What are Commons and Common-Pool Resources?

Common-pool resources (sometimes designated "common property") such as land, fish, and water can be identified and quantified, while the commons is a broader concept that includes the context in which common-pool resources exist and the property system embracing them. Indeed, the switch from discussing the commons to analyzing common-pool resources and property rights illustrates the disagreement many biological and social scientists have with Hardin's original thesis.

In the broad sense, a commons includes the resources held in common by a group of people, all of whom have access and who derive benefit with increasing access. Access may be equal or unequal, and control may be democratic or not. There is some disagreement as to what constitutes a common-pool resource. The term is often restricted to land, grass, wildlife, fish, forests, and water. The concept can also be applied to non-natural resources such as national treasuries, medical care, and the Internet,[8] but the focus of this article will be on the more traditional commons issues of fisheries, recreational areas, public land, and air quality (although atmosphere has been a highly disputed commons with unique qualities to be discussed later). Once these resources could be held in common by small tribes or villages, communities could limit both access to the resources and the amount extracted. Limitation often involved aggression against would-be usurpers.

In many places, this system still exists. In the dry desert lands of northern Namibia and southern Angola, tribal councils control large blocks of land and tribe members are free to build their houses and farm wherever they choose. The councils can mediate disputes, limit intruders, and impose sanctions. While the primary resource held in common is land for farming and grazing, another very important resource in these arid lands is water for people and livestock. Thus, even where overall population density is low, land is not equally desirable and people congregate near the rivers and marshes, potentially leading to overexploitation of these lands and depletion and fouling of water.

Categories of Commons and Property Rights

Current reexamination of the applicability of common-pool resource management is fitting because the use of many resources has become truly globalized, requiring new and more global solutions. International attention has now focused on various aspects of sustainability. Global economies, multinational corporations, international trade agreements, and international commissions have created an institutional framework in which resource sustainability is one prevailing theme among a virtual cacophony of others.[9] As Elinor Ostrom, professor of political theory and policy analysis at Indiana University, points out, it is unclear whether existing international cooperative efforts are adequate to protect essential resources.[10] Rates of population growth and resource consumption vary among regions. The gap in who has access to resources is not narrowing, and there is rapid emergence of new technologies that allow even more efficient exploitation of resources. At the same time, improved communication has heightened expectations of a higher standard of living, even in remote re-

HARDIN'S TRAGEDY OF THE COMMONS THESIS

Hardin based his thesis of the tragedy of the commons on earlier studies written during the late 18th and early 19th centuries. In 1798, Thomas Robert Malthus wrote that human population could grow exponentially, unmatched by resource growth.[1] Charles Darwin's theory of evolution predicted that the characteristics of people who produced more children than others would increase over time. These observations are even more true today given medical care and social systems to protect the children of those who cannot support them. For most of human history, the world seemed like an infinite space with unlimited resources (forests, oceans, wildlife) available for the taking because in nearly every part of the globe there were sufficient resources for the existing, low-density populations. In the past century, however, human population has increased almost everywhere, demonstrating that demand can more than match even very abundant resources.[2]

In 1968, Hardin predicted that with increasing population the eventual fate of all common resources was overexploitation and degradation.[3] His credo, "Freedom in a Commons brings ruin to all," became a universal cry. Others made the same point, although with less flare and consequently less effect. Hardin's concerns focused people's attention on the relationship between individual behavior and resource sustainability.[4] The underlying tenet of his thesis, however, was that populations were increasing beyond the ability of the Earth's resources to support them at a sufficiently high standard of living.

Hardin used William Foster Lloyd's example of herdsmen sharing village lands to graze cattle.[5] Each herdsman derives full benefit from each cow he adds to his herd, while the depletion of grass attributable to that cow is shared among all users. Thus, at each decision point, Hardin argued, each herdsman would choose to add a cow rather than maintain status quo. This leads to each herdsman increasing his herd without limit and to ultimate and inevitable ruin for all. Hardin made several assumptions, including that the world and its resources are finite, human populations will continue to increase, and every person will want to use an increasing share of the resources. Hardin's solution was to have government controls to limit access to the commons or to privatize common-pool resources and, above all, to limit population, even through coercion. Recently Hardin has reaffirmed his predictions, noting that expanding cities must control traffic and parking, nations seek to limit air pollution, and the freedom of the seas is being constrained.[6]

1. T. R. Malthus, *Population: The First Essay* (Ann Arbor, Mich.; University of Michigan Press, 1959).

2. J. Cohen, *How Many People Can the Earth Support?* (New York: W. W. Norton, 1995).

3. G. Hardin, "The Tragedy of the Commons," *Science*, 13 December 1968, 1,243–48.

4. G. Hardin and J. Baden, eds., *Managing the Commons* (New York; W. H. Freeman, 1977).

5. W. F. Lloyd, "Population," "Value," "Poor-laws," and "Rent," in *Reprints of Economic Classics* (New York: Kelley, 1968).

6. G. Hardin, "Extensions of 'The Tragedy of the Commons'," *Science*, 1 May 1998, 682–83.

gions of developing nations. The rate of these changes is also accelerating.

The following examples of commons challenges are drawn from fisheries, public land use, and air quality. Each of these represent similar themes but different scales and solutions. Ostrom identifies four properties of these resources that facilitate cooperative management: the resource has not already been depleted beyond hope of recovery; there are reliable indicators of resource condition; the resource is sufficiently predictable; and the distribution of the resource is sufficiently localized to be studied and controlled by the political entity.

There are also four general categories of property rights: open or uncon-

trolled access, communal, state, and private[11] (see Table 1). Access refers to who controls access or who has access to the resources under what conditions or for what time period (while subtractability refers to the ability of one user to subtract from the welfare of the others).[12] These categories are not discrete but intergrade, and some common-pool resources can be managed under more than one category.[13] For some fisheries, such as shellfish, the government enforces regulations on seasons, size limits, and overall take, but the local shell-fishermen may claim traditional rights or ownership of particular clam beds that they seed with young shellfish, waiting for them to mature to a marketable size. Infringement of these beds can often lead to violence, as has happened in Maine when interlopers tried to fish for lobster in a territory claimed by someone else.[14]

Different modes of property rights may compete—for example federal versus local government or privatization versus community control. Even where a community or state maintains ownership, restricted access for exploitation of certain resources may be granted by concession. There may be a dissociation between resources: One may own land and trees privately whereas wildlife is communal property, or individuals may own rights to certain trees on communal land.

Basically, there is the question of how access can be controlled or managed, and who wins and who loses. Access can be managed by agreed-upon rights and rules, which are uniformly adhered to or enforced.[15] For example, in a small fishing community without outsiders, fishermen can agree to fish only in certain zones or only at certain times, catching a prescribed amount of fish. As long as everyone follows the rules, and the community governs wisely, the fishermen's extraction would not exceed the carrying capacity or regeneration rate of the fish stocks, the resource would not be depleted, and the situation would be considered sustainable. The failure of someone to follow the agreed-upon rules necessitates sanctions, which must also be agreed upon by the users or commoners.

One difficulty in protecting common-pool resources is that there is often an incongruence between the distribution of the resource and property regimes. Fisheries provide many examples of such disparities; many commercial fish are migratory, making property rights, even those as broad as the 200-mile exclusive economic zones, effective for only part of the year.

Local to Global: The Commons Comes of Age

The greatest changes that have come about since Hardin proposed the tragedy of the commons have been an increase in human populations worldwide, shrinking resources, and the globalization of economies. The focus on commons management as a discrete area of social and economic challenge overlaps broadly with the focus on sustainability of resources. Indeed, those concerned with the sustainability of resources should examine whether a commons represents an appropriate avenue for developing sustainable management, and if so, at what spatial scale.

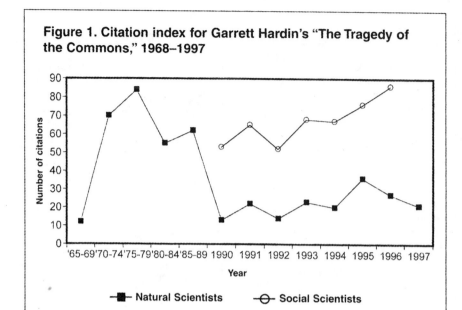

Figure 1. Citation index for Garrett Hardin's "The Tragedy of the Commons," 1968–1997

Temporal scales are important as well, for there are intergenerational aspects that need consideration.[16] The wise or unwise use of resources today directly affects the health and well-being of the next generation. We are borrowing from the next generation at a time when our resources are not only decreasing but population, and thus resource needs, is increasing.

Understanding the use of common-pool resources has been greatly enhanced by two developments: a new economic theory of cooperation that suggests cooperation could lead to the wise and sustained use of common-pool resources; and very detailed empirical work on commons issues that were well founded in theory.[17] Both were essential for the field to move forward. The traditional model of examining commons assumed that each person acted only in his or her own immediate best interest. More recently, economic theory suggests that by cooperating with others (even if there is an initial cost), common users can not only protect the resource but keep it sustainable as well. By examining case studies, researchers found that cooperation can lead to sustainable use and economic viability at least on small regional and short temporal scales.

There are many examples of both failed and successful attempts to manage common resources on local levels. Bonnie McCay, professor of human ecology at Rutgers University, and Fikret Berkes, professor at the Natural Resources Institute at the University of Manitoba, have been instrumental in providing examples where fishermen, hunters, and foresters have established norms, rules, and institutions to successfully extract resources without overexploitation.[18] They argue persuasively that although the tragedy of the commons has been accorded the status of a scientific law, much more detailed study of common-property resource management is needed. There are good examples of self-regulation, including Maine lobstermen and New Jersey fisherman who maintain yields and thus stable or even raise prices.

Three major categories of environmental problems that are useful in understanding methods of dealing with common-pool resources are fisheries, public lands, and air pollution. They are interesting because they illustrate the two main abuses of the commons: resource depletion and capacity depletion (due to pollution). Fish are a traditional, depletable common-pool resource, public land can fall under either category, and air pollution is a more global issue that affects the world at large. Traditional approaches to air pollution are being complemented by viewing the atmosphere's capacity to absorb pollutants as a commons issue. Although at first air seems markedly different from renewable resources such as fish, firewood, and lumber, viewing the atmosphere in terms of limited capacity means that everyone who introduces pollutants, even though he or she does not consume air, subtracts from its use by others.

Fisheries and Water-Related Issues

Fisheries offer classic examples of commons issues that can be local (a stream or lake), regional (North Atlantic), or global, depending upon the fishery. Fisheries also provide examples of the best and worst management schemes concerning common resources. They are instructive because there are examples of local and institutional control that have effectively protected a resource and a livelihood in a sustainable manner, as well as examples of massive overexploitation that threaten not only the commercial but the biological viability of some species.[19]

Although the major fisheries challenges involve oceanic species and a system of uncontrolled, open access on the high seas, shrimp farming along the tropical coastlines is one important example of a land-margin commons issue where a combination of private, state, and communal property rights prevail. Increasing market demand for shrimp provides an incentive for the conversion of otherwise valuable mangrove habitats—where many fish species spawn—to shrimp farms, often at the expense rather than enrichment of fishing communities. The shrimp farms are often owned by corporations with the capital for transforming the habitat rather than being managed by local cooperatives. These farms are economically viable for only a few years, after which they are often abandoned.

In the United States, inshore marine resources are managed by state governments as a trust for all citizens. However, even within this government-regulated system fishermen can cooperate locally to preserve a common-pool resource. Lobster, for example, are a common-pool resource that can be easily overexploited if everyone has a right to harvest and if there are no limits on the number of lobster each fisherman takes. Additionally, if fishermen can come in from outside the region the problem increases. In Maine, the state government does not limit the number of licenses, but the lobstermen practice exclusion through a system of traditional fishing rights. Acceptance into the lobster fishing community is essential before someone can fish, and thereafter one can extract lobsters only in the territory held by that community. The end result has been a sustainable harvest and higher catches of larger, commercially valuable lobsters by fishermen in the exclusion communities.[20]

McCay and colleagues have also shown that in a trawl fishery in the New York Bight, the fishermen belong to a cooperative that has maintained relatively high prices and sustainability by limiting entry into the local fishery and establishing catch quotas among members. In this case, self-regulation is both flexible and effective.[21] The trawl fishery illustrates another critical point about commons resources: It is not only important to have sufficient fish to catch and fish populations that are stable but the price must be maintained or the industry will not be viable for the users.

In both of these examples, success has been achieved by local management of a local fishery. The fishermen were effective in excluding outsiders and in limiting the fishing rights of insiders so that fish stocks were sustainable, prices were maintained, and the fishery was viable. Management of fisheries on a regional or global scale is far more problematic because of the difficulty of exclusion or of monitoring catches.

At the opposite end of the spatial spectrum are cases where fisheries management has been ineffective, with declines in the fish stocks so serious that both the fishery and the fish are threatened with extinction.[22] Many of the examples where fish stocks have declined precipitously involve marine fish with wide geographical ranges. Swordfish and bluefin tuna are classic illustrations of Hardin's thesis. They are a common-pool resource that have suffered overexploitation because of the difficulty of exclusion and the pressure from fishermen of many countries who, stimulated by high market value, push for the maximum catch possible.

Carl Safina, conservation writer and director of the Living Oceans Program of the National Audubon Society, has highlighted the plight of bluefin tuna, one of the largest, fastest, and most wide-ranging fish in the ocean.[23] Its west Atlantic breeding population has declined by 90 percent since 1975 (see Figure 2).[24] The International Commission for the Conservation of Atlantic Tunas, responsible for stewardship of these tuna, is made up of members from 20 countries, many of whom are major tuna users. Although the commission's scientific advisory committees repeatedly presented them with data showing drastic declines, the commission continued to allow catches that exceeded the maximum sustainable yield. In this case, the problem has multiple facets: Some countries that catch tuna are not members of the commission and are thus not regulated; some countries that belong to the commission fish under flags from noncommission countries so they elude the regulations; the fishery is pelagic and global, making enforcement of regulation difficult if not impossible; and last-ditch efforts to place the species on Convention on International Trade in Endangered Species of Wild Flora and Fauna (CITES) lists, which would limit fishing, have so far failed due to pressure from user nations. Even aggressive efforts by the conservation community have been unable to prevent the continued destruction of bluefin tuna.

On a national scale, the United States successfully excluded foreign fishing fleets from its exclusive economic zone and dominated these waters. Even where outsiders have been excluded, however, the commons problem remains because without institutional controls, insiders are free to exploit resources. As Safina pointed out, excluding outsiders did not prevent U.S. fishermen from overexploiting fish resources despite the establishment of agencies and commissions nominally charged with protecting these resources. The problem partly lies with the membership on such commissions; many members represent fishermen who want to get their share (or more) rather than people who are charged with protecting the fish stocks regardless of the economic pressures.

Although successful management of common-pool resources for sustainability is desirable, there are other approaches that do not incorporate sustainability. Some marine fisheries exemplify an alternative approach involving overcapitalization of fleets, rapid sequential elimination of one common-pool fishery resource after another, and shifting to new resources.[25] This allows the industry to perpetuate itself in the short term with little attention to sustainability of a specific resource. When local resources are exhausted, fishermen must exploit more distant sources or sell their fleets and make other investments.

Moreover, understanding of traditional common-pool resources in fisheries has been expanded to include other coastal resources. Two examples illustrate this point: the serious reduction in horseshoe crabs and shorebirds; and personal watercraft users versus fishermen and other water users.

Horseshoe Crabs and Shorebirds

Since 1990, a directed fishery for horseshoe crabs has developed along the East Coast of North America to fulfill the demand for bait for eels, conch, and other fish. This had led to overexploitation and the reduction in the number of horseshoe crabs spawning in many regions. While this problem may once have been considered a fisheries issue only, it is compounded by the fact that several species of migratory shorebirds are threatened by the massive reduction in horseshoe crab eggs, their major food source on Delaware Bay and other stopover places during their northward migration.[26] Although the animals themselves are transitory, the phenomenon

Table 1. Types of property rights regimes		
Type	Description	Example
Open Access	Absence of any well-defined property rights; completely open access to resources that are free to everyone	Recreational fishing in open ocean. Bison and passenger pigeon overharvested leading to decline and even extinction
Common Property	Resource held by community of users who may apportion or regulate access by members and may exclude non-members	Small fishing village that regulates fishing rights among users
State Property	The resource is held by government, which may regulate or exploit the resource or grant public access; government can enforce, sanction, or subsidize the use by some people	Public lands such as national forests or parks where grazing, lumber, or recreational rights are granted by government
Private Property	Individual owns property and has the right to exclude others from use as well as sell or rent the property rights	Private ownership of a woods where owner can sell or rent the land, cut or sell the trees.

occurs annually and predictably on the same beaches.

Apart from the fishermen, several local communities depend on the tourist income generated by the attraction of huge concentrations of migratory shorebirds and breeding horseshoe crabs. Having large populations of both species available for viewing is a commons resource. The fishermen's direct extraction of crabs and the indirect extraction of birds reduces the resource attractive to the tourists, decreasing their pleasure and ultimately their visits. Although less conspicuous from an ecotourism viewpoint, other species, such as green sea turtles, also depend on horseshoe crabs for food, while a medical industry relies on horseshoe crabs for the production of an important laboratory reagent. Thus, a traditional commons fisheries problem has now emerged as a multispecies conservation problem involving other vertebrates as well as economic issues that affect not only the conservation community but fishing, industry, and tourism. A further complication is that the demand for the eels (for which ground up female horseshoe crab is the only bait) emanates from around the globe. Japan, having depleted its own eel populations and those of nearby Asian countries, now offers extremely high prices for American eels, rendering both eel trapping and horseshoe crabbing economically lucrative. Rapid extraction and rapid financial remuneration is the apparent priority rather than sustainability of the resource. While the Atlantic Coastal Marine Fisheries Commission is responsible for maintaining sustainable horseshoe crab populations, the protection of the shorebirds falls under the U.S. Fish and Wildlife Service, which has a conflicting prerogative.

Personal Watercraft versus Fishermen

Recreational use of public land and waterways, one of the commons resources that Hardin mentioned in his classic paper, has received little attention or rigorous analysis. In this example, the massive increase in the use of personal watercraft (often called "jet skis") threatens a number of common-pool resources: the safe nesting of estuarine birds and other animals, the quality of aquatic vegetation so essential to the production of fish and shellfish, the peace and quiet of residents in shore communities, the physical safety of others using aquatic environments, and the undisturbed fishing of both recreational and commercial fishermen.[27] Fast, noisy, and numerous, these craft speed through habitats inaccessible to boats and are not yet regulated in most areas. However, the U.S. National Park Service is in the process of restricting or eliminating their use.

In many estuaries, commercial fishermen are already reporting decreases in catch because of the physical disturbance caused by personal watercraft, while other users report a serious reduction in aesthetic values such as "peace and quiet." This problem is not limited to coastal environments but threatens inland waterways as well. At issue is the freedom of personal watercraft users to take over aquatic environments where their open access subtracts ecological, aesthetic, and commercial benefits long sought by others.[28] Regulation of their use is in its infancy. Ultimately it may be the fatalities they cause, rather than aesthetic or economic impacts on the commons, that leads to further regulation and exclusion.

Public Land

One commons resource currently under discussion in the United States is the huge tracts of public land used for nuclear weapons production by the U.S. Department of Energy during the Cold War.[29] These are now being considered for transfer back to regional, local, or even private ownership, with the inherent problems of determining access and subtractability. For 50 years the federal government excluded all other users from these lands, which in the future could become commons for recreational, industrial, or agricultural use. Which option will be chosen remains undetermined and is likely to vary from site to site.

The Department of Energy's Savannah River Site in South Carolina is a good example. The site is composed of 800 square kilometers of land alongside the river. It includes habitat for a number of endangered species, such as the red-cockaded woodpecker, the wood stork, and the bald eagle, as well as some of the only remaining pristine Carolina Bay habitats. It also offers excellent hunting, fishing, and forestry opportunities. Only a small portion of the area contains industrial facilities and converting these to alternative industrial applications could be accomplished without detracting from the recreational and other uses of the site. Deciding how these lands will be used is a commons issue because the use by one group of people (expanded industrial development, agriculture, or forestry) could detract from the use of others. There are many users with conflicting ideas and stakes in how these public lands should be used, and the question of winners and losers is not only one of human values but of ecological values as well.

Air Quality as a Common-Pool Concern

Traditional approaches to the commons have usually not dealt with air quality or air pollution. Although the atmosphere is a common-pool resource, it is in its use for waste disposal—where unequal access is very difficult to control—that the resource suffers degradation. Studies of global atmospheric transport reveal that air pollutants travel around the globe, to be deposited thousands of miles away from the source. (See "Atmosphere as a Global Commons" in the March 1998 issue of *Environment* for more on transboundary air pollution problems.)

Air pollution from power plants has long been of concern to downwind receptors. While the downwind states and provinces in northeastern North America were encouraging more stringent air pollution standards to control emissions of acid gases and toxic air pollutants, a serious countervailing force arose in the form of energy deregulation. By requiring states to allow the importation of electric power from any producer, the production of cheaper electricity from more polluting plants might actually be increased through demand from users within the downwind states (who will both provide the incentive for and suffer the consequences of increased energy production). Current legislation in the 12 states that have already deregulated electricity includes a variety of incentives for producers (both in-state and out-of-state) to reduce emissions, including disclosure portfolios that would allow consumers to know the emission characteristics of their vendors. In this example, there is no community of producers or consumers of electricity, but there is a clear community of users of air quality, who may have little prerogative for controlling the quality of their air. The states, which would normally be responsible for protecting their residents' health, are clearly not sufficient, and even the regional or multistate consortiums that have formed may be inadequate to protect this common environment. Moreover, the prerogative of the responsible federal agency, the U.S. Environmental Protection Agency, is in jeopardy unless the final deregulation legislation empowers that agency to improve air pollution standards nationwide (even if the lowered cost of electricity is compromised).

Common-Pool Resources and Conflict Resolution

Our understanding of both common-pool resources and the institutions governing their use comes from a number of case studies of resource management in a variety of cultures. Some of the most enlightening case studies deal with the use of public lands, fisheries, agriculture and irrigation systems, groundwater, and contamination of the air. Conflicts inevitably arise and are resolved differently under different property access systems. By examining what systems have worked as well as which ones have allowed or even accelerated resource depletion and habitat degradation, it is possible to begin to understand the rights, rules, and institutions that govern the wise and sustainable use of common-pool resources. This is the legacy of Hardin's initial article, and the responsibility falls on a wide range of disciplines to accomplish it.

The management of common-pool resources is in various stages of development. Recreational and agricultural lands and forests remain as commons in some regions of the world but are privately or governmentally owned in others. Other common-pool resources, such as clean air and water, are clearly regional or global concerns requiring cooperation among widely dispersed people and governments. In many cases, existing national governments are not presently able to manage them effectively at the national, much less at the global, scale.

The oceans may be in transition from being nationally managed to being regionally or globally controlled, as reflected in the increasing reliance on international treaties to establish exclusive economic zones and international commissions to set quotas, close certain fisheries, and maintain catch statistics. Likewise, there are attempts through international conventions to protect major regional airsheds and even the global atmosphere and ultimately global climate. The Montreal protocol offers an example of a partially successful attempt to retard ozone depletion on a global scale by limiting the use of chlorofluorocarbons.

Social Policy Meets Ecology

Our understanding of common-pool resources is entering a new era of more global influence over resource use and pollution abatement coupled with local institutions managing the resources within their own domains. Ostrom argues that international treaty practices are in a position to take commons management actions on a global scale. But this will require hard decisions and long-term considerations on the part of user nations. Such decisions are often difficult to make in light of short-term domestic economic constraints influenced by the multinational nature of corporations wielding power and the potential for blackmail by user nations.[30]

The United Nations is the logical forum for developing commons approaches, but its potential is yet to be realized and it seems to be dismissed or ignored in most discussions. Nonetheless, under its aegis a number of attempts are being made. These include the Kyoto protocol on climate change and the Convention on Long-Range Transboundary Air Pollution, which together call for at least a 50 percent reduction in metal emissions and cover basic obligations, cooperative research, reporting, monitoring, compliance, and dispute resolution.[31]

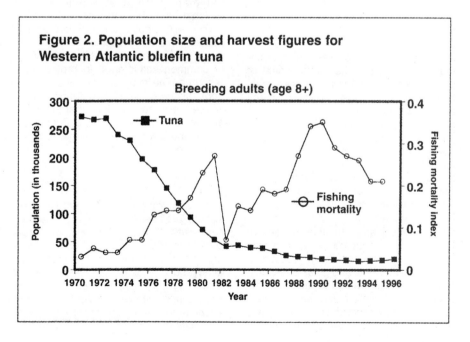

Figure 2. Population size and harvest figures for Western Atlantic bluefin tuna

It has become increasingly clear that there are great differences among individuals' access and use of resources, both locally and globally. We will not find one set of rights, rules, and regulations that will fit all common-pool resources. It is also apparent that the social institutions of an ethnically homogeneous, interrelated tribe cooperatively managing a local fishery or plot of land are not a complete model for cooperation among diverse nations managing a global resource.

Social scientists have established a framework for evaluating characteristics of the user. Commons management is more likely to succeed when the users depend on the resource and share a common understanding of it; when there are grounds for trust; when the users can form an autonomous controlling body; and when they have prior experience with successful management.[32]

Increasing our understanding of how to make the Earth sustainable will require more detailed knowledge about the biology of resources, the social and cultural systems that depend upon these resources, and the economic pressures that govern them. Discussions of sustainability must be based on an understanding of common-pool resource management. The reliance of Western civilization on technological solutions will not solve many of the issues raised by common-pool resources.[33] This is particularly true given the globalization of resource extraction, where the parties that benefit and those that incur the costs are separated geographically and economically and where the benefits are derived by one generation but the costs are incurred by future generations.

Conclusions

Hardin's thesis was seminal in defining the problem of the management of commons and common-pool resources. It met with immediate support by many resource managers who noted that a variety of species had declined dramatically because of overexploitation. Thereafter, social scientists began to note exceptions to his tragedy scenarios, arguing that his thesis was oversimplified and providing examples where institutions allowed people to manage resources sustainably.

Theoretical and empirical studies of common-pool resources have centered on two areas: depletable resources (such as fish, trees, and grasslands) and the depletable waste capacity of resources (such as air and water). While concepts of the commons are equally applicable to both types and remarkable headway has been made in the study of depletable resources, understanding the use of air and water as waste sinks still lags behind. This difference may reflect the more global nature of air and water resources, where the benefits are accrued in one place and the costs are borne by users many thousands of miles away.

The major research thrusts for the future will be in understanding the management of common-pool resources on different temporal and spatial scales, with a view to applying the lessons learned and expanding their applicability. Many of the examples of wise management of common-pool resources have been managed effectively, such as the successful recovery of striped bass along the Atlantic Coast, which required a combination of governmental intervention and user cooperation. The management of global commons resources, particularly fisheries, forests, and wildlife has received considerable attention over the last 10 years and will be an important global issue for many years to come. Although we are making some headway with international treaties to manage global resources (the Montreal protocol and CITES are good examples), other attempts have been disastrous (for example, bluefin tuna). Yet institutions must be developed to deal with these resources or the species are doomed, along with their fisheries.

The difficulty of managing global resources is partly one of attempting to create global treaties and other institutions where there is no global government and few global sanctions. Privatization or long-term governmental stewardship offer alternatives to communal management at the local and regional level, but international conservation will require cooperation across nations where principles of the commons can be invoked to take advantage of common self-interest in protecting a resource. The management of common-pool resources seems to function best where there are sanctions that everyone agrees to and that can be enforced and where the benefits of management are widely recognized.

Joanna Burger is a distinguished research professor of biology at Rutgers University. Michael Gochfeld is clinical professor of environmental and community medicine at the University of Medicine and Dentistry of New Jersey's Robert Wood Johnson Medical School. The authors can be reached at the Environmental and Occupational Health Sciences Institute, Piscataway, NJ 08854 (Burger's telephone: (732) 445-4318, e-mail: burger@biology.rutgers.edu; Gochfeld's telephone: (732) 445-0123, ext. 627, e-mail: gochfeld@ eohsi.rutgers.edu).

The authors would like to thank several people for comments on the manuscript, including C. Safina, B. McCay, C. Powers, and B. Goldstein. Alana Darnell extracted data from citation abstracts and Robert Ramos prepared the illustrations. Some of the research discussed herein was funded by the New Jersey Department of Environmental Protection, the U.S. Environmental Protection Agency, the Trust for Public Lands, the Department of Energy (cooperative agreement with the Consortium for Risk Evaluation with Stakeholder Participation, DE-FC01-95EW55084) and the National Institute for Environmental Health Sciences (ES05022).

Notes

1. G. Hardin, "The Tragedy of the Commons," *Science*, 13 December 1968, 1,243–48.

2. Ibid.

3. G. Hardin, "Extensions of 'The Tragedy of the Commons'," Science, 1 May 1998, 682–83.

4. P. Ehrlich and J. P. Holdren, "Impact of Population Growth," *Science*, March 1971, 1,212–17.

5. B. Commoner, *The Closing Circle* (New York: A. Knopf, 1971); and G. C. Daily and P. R. Ehrlich, "Population, Sustainability, and Earth's Carrying Capacity," *BioScience* 42 (1992): 761–71.

6. D. Feeny, F. Berkes, B. J. McCay, and J. M. Acheson, "The Tragedy of the Commons: Twenty-Two Years Later," *Human Ecology* 18 (1990): 1–19; and F. Berkes, ed., *Common Property Resources: Ecology and Community-Based Sustainable Development* (London; Belhaven Press, 1989).

7. E. Ostrom, "Self-Governance of Common-Pool Resources," in P. Newman, ed., *The New Palgrave Dictionary of Economics and the Law* (London: MacMillan, in press).

8. C. Hess, "Untangling the Web: The Internet as a Commons" (unpublished manuscript presented at the workshop Reinventing the Commons, Transnational Institute, Bonn, Germany, 4–5 November 1995).

9. M. McGinnis and E. Ostrom, "Design Principles for Local and Global Commons," in O. R. Young et al, eds., *International Political Economy and International Institutions* 11 (Cheltenham, U.K.: Edward Elgar Publications, 1996).

10. Ostrom, note 7 above.

11. S. Hanna and M. Monasinghe, eds., *Property Rights and the Environment:*

Social and Ecological Issues (Washington, D.C.: The Beijer Institute of Ecological Economics and the World Bank, 1995); and National Research Council, *Proceedings of the Conference on Common Property Resource Management* (Washington, D.C.: National Academy Press, 1986).

12. F. Berkes, D. Feeny, B. J. McCay, and J. M. Acheson, "The Benefits of the Commons," *Nature,* July 1989, 91–93.

13. Feeny et al., note 6 above.

14. J. M. Acheson, *The Lobster Gangs of Maine* (Hanover, N.H.: University Press of New England, 1988).

15. E. Ostrom, *Governing the Commons: The Evolution of Institutions for Collective Action* (New York: Cambridge University Press, 1990).

16. R. B. Norgaard, "Intergenerational Commons, Globalization, Economism, and Unsustainable Development," *Advances in Human Ecology* 4 (1995): 141–71.

17. Ostrom, note 7 above.

18. B. J. McCay, "Muddling through the Clam Beds: Cooperative Management of New Jersey's Hard Clam Spawner Sanctuaries," *Journal of Shellfish Research* 7 (1988): 327–40; and B. J. McCay and J. M. Acheson, eds., *The Question of the Commons: The Culture and Ecology of Communal Resources* (Tucson, Ariz.: University of Arizona Press, 1987).

19. C. Safina, *Song for the Blue Ocean* (New York: Henry Holt & Co., 1997).

20. Ostrom, note 15 above.

21. Norgaard, note 16 above.

22. C. Safina, "Where Have All the Fishes Gone?," *Issues in Science and Technology* 10 (1994): 37–43.

23. C. Safina, "Bluefin Tuna in the West Atlantic: Negligent Management and the Making of an Endangered Species," *Conservation Biology* 7 (1993): 229–34.

24. International Commission for the Conservation of Atlantic Tunas, *Report of the Standing Committee on Research and Statistics* (Genoa, Italy, 1996).

25. Safina, note 19 above.

26. J. Burger, *A. Naturalist along the Jersey Shore* (New Brunswick, N.J.: Rutgers University Press, 1996).

27. J. Burger, "Effects of Motorboats and Personal Watercraft on Flight Behavior over a Colony of Common Terns," *Condor* 105 (1998): 528–34.

28. J. Burger, "Attitudes about Recreation, Environmental Problems, and Estuarine Health along the New Jersey Shore, U.S.A." *Environmental Management* 22 (1998): 889–96.

29. Commission on Risk Assessment and Risk Management, *Report of the Commission on Risk Assessment and Risk Management* (Washington, D.C.: U.S. Congress, 1997); Department of Energy, *Charting the Course: the Future Use Report,* DOE/EM-0283 (Washington, D.C., 1996); and J. Burger, J. Sanchez, J. W. Gibbons, and M. Gochfeld, "Risk Perception, Federal Spending, and the Savannah River Site: Attitudes of Hunters and Fishermen," *Risk Analysis* 17 (1997): 313–20.

30. Safina, note 22 above.

31. H. E. Ott, "The Kyoto Protocol; Unfinished Business," *Environment,* July/August 1998, 17; Economic and Social Council, "Convention on the Long-Range Transboundary Air Pollution of Heavy Metals" (Aarhus, Denmark: United Nations, 1998).

32. Ostrom, note 7 above.

33. Hardin, note 1 above.

Where Have All the Farmers Gone?

The globalization of industry and trade is bringing more and more uniformity to the management of the world's land, and a spreading threat to the diversity of crops, ecosystems, and cultures. As Big Ag takes over, farmers who have a stake in their land—and who often are the most knowledgeable stewards of the land—are being forced into servitude or driven out.

by Brian Halweil

Since 1992, the U.S. Army Corps of Engineers has been developing plans to expand the network of locks and dams along the Mississippi River. The Mississippi is the primary conduit for shipping American soybeans into global commerce—about 35,000 tons a day. The Corps' plan would mean hauling in up to 1.2 million metric tons of concrete to lengthen ten of the locks from 180 meters to 360 meters each, as well as to bolster several major wing dams which narrow the river to keep the soybean barges moving and the sediment from settling. This construction would supplement the existing dredges which are already sucking 85 million cubic meters of sand and mud from the river's bank and bottom each year. Several different levels of "upgrade" for the river have been considered, but the most ambitious of them would purportedly reduce the cost of shipping soybeans by 4 to 8 cents per bushel. Some independent analysts think this is a pipe dream.

Around the same time the Mississippi plan was announced, the five governments of South America's La Plata Basin—Bolivia, Brazil, Paraguay, Argentina, and Uruguay—announced plans to dredge 13 million cubic meters of sand, mud, and rock from 233 sites along the Paraguay-Paraná River. That would be enough to fill a convoy of dump trucks 10,000 miles long. Here, the plan is to straighten natural river meanders in at least seven places, build dozens of locks, and construct a major port in the heart of the Pantanal—the world's largest wetland. The Paraguay-Paraná flows through the center of Brazil's burgeoning soybean heartland—second only to the United States in production and exports. According to statements from the Brazilian State of Mato Grosso, this "Hidrovía" (water highway) will give a further boost to the region's soybean export capacity.

Lobbyists for both these projects argue that expanding the barge capacity of these rivers is necessary in order to improve competitiveness, grab world market share, and rescue farmers (either U.S. or Brazilian, depending on whom the lobbyists are addressing) from their worst financial crisis since the Great Depression. Chris Brescia, president of the Midwest River Coalition 2000, an alliance of commodity shippers that forms the primary lobbying force for the Mississippi plan, says, "The sooner we provide the waterway infrastructure, the sooner our family farmers will benefit." Some of his fellow lobbyists have even argued that these projects are essential to feeding the world (since the barges can then more easily speed the soybeans to the world's hungry masses) and to saving the environment (since the hungry masses will not have to clear rainforest to scratch out their own subsistence).

Probably very few people have had an opportunity to hear both pitches and compare them. But anyone who has may find something amiss with the argument that U.S. farmers will become more competitive versus their Brazilian counterparts, at the same time that Brazilian farmers will, for the same reasons, become more competitive with their U.S. counterparts. A more likely outcome is that farmers of these two nations will be pitted against each other in a costly race to maximize production, resulting in short-cut practices that essentially strip-mine their soil and throw long-term investments in the land to the wind. Farmers in Iowa will have stronger incentives to plow up land along stream banks, triggering faster erosion of topsoil. Their brethren in Brazil will find themselves needing to cut deeper into the savanna, also accelerating erosion. That will increase the flow of soybeans, all right—both north and south. But it will also further depress prices, so that even as the farmers are shipping more, they're getting less income per ton shipped. And in any case, increasing volume can't help the farmers survive in the long run, because sooner or later they will be swallowed by larger, corporate, farms that can make up for the smaller per-ton margins by producing even larger volumes.

So, how can the supporters of these river projects, who profess to be acting in the farmer's best interests, not notice the illogic of this form of competition? One explanation is that from the advocates' (as opposed to the farmers') standpoint, this competition isn't illogical at all—because the lobbyists aren't really representing farmers. They're working for the commodity processing, shipping, and trading firms who want the price of soybeans to fall, because these are the firms that buy the crops from the farmers. In fact, it is the same three agribusiness conglomerates—Archer Daniels Midland (ADM), Cargill, and Bunge—that are the top soybean processors and traders along both rivers.

Welcome to the global economy. The more brutally the U.S. and Brazilian farmers can batter each-other's prices (and standards of living) down, the greater the margin of profit these three giants gain. Meanwhile, another handful of companies controls the markets for genetically modified seeds, fertilizers, and herbicides used by the farmers—charging oligopolistically high prices both north and south of the equator.

In assessing what this proposed digging-up and reconfiguring of two of the world's great river basins really means, keep in mind that these projects will not be the activities of private businesses operating inside their own private property. These are proposed public works, to be undertaken at huge public expense. The motive is neither the plight of the family farmer nor any moral obligation to feed the world, but the opportunity to exploit poorly informed public sentiments about farmers' plights or hungry masses as a means of usurping public policies to benefit private interests. What gets thoroughly Big Muddied, in this usurping process, is that in addition to subjecting farmers to a gladiator-like attrition, these projects will likely bring a cascade of damaging economic, social, and ecological impacts to the very river basins being so expensively remodeled.

What's likely to happen if the lock and dam system along the Mississippi is expanded as proposed? The most obvious effect will be increased barge traffic, which will accelerate a less obvious cascade of events that has been underway for some time, according to Mike Davis of the Minnesota Department of Natural Resources. Much of the Mississippi River ecosystem involves aquatic rooted plants, like bullrush, arrowhead, and wild celery. Increased barge traffic will kick up more sediment, obscuring sunlight and reducing the depth to which plants can survive. Already, since the 1970s, the number of aquatic plant species found in some of the river has been cut from 23 to about half that, with just a handful thriving under the cloudier conditions. "Areas of the river have reached an ecological turning point," warns Davis. "This decline in plant diversity has triggered a drop in the invertebrate communities that live on these plants, as well as a drop in the fish, mollusk, and bird communities that depend on the diversity of insects and plants." On May 18, 2000, the U.S. Fish and Wildlife Service released a study saying that the Corps of Engineers project would threaten the 300 species of migratory birds and 12 species of fish in the Mississippi watershed, and could ultimately push some into extinction. "The least tern, the pallid sturgeon, and other species that evolved with the ebbs and flows, sandbars and depths, of the river are progressively eliminated or forced away as the diversity of the river's natural habitats is removed to maximize the barge habitat," says Davis.

The outlook for the Hidrovía project is similar. Mark Robbins, an ornithologist at the Natural History Museum at the University of Kansas, calls it "a key step in creating a Florida Everglades-like scenario of destruction in the Pantanal, and an American Great Plains-like scenario in the Cerrado in southern Brazil." The Paraguay-Paraná feeds the Pantanal wetlands, one of the most diverse habitats on the planet, with its populations of woodstorks, snailkites, limpkins, jabirus, and more than 650 other species of birds, as well as more than 400 species of fish and hundreds of other less-studied plants, mussels, and marshland organisms. As the river is dredged and the banks are built up to funnel the surrounding wetlands water into the navigation path, bird nesting habitat and fish spawning grounds will be eliminated, damaging the indigenous and other traditional societies that depend on these resources. Increased barge traffic will suppress river species here just as it will on the Mississippi. Meanwhile, herbicide-intensive soybean monocultures—on farms so enormous that they dwarf even the biggest operations in the U.S. Midwest—are rapidly replacing diverse grasslands in the fragile Cerrado. The heavy plowing and periodic absence of ground cover associated with such farming erodes 100 million tons of soil per year. Robbins notes that "compared to the Mississippi, this southern river system and surrounding grassland is several orders of magnitude more diverse and has suffered considerably less, so there is much more at stake."

Supporters of such massive disruption argue that it is justified because it is the most "efficient" way to do business. The perceived efficiency of such farming might be compared to the perceived efficiency of an energy system based on coal. Burning coal looks very efficient if you ignore its long-term impact on air quality and climate stability. Similarly, large farms look more efficient than small farms if you don't count some of their largest costs—the loss of the genetic diversity that underpins agriculture, the pollution caused by agro-chemicals, and the dislocation of rural cultures. The simultaneous demise of small, independent farmers and rise of multinational food giants is troubling not just for those who empathize with dislocated farmers, but for anyone who eats.

An Endangered Species

Nowadays most of us in the industrialized countries don't farm, so we may no longer really understand that way of life. I was born in the apple orchard and dairy country of Dutchess County, New York, but since age five have spent most of my life in New York City—while most of the farms back in Dutchess County have given way to spreading subdivisions. It's also hard for those of us who get our food from supermarket shelves or drive-thru windows to know how dependent we are on the viability of rural communities.

Whether in the industrial world, where farm communities are growing older and emptier, or in developing nations where population growth is pushing the number of farmers continually higher and each generation is inheriting smaller family plots, it is becoming harder and harder to make a living as a farmer. A combination of falling incomes, rising debt, and worsening rural poverty is forcing more people to either abandon farming as their primary activity or to leave the countryside altogether—a bewildering juncture, considering that farmers produce perhaps the only good that the human race cannot do without.

Since 1950, the number of people employed in agriculture has plummeted in all industrial nations, in some regions by more than 80 percent. Look at the numbers, and you might think farmers are being singled out by some kind of virus:

- In Japan, more than half of all farmers are over 65 years old; in the United States, farmers over 65 outnumber those under 35 by three to one. (Upon retirement or death, many will pass the farm on to children who live in the city and have no interest in farming themselves.)
- In New Zealand, officials estimate that up to 6,000 dairy farms will disappear during the next 10 to 15 years—dropping the total number by nearly 40 percent.
- In Poland, 1.8 million farms could disappear as the country is absorbed into the European Union—dropping the total number by 90 percent.
- In Sweden, the number of farms going out of business in the next decade is expected to reach about 50 percent.
- In the Philippines, Oxfam estimates that over the next few years the number of farm households in the corn producing region of Mindanao could fall by some 500,000—a 50 percent loss.
- In the United States, where the vast majority of people were farmers at the time of the American Revolution, fewer people are now full-time farmers (less than 1 percent of the population) than are full-time prisoners.
- In the U.S. states of Nebraska and Iowa, between a fifth and a third of farmers are expected to be out of business within two years.

Of course, the declining numbers of farmers in industrial nations does not imply a decline in the importance of the farming sector. The world still has to eat (and 80 million more mouths to feed each year than the year before), so smaller numbers of farmers mean larger farms and greater concentration of ownership. Despite a precipitous plunge in the number of people employed in farming in North America, Europe, and East Asia, half the world's people still make their living from the land. In sub-Saharan Africa and South Asia, more than 70 percent do. In these regions, agriculture accounts, on average, for half of total economic activity.

Some might argue that the decline of farmers is harmless, even a blessing, particularly for less developed nations that have not yet experienced the modernization that moves peasants out of backwater rural areas into the more advanced economies of the cities. For most of the past two centuries, the shift toward fewer farmers has generally been assumed to be a kind of progress. The substitution of high-powered diesel tractors for slow-moving women and men with hoes, or of large mechanized industrial farms for clusters of small "old fashioned" farms, is typically seen as the way to a more abundant and affordable food supply. Our urban-centered society has even come to view rural life, especially in the form of small family-owned businesses, as backwards or boring, fit only for people who wear overalls and go to bed early—far from the sophistication and dynamism of the city.

Urban life does offer a wide array of opportunities, attractions, and hopes—some of them falsely created by urban-oriented commercial media—that many farm families decide to pursue willingly. But city life often turns out to be a disappointment, as displaced farmers find themselves lodged in crowded slums, where unemployment and ill-health are the norm and where they are worse off than they were back home. Much evidence suggests that farmers aren't so much being lured to the city as they are being driven off their farms by a variety of structural changes in the way the global food chain operates. Bob Long, a rancher in McPherson County, Nebraska, stated in a recent *New York Times* article that passing the farm onto his son would be nothing less than "child abuse."

As long as cities are under the pressure of population growth (a situation expected to continue at least for the next three or four decades), there will always be pressure for a large share of humanity to subsist in the countryside. Even in highly urbanized North America and Europe, roughly 25 percent of the population—275 million people—still reside in rural areas. Meanwhile, for the 3 billion Africans, Asians, and Latin Americans who remain in the countryside—and who will be there for the foreseeable future—the marginalization of farmers has set up a vicious cycle of low educational achievement, rising infant mortality, and deepening mental distress.

Hired Hands on Their Own Land

In the 18th and 19th centuries, farmers weren't so trapped. Most weren't wealthy, but they generally enjoyed stable incomes and strong community ties. Diversified farms yielded a range of raw and processed goods that the farmer could typically sell in a local market. Production costs tended to be much lower than now, as many of the needed inputs were home-grown: the farmer planted seed that he or she had saved from the previous year, the farm's cows or pigs provided fertilizer, and the diversity of crops—usually a large range of grains, tubers, vegetables, herbs, flowers, and fruits for home use as well as for sale—effectively functioned as pest control.

Things have changed, especially in the past half-century, according to Iowa State agricultural economist Mike Duffy. "The end of World War II was a watershed period," he says. "The widespread introduction of chemical fertilizers and synthetic pesticides, produced as part of the war effort, set in motion dramatic changes in how we farm—and a dramatic decline in the number of farmers." In the post-war period, along with increasing mechanization, there was an increasing tendency to "outsource" pieces of the work that the farmers had previously done themselves—from producing their own fertilizer to cleaning and packaging their harvest. That outsourcing, which may have seemed like a welcome convenience at the time, eventually boomeranged: at first it enabled the farmer to increase output, and thus profits, but when all the other farmers were doing it too, crop prices began to fall.

Before long, the processing and packaging businesses were adding more "value" to the purchased product than the farmer, and it was those businesses that became the dominant players in the food industry. Instead of farmers outsourcing to contractors, it became a matter of large food processors buying raw materials from farmers, on the processors' terms. Today, most of the money is in the work the farmer no longer does—or even controls. In the United States, the share of the consumer's food dollar that trickles

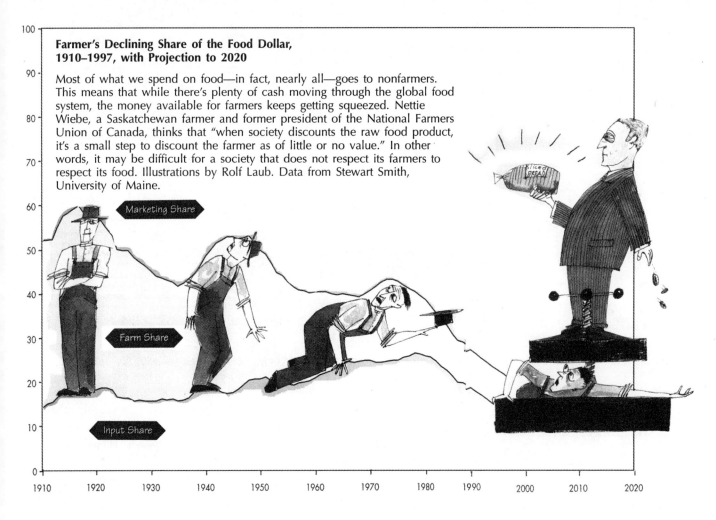

Farmer's Declining Share of the Food Dollar, 1910–1997, with Projection to 2020

Most of what we spend on food—in fact, nearly all—goes to nonfarmers. This means that while there's plenty of cash moving through the global food system, the money available for farmers keeps getting squeezed. Nettie Wiebe, a Saskatchewan farmer and former president of the National Farmers Union of Canada, thinks that "when society discounts the raw food product, it's a small step to discount the farmer as of little or no value." In other words, it may be difficult for a society that does not respect its farmers to respect its food. Illustrations by Rolf Laub. Data from Stewart Smith, University of Maine.

5 ❖ RESOURCES: Land

ConAgra: *Vertical Integration, Horizontal Concentration, Global Omnipresence*

Three conglomerates (ConAgra/DuPont, Cargill/Monsanto, and Novartis/ADM) dominate virtually every link in the North American (and increasingly, the global) food chain. Here's a simplified diagram of one conglomerate.

KEY: ⬇ Vertical integration of production links, from seed to supermarket   ⬌ Concentration within a link

INPUTS
Distribution of farm chemicals, machinery, fertilizer, and seed ⬌
- 3 companies dominate North American farm machinery sector
- 6 companies control 63% of global pesticide market
- 4 companies control 69% of North American seed corn market
- 3 companies control 71% of Canadian nitrogen fertilizer capacity
- **ConAgra** distributes all of these inputs, and is in a joint venture with DuPont to distribute DuPont's transgenic high-oil corn seed.

⬇

FARMS ⬌ The farm sector is rapidly consolidating in the industrial world, as farms "get big or get out." Many go under contract with **ConAgra** and other conglomerates; others just go under. In the past 50 years, the number of farmers has declined by 86% in Germany, 85% in France, 85% in Japan, 64% in the U.S., 59% in South Korea, and 59% in the U.K.

⬇

GRAIN COLLECTION ⬌ A proposed merger of Cargill and Continental Grain will control half of the global grain trade; **ConAgra** has about one-quarter.

⬇

GRAIN MILLING ⬌ **ConAgra** and 3 other companies account for 62% of the North American market.

⬇

PRODUCTION OF BEEF, PORK, TURKEY, CHICKEN, AND SEAFOOD
⬌ **ConAgra** ranks 3rd in cattle feeding and 5th in broiler production.
⬌ **ConAgra** Poultry, Tyson Foods, Perdue, and 3 other companies control 60% of U.S. chicken production

⬇

PROCESSING OF BEEF, PORK, TURKEY, CHICKEN, AND SEAFOOD
⬌ IBP, **ConAgra,** Cargill, and Farmland control 80% of U.S. beef packing
⬌ Smithfield, **ConAgra**, and 3 other companies control 75% of U.S. pork packing

⬇

SUPERMARKETS ⬌ **ConAgra** divisions own Wesson oil, Butterball turkeys, Swift Premium meats, Peter Pan peanut butter, Healthy Choice diet foods, Hunt's tomato sauce, and about 75 other major brands.

20. Where Have All the Farmers Gone?

back to the farmer has plunged from nearly 40 cents in 1910 to just above 7 cents in 1997, while the shares going to input (machinery, agrochemicals, and seeds) and marketing (processing, shipping, brokerage, advertising, and retailing) firms have continued to expand. (See graph "Farmer's Declining Share of the Food Dollar") The typical U.S. wheat farmer, for instance, gets just 6 cents of the dollar spent on a loaf of bread—so when you buy that loaf, you're paying about as much for the wrapper as for the wheat.

Ironically, then, as U.S. farms became more mechanized and more "productive," a self-destructive feedback loop was set in motion: over-supply and declining crop prices cut into farmers' profits, fueling a demand for more technology aimed at making up for shrinking margins by increasing volume still more. Output increased dramatically, but expenses (for tractors, combines, fertilizer, and seed) also ballooned—while the commodity prices stagnated or declined. Even as they were looking more and more modernized, the farmers were becoming less and less the masters of their own domain.

On the typical Iowa farm, the farmer's profit margin has dropped from 35 percent in 1950 to 9 percent today. In order to generate the same income, this farm would need to be roughly four times as large today as in 1950—or the farmer would need to get a night job. And that's precisely what we've seen in most industrialized nations: fewer farmers on bigger tracts of land producing a greater share of the total food supply. The farmer with declining margins buys out his neighbor and expands or risks being cannibalized himself.

There is an alternative to this huge scaling up, which is to buck the trend and bring some of the input-supplying and post-harvest processing—and the related profits—back onto the farm. But more self-sufficient farming would be highly unpopular with the industries that now make lucrative profits from inputs and processing. And since these industries have much more political clout than the farmers do, there is little support for rescuing farmers from their increasingly servile condition—and the idea has been largely forgotten. Farmers continue to get the message that the only way to succeed is to get big.

The traditional explanation for this constant pressure to "get big or get out" has been that it improves the efficiency of the food system—bigger farms replace smaller farms, because the bigger farms operate at lower costs. In some respects, this is quite true. Scaling up may allow a farmer to spread a tractor's cost over greater acreage, for example. Greater size also means greater leverage in purchasing inputs or negotiating loan rates—increasingly important as satellite-guided combines and other equipment make farming more and more capital-intensive. But these economies of scale typically level off. Data for a wide range of crops produced in the United States show that the lowest production costs are generally achieved on farms that are much smaller than the typical farm now is. But large farms can tolerate lower margins, so while they may not *produce* at lower cost, they can afford to *sell* their crops at lower cost, if forced to do so—as indeed they are by the food processors who buy from them. In short, to the extent that a giant farm has a financial benefit over a small one, it's a benefit that goes only to the processor—not to the farmer, the farm community, or the environment.

This shift of the food dollar away from farmers is compounded by intense concentration in every link of the food chain—from seeds and herbicides to farm finance and retailing. In Canada, for example, just three companies control over 70 percent of fertilizer sales, five banks provide the vast majority of agricultural credit, two companies control over 70 percent of beef packing, and five companies dominate food retailing. The merger of Philip Morris and Nabisco will create an empire that collects nearly 10 cents of every dollar a U.S. consumer spends on food, according to a company spokesperson. Such high concentration can be deadly for the bottom line, allowing agribusiness firms to extract higher prices for the products farmers buy from them, while offering lower prices for the crop they buy from the farmers.

An even more worrisome form of concentration, according to Bill Heffernan, a rural sociologist at the University of Missouri, is the emergence of several clusters of firms that—through mergers, takeovers, and alliances with other links in the food chain—now possess "a seamless and fully vertically integrated control of the food system from gene to supermarket shelf." (See diagram "ConAgra") Consider the recent partnership between Monsanto and Cargill, which controls seeds, fertilizers, pesticides, farm finance, grain collection, grain processing, livestock feed processing, livestock production, and slaughtering, as well as some well-known processed food brands. From the standpoint of a company like Cargill, such alliances yield tremendous control over costs and can therefore be extremely profitable.

But suppose you're the farmer. Want to buy seed to grow corn? If Cargill is the only buyer of corn in a hundred mile radius, and Cargill is only buying a particular Monsanto corn variety for its mills or elevators or feedlots, then if you don't plant Monsanto's seed you won't have a market for your corn. Need a loan to buy the seed? Go to Cargill-owned Bank of Ellsworth, but be sure to let them know which seed you'll be buying. Also mention that you'll be buying Cargill's Saskferco brand fertilizer. OK, but once the corn is grown, you don't like the idea of having to sell to Cargill at the prices it dictates? Well, maybe you'll feed the corn to your pigs, then, and sell them to the highest bidder. No problem—Cargill's Excel Corporation buys pigs, too. OK, you're moving to the city, and renouncing the farm life! No more home-made grits for breakfast, you're buying corn flakes. Well, good news: Cargill Foods supplies corn flour to the top cereal makers. You'll notice, though, that all the big brands of corn flakes seem to have pretty much the same hefty price per ounce. After all, they're all made by the agricultural oligopoly.

As these vertical food conglomerates consolidate, Heffernan warns, "there is little room left in the global food system for independent farmers"—the farmers being increasingly left with "take it or leave it" contracts from the remaining conglomerates. In the last two decades, for example, the share of American agricultural output produced under contract has more than tripled, from 10 percent to 35 percent—and this doesn't include the contracts that farmers must sign to plant genetically engineered seed. Such centralized control of the food system, in which farmers are in effect reduced to hired hands on their own land, reminds Heffernan of the Soviet-style state farms, but with the Big Brother role now being played by agribusiness executives. It is also reminiscent of the "company store" which once dominated small American mining or factory towns, except that if you move out of town now, the store is still with you. The company store has gone global.

With the conglomerates who own the food dollar also owning the political clout, it's no surprise that agricultural policies—including subsidies, tax breaks, and environmental legislation at both the national and international levels—do not generally favor the farms. For example, the conglomerates command growing influence over both private and public agricultural research priorities, which might explain why the U.S. Department of Agriculture (USDA), an agency ostensibly beholden to farmers, would help to develop the seed-sterilizing Terminator technology—a biotechnology that offers farmers only greater dependence on seed companies. In some cases the influence is indirect, as manifested in government funding decisions, while in others it is more blatant. When Novartis provided $25 million to fund a research partnership with the plant biology department of the University of California at Berkeley, one of the conditions was that Novartis has the first right of refusal for any patentable inventions. Under those circumstances, of course, the UC officials—mindful of where their funding comes from—have strong incentives to give more attention to technologies like the Terminator seed, which shifts profit away from the farmer, than to technologies that directly benefit the farmer or the public at large.

Even policies that are touted to be in the best interest of farmers, like liberalized trade in agricultural products, are increasingly shaped by non-farmers. Food traders, processors, and distributors, for example, were some of the principal architects of recent revisions to the General Agreement on Trade and Tariffs (GATT)—the World Trade Organization's predecessor—that paved the way for greater trade flows in agricultural commodities. Before these revisions, many countries had mechanisms for assuring that their farmers wouldn't be driven out of their own domestic markets by predatory global traders. The traders, however, were able to do away with those protections.

The ability of agribusiness to slide around the planet, buying at the lowest possible price and selling at the highest, has tended to tighten the squeeze already put in place by economic marginalization, throwing every farmer on the planet into direct competition with every other farmer.

A recent UN Food and Agriculture Organization assessment of the experience of 16 developing nations in implementing the latest phase of the GATT concluded that "a common reported concern was with a general trend towards the concentration of farms," a process that tends to further marginalize small producers and exacerbate rural poverty and unemployment. The sad irony, according to Thomas Reardon, of Michigan State University, is that while small farmers in all reaches of the world are increasingly affected by cheap, heavily subsidized imports of foods from outside of their traditional rural markets, they are nonetheless often excluded from opportunities to participate in food exports themselves. To keep down transaction costs and to keep processing standardized, exporters and other downstream players prefer to buy from a few large producers, as opposed to many small producers.

As the global food system becomes increasingly dominated by a handful of vertically integrated, international corporations, the servitude of the farmer points to a broader society-wide servitude that OPEC-like food cartels could impose, through their control over food prices and food quality. Agricultural economists have already noted that the widening gap between retail food prices and farm prices in the 1990s was due almost exclusively to exploitation of market power, and not to extra services provided by processors and retailers. It's questionable whether we should pay as much for a bread wrapper as we do for the nutrients it contains. But beyond this, there's a more fundamental question. Farmers are professionals, with extensive knowledge of their local soils, weather, native plants, sources of fertilizer or mulch, native pollinators, ecology, and community. If we are to have a world where the land is no longer managed by such professionals, but is instead managed by distant corporate bureaucracies interested in extracting maximum output at minimum cost, what kind of food will we have, and at what price?

Agrarian Services

No question, large industrial farms can produce lots of food. Indeed, they're designed to maximize quantity. But when the farmer becomes little more than the lowest-cost producer of raw materials, more than his own welfare will suffer. Though the farm sector has lost power and profit, it is still the one link in the agrifood chain accounting for the largest share of agriculture's public goods—including half the world's jobs, many of its most vital communities, and many of its most diverse landscapes. And in providing many of these goods, small farms clearly have the advantage.

Local economic and social stability: Over half a century ago, William Goldschmidt, an anthropologist working at the USDA, tried to assess how farm structure and size affect the health of rural communities. In California's San Joaquin Valley, a region then considered to be at the cutting edge of agricultural industrialization, he identified two small towns that were alike in all basic economic and geographic dimensions, including value of agricultural production, except in farm size. Comparing the two, he found an inverse correlation between the sizes of the farms and the well-being of the communities they were a part of.

The small-farm community, Dinuba, supported about 20 percent more people, and at a considerably higher level of living—including lower poverty rates, lower levels of economic and social class distinctions, and a lower crime rate—than the large-farm community of Arvin. The majority of Dinuba's residents were independent entrepreneurs, whereas fewer than 20 percent of Arvin's residents were—most of the others being agricultural laborers. Dinuba had twice as many business establishments as Arvin, and did 61 percent more retail business. It had more schools, parks, newspapers, civic organizations, and churches, as well as better physical infrastructure—paved streets, sidewalks, garbage disposal, sewage disposal and other public services. Dinuba also had more institutions for democratic decision making, and a much broader participation by its citizens. Political scientists have long recognized that a broad base of independent entrepreneurs and property owners is one of the keys to a healthy democracy.

The distinctions between Dinuba and Arvin suggest that industrial agriculture may be limited in what it can do for a community. Fewer (and less meaningful) jobs, less local spending, and a hemorrhagic flow of profits to absentee landowners and distant suppliers means that industrial farms can actually be a net drain on the local economy. That hypothesis has been corroborated by Dick Levins, an agricultural economist at the University of Minnesota. Levins studied the economic receipts from Swift County, Iowa, a typical Midwestern corn and soybean community, and found that although total farm sales are near an all-time high, farm income there has been dismally low—and that many of those who were once the financial stalwarts of the community are now deeply in debt. "Most of the U.S. Corn Belt, like Swift County, is a colony, owned and operated by people who don't live there and for the benefit of those who don't live there," says Levin. In fact, most of the land in Swift County is rented, much of it from absentee landlords.

This new calculus of farming may be eliminating the traditional role of small farms in anchoring rural economies—the kind of tradition, for example, that we saw in the emphasis given to the support of small farms by Japan, South Korea, and Taiwan following World War II. That emphasis, which brought radical land reforms and targeted investment in rural areas, is widely cited as having been a major stimulus to the dramatic economic boom those countries enjoyed.

Not surprisingly, when the economic prospects of small farms decline, the social fabric of rural communities begins to tear. In the United States, farming families are more than twice as likely as others to live in poverty. They have less education and lower rates of medical protection, along with higher rates of infant mortality, alcoholism, child abuse, spousal abuse, and mental stress. Across Europe, a similar pattern is evident. And in sub-Saharan Africa, sociologist Deborah Bryceson of the Netherlands-based African Studies Centre has studied the dislocation of small farmers and found that "as de-agrarianization proceeds, signs of social dysfunction associated with urban areas [including petty crime and breakdowns of family ties] are surfacing in villages."

People without meaningful work often become frustrated, but farmers may be a special case. "More so than other occupations, farming represents a way of life and defines who you are," says Mike Rosemann, a psychologist who runs a farmer counseling network in Iowa. "Losing the family farm, or the prospect of losing the family farm, can generate tremendous guilt and anxiety, as if one has failed to protect the heritage that his ancestors worked to hold onto." One measure of the despair has been a worldwide surge in the number of farmers committing suicide. In 1998, over 300 cotton farmers in Andhra Pradesh, India, took their lives by swallowing pesticides that they had gone into debt to purchase but that had nonetheless failed to save their crops. In Britain, farm workers are two-and-a-half times more likely to commit suicide than the rest of the population. In the United States, official statistics say farmers are now five times as likely to commit suicide as to die from farm accidents, which have been traditionally the most frequent cause of unnatural death for them. The true number may be even higher, as suicide hotlines report that they often receive calls from farmers who want to know which sorts of accidents (Falling into the blades of a combine? Getting shot while hunting?) are least likely to be investigated by insurance companies that don't pay claims for suicides.

Whether from despair or from anger, farmers seem increasingly ready to rise up, sometimes violently, against government, wealthy landholders, or agribusiness giants. In recent years we've witnessed the Zapatista revolution in Chiapas, the seizing of white-owned farms by landless blacks in Zimbabwe, and the attacks of European farmers on warehouses storing genetically engineered seed. In the book *Harvest of Rage,* journalist Joel Dyer links the 1995 Oklahoma City bombing that killed nearly 200 people—as well as the rise of radical right and antigovernment militias in the U.S. heartland—to a spreading despair and anger stemming from the ongoing farm crisis. Thomas Homer-Dixon, director of the Project on Environment, Population, and Security at the University of Toronto, regards farmer dislocation, and the resulting rural unemployment and poverty, as one of the major security threats for the coming decades. Such dislocation is responsible for roughly half of the growth of urban populations across the Third World, and such growth often occurs in volatile shantytowns that are already straining to meet the basic needs of their residents. "What was an extremely traumatic transition for Europe and North America from a rural society to an urban one is now proceeding at two to three times that

In the Developing World, an Even Deeper Farm Crisis

"One would have to multiply the threats facing family farmers in the United States or Europe five, ten, or twenty times to get a sense of the handicaps of peasant farmers in less developed nations," says Deborah Bryceson, a senior research fellow at the African Studies Centre in the Netherlands. Those handicaps include insufficient access to credit and financing, lack of roads and other infrastructure in rural areas, insecure land tenure, and land shortages where population is dense.

Three forces stand out as particularly challenging to these peasant farmers:

Structural adjustment requirements, imposed on indebted nations by international lending institutions, have led to privatization of "public commodity procurement boards" that were responsible for providing public protections for rural economies. "The newly privatized entities are under no obligation to service marginal rural areas," says Rafael Mariano, chairman of a Filipino farmers' union. Under the new rules, state protections against such practices as dumping of cheap imported goods (with which local farmers can't compete) were abandoned at the same time that state provision of health care, education, and other social services was being reduced.

Trade liberalization policies associated with structural adjustment have reduced the ability of nations to protect their agricultural economies even if they want to. For example, the World Trade Organization's Agreement on Agriculture will forbid domestic price support mechanisms and tariffs on imported goods—some of the primary means by which a country can shield its own farmers from overproduction and foreign competition.

The growing emphasis on agricultural grades and standards—the standardizing of crops and products so they can be processed and marketed more "efficiently"—has tended to favor large producers, and to marginalize smaller ones. Food manufacturers and supermarkets have emerged as the dominant entities in the global agri-food chain, and with their focus on brand consistency, ingredient uniformity, and high volume, smaller producers often are unable to deliver—or aren't even invited to bid.

Despite these daunting conditions, many peasant farmers tend to hold on long after it has become clear that they can't compete. One reason, says Peter Rosset of the Institute for Food and Development Policy, is that "even when it gets really bad, they will cling to agriculture because of the fact that it at least offers some degree of food security—that you can feed yourself." But with the pressures now mounting, particularly as export crop production swallows more land, even that fallback is lost.

---

speed in developing nations," says Homer-Dixon. And, these nations have considerably less industrialization to absorb the labor. Such an accelerated transition poses enormous adjustment challenges for India and China, where perhaps a billion and a half people still make their living from the land.

Ecological stability: In the Andean highlands, a single farm may include as many as 30 to 40 distinct varieties of potato (along with numerous other native plants), each having slightly different optimal soil, water, light, and temperature regimes, which the farmer—given enough time—can manage. (In comparison, in the United States, just four closely related varieties account for about 99 percent of all the potatoes produced.) But, according to Karl Zimmerer, a University of Wisconsin sociologist, declining farm incomes in the Andes force more and more growers into migrant labor forces for part of the year, with serious effects on farm ecology. As time becomes constrained, the farmer manages the system more homogenously—cutting back on the number of traditional varieties (a small home garden of favorite culinary varieties may be the last refuge of diversity), and scaling up production of a few commercial varieties. Much of the traditional crop diversity is lost.

Complex farm systems require a highly sophisticated and intimate knowledge of the land—something small-scale, full-time farmers are more able to provide. Two or three different crops that have different root depths, for example, can often be planted on the same piece of land, or crops requiring different drainage can be planted in close proximity on a tract that has variegated topography. But these kinds of cultivation can't be done with heavy tractors moving at high speed. Highly site-specific and management-intensive cultivation demands ingenuity and awareness of local ecology, and can't be achieved by heavy equipment and heavy applications of agrochemicals. That isn't to say that being small is always sufficient to ensure ecologically sound food production, because economic adversity can drive small farms, as well as big ones, to compromise sustainable food production by transmogrifying the craft of land stewardship into the crude labor of commodity production. But a large-scale, highly mechanized farm is simply not equipped to preserve landscape complexity. Instead, its normal modus is to use blunt management tools, like crops that have been genetically engineered to churn out insecticides, which obviate the need to scout the field to see if spraying is necessary at all.

In the U.S. Midwest, as farm size has increased, cropping systems have gotten more simplified. Since 1972, the number of counties with more than 55 percent of their acreage planted in corn and soybeans has nearly tripled, from 97 to 267. As farms scaled up, the great simplicity of managing the corn-soybean rotation—an 800 acre farm, for instance, may require no more than a couple of weeks planting in the spring and a few weeks harvesting in the fall—became its big selling point. The various arms of the agricultural economy in the region, from extension services to grain elevators to seed suppliers, began to solidify around this corn-soybean rotation, reinforcing the farmers' movement away from other crops. Fewer and fewer farmers kept livestock, as beef and hog production became "economical" only in other parts of the country where it was becoming more concentrated. Giving up livestock meant eliminating clover, pasture mixtures, and a key source of fertilizer in the Midwest, while creating tremendous manure concentrations in other places.

But the corn and soybean rotation—one monoculture followed by another—is extremely inefficient or "leaky" in its use of applied fertilizer, since low levels of biodiversity tend to leave a range of vacant niches in the field, including different root depths and different nutrient preferences. Moreover, the Midwest's shift to monoculture has subjected the country to a double hit of nitrogen pollution, since not only does the removal and concentration of livestock tend to dump inordinate amounts of feces in the places (such as Utah and North Carolina) where the livestock operations are now located, but the monocultures that remain in the Midwest have much poorer nitrogen retention than they would if their cropping were more complex. (The addition of just a winter rye crop to the corn-soy rotation has been shown to reduce nitrogen runoff by nearly 50 percent.) And maybe this disaster-in-the-making should really be regarded as a triple hit, because in addition to contaminating Midwestern water supplies, the runoff ends up in the Gulf of Mexico, where the nitrogen feeds massive algae blooms. When the algae die, they are decomposed by bacteria, whose respiration depletes the water's oxygen—suffocating fish, shellfish, and all other life that doesn't escape. This process periodically leaves 20,000 square kilometers of water off the coast of Louisiana biologically dead. Thus the act of simplifying the ecology of a field in Iowa can contribute to severe pollution in Utah, North Carolina, Louisiana, *and* Iowa.

The world's agricultural biodiversity—the ultimate insurance policy against climate variations, pest outbreaks, and other unforeseen threats to food security—depends largely on the millions of small farmers who use this diversity in their local growing environments. But the marginalization of farmers who have developed or inherited complex farming systems over generations means more than just the loss of specific crop varieties and the knowledge of how they best grow. "We forever lose the best available knowledge and experience of place, including what to do with marginal lands not suited for industrial production," says Steve Gleissman, an agroecologist at the University of California at Santa Cruz. The 12 million hogs produced by Smithfield Foods Inc., the largest hog producer and processor in the world and a pioneer in vertical integration, are nearly genetically identical and raised under identical conditions—regardless of whether they are in a Smithfield feedlot in Virginia or Mexico.

As farmers become increasingly integrated into the agribusiness food chain, they have fewer and fewer controls over the totality of the production process—shifting more and more to the role of "technology applicators," as opposed to managers making informed and independent decisions. Recent USDA surveys of contract poultry farmers in the United States found that in seeking outside advice on their operations, these farmers now turn first to bankers and then to the corporations that hold their contracts. If the contracting corporation is also the same company that is selling the farm its seed and fertilizer, as is often the case, there's a strong likelihood that the company's procedures will be followed. That corporation, as a global enterprise with no compelling local ties, is also less likely to be concerned about the pollution and resource degradation created by those procedures, at least compared with a farmer who is rooted in that community. Grower contracts generally disavow any environmental liability.

And then there is the ecological fallout unique to large-scale, industrial agriculture. Colossal confined animal feeding operations (CAFOs)—those "other places" where livestock are concentrated when they are no longer present on Midwestern soy/corn farms—constitute perhaps the most egregious example of agriculture that has, like a garbage barge in a goldfish pond, overwhelmed the scale at which an ecosystem can cope. CAFOs are increasingly the norm in livestock production, because, like crop monocultures, they allow the production of huge populations of animals which can be slaughtered and marketed at rock-bottom costs. But the disconnection between the livestock and the land used to produce their feed means that such CAFOs generate gargantuan amounts of waste, which the surrounding soil cannot possibly absorb. (One farm in Utah will raise over five million hogs in a year, producing as much waste each day as the city of Los Angeles.) The waste is generally stored in large lagoons, which are prone to leak and even spill over during heavy storms. From North Carolina to South Korea, the overwhelming stench of these lagoons—a combination of hydrogen sulfide, ammonia, and methane gas that smells like rotten eggs—renders miles of surrounding land uninhabitable.

A different form of ecological disruption results from the conditions under which these animals are raised. Because massive numbers of closely confined livestock are highly susceptible to infection, and because a steady diet of antibiotics can modestly boost animal growth, overuse of antibiotics has become the norm in industrial animal production. In recent months, both the Centers for Disease Control and Prevention in the United States and the World Health Organization have identified such industrial feeding operations as principal causes of the growing antibiotic resistance in food-borne bacteria like *salmonella* and *campylobacter*. And as decisionmaking in the food chain grows ever more concentrated—confined behind fewer corporate doors—there may be other food safety issues that you won't even hear about, particularly in the burgeoning field of genetically modified organisms (GMOs). In reaction to growing public concern over GMOs, a coalition that ingeniously calls itself the "Alliance for Better Foods"—actually made up of large food retailers, food processors, biotech companies and corporate-financed farm organizations—has launched a $50 million public "educational" campaign, in addition to giving over $676,000 to U.S. lawmakers and political parties in 1999, to head off the mandatory labeling of such foods.

5 ❖ RESOURCES: Land

Perhaps most surprising, to people who have only casually followed the debate about small-farm values versus factory-farm "efficiency," is the fact that a wide body of evidence shows that small farms are actually more productive than large ones—by as much as 200 to 1,000 percent greater output per unit of area. How does this jive with the often-mentioned productivity advantages of large-scale mechanized operations? The answer is simply that those big-farm advantages are always calculated on the basis of how much *of one crop* the land will yield per acre. The greater productivity of a smaller, more complex farm, however, is calculated on the basis of how much food *overall* is produced per acre. The smaller farm can grow several crops utilizing different root depths, plant heights, or nutrients, on the same piece of land simultaneously. It is this "polyculture" that offers the small farm's productivity advantage.

To illustrate the difference between these two kinds of measurement, consider a large Midwestern corn farm. That farm may produce more corn per acre than a small farm in which the corn is grown as part of a polyculture that also includes beans, squash, potato, and "weeds" that serve as fodder. But in overall output, the polycrop—under close supervision by a knowledgeable farmer—produces much more food overall, whether you measure in weight, volume, bushels, calories, or dollars.

The inverse relationship between farm size and output can be attributed to the more efficient use of land, water, and other agricultural resources that small operations afford, including the efficiencies of intercropping various plants in the same field, planting multiple times during the year, targeting irrigation, and integrating crops and livestock. So in terms of converting inputs into outputs, society would be better off with small-scale farmers. And as population continues to grow in many nations, and the agricultural resources per person continue to shrink, a small farm structure for agriculture may be central to meeting future food needs.

Rebuilding Foodsheds

Look at the range of pressures squeezing farmers, and it's not hard to understand the growing desperation. The situation has become explosive, and if stabilizing the erosion of farm culture and ecology is now critical not just to farmers but to everyone who eats, there's still a very challenging question as to what strategy can work. The agribusiness giants are deeply entrenched now, and scattered protests could have as little effect on them as a mosquito bite on a tractor. The prospects for farmers gaining political strength on their own seem dim, as their numbers—at least in the industrial countries—continue to shrink.

A much greater hope for change may lie in a joining of forces between farmers and the much larger numbers of other segments of society that now see the dangers, to their own particular interests, of continued restructuring of the countryside. There are a couple of prominent models for such coalitions, in the constituencies that have joined forces to fight the Mississippi River Barge Capacity and Hidrovía Barge Capacity projects being pushed forward in the name of global soybean productivity.

The American group has brought together at least the following riverbedfellows:

- National environmental groups, including the Sierra Club and National Audubon Society, which are alarmed at the prospect of a public commons being damaged for the profit of a small commercial interest group;
- Farmers and farmer advocacy organizations, concerned about the inordinate power being wielded by the agribusiness oligopoly;
- Taxpayer groups outraged at the prospect of a corporate welfare payout that will drain more than $1 billion from public coffers;
- Hunters and fishermen worried about the loss of habitat;
- Biologists, ecologists, and birders concerned about the numerous threatened species of birds, fish, amphibians, and plants;
- Local-empowerment groups concerned about the impacts of economic globalization on communities;
- Agricultural economists concerned that the project will further entrench farmers in a dependence on the export of low-cost, bulk commodities, thereby missing valuable opportunities to keep money in the community through local milling, canning, baking, and processing.

A parallel coalition of environmental groups and farmer advocates has formed in the Southern hemisphere to resist the Hidrovía expansion. There too, the river campaign is part of a larger campaign to challenge the hegemony of industrial agriculture. For example, a coalition has formed around the Landless Workers Movement, a grassroots organization in Brazil that helps landless laborers to organize occupations of idle land belonging to wealthy landlords. This coalition includes 57 farm advocacy organizations based in 23 nations. It has also brought together environmental groups in Latin America concerned about the related ventures of logging and cattle ranching favored by large landlords; the mayors of rural towns who appreciate the boost that farmers can give to local economies; and organizations working on social welfare in Brazil's cities, who see land occupation as an alternative to shantytowns.

The Mississippi and Hidrovía projects, huge as they are, still constitute only two of the hundreds of agro-industrial developments being challenged around the world. But the coalitions that have formed around them represent the kind of focused response that seems most likely to slow the juggernaut, in part because the solutions these coalitions propose are not vague or quixotic expressions of

20. Where Have All the Farmers Gone?

Past and Future: Connecting the Dots

Given the direction and speed of prevailing trends, how far can the decline in farmers go? The lead editorial in the September 13, 1999 issue of *Feedstuffs,* an agribusiness trade journal, notes that "Based on the best estimates of analysts, economists and other sources interviewed for this publication, American agriculture must now quickly consolidate all farmers and livestock producers into about 50 production systems . . . each with its own brands," in order to maintain competitiveness. Ostensibly, other nations will have to do the same in order to keep up.

To put that in perspective, consider that in traditional agriculture, each farm is an independent production system. In this map of Ireland's farms circa 1930, each dot represents 100 farms, so the country as a whole had many thousands of independent production systems. But if the *Feedstuffs* prognosis were to come to pass, this map would be reduced to a single dot. And even an identically keyed map of the much larger United States would show the country's agriculture reduced to just one dot.

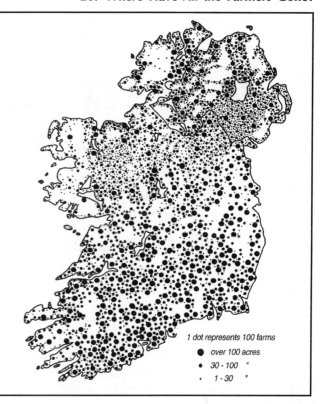

1 dot represents 100 farms
● over 100 acres
• 30 - 100 "
· 1 - 30 "

idealism, but are site-specific and practical. In the case of the alliance forming around the Mississippi River project, the coalition's work has included questioning the assumptions of the Corps of Engineers analysis, lobbying for stronger antitrust examination of agribusiness monopolies, and calling for modification of existing U.S. farm subsidies, which go disproportionately to large farmers. Environmental groups are working to re-establish a balance between use of the Mississippi as a barge mover and as an intact watershed. Sympathetic agricultural extensionists are promoting alternatives to the standard corn-soybean rotation, including certified organic crop production, which can simultaneously bring down input costs and garner a premium for the final product, and reduce nitrogen pollution.

The United States and Brazil may have made costly mistakes in giving agribusiness such power to reshape the rivers and land to its own use. But the strategy of interlinked coalitions may be mobilizing in time to save much of the world's agricultural health before it is too late. Dave Brubaker, head of the Spira/GRACE Project on Industrial Animal Production at the Johns Hopkins University School of Public Health, sees these diverse coalitions as "the beginning of a revolution in the way we look at the food system, tying in food production with social welfare, human health, and the environment." Brubaker's project brings together public health officials focused on antibiotic overuse and water contamination resulting from hog waste; farmers and local communities who oppose the spread of new factory farms or want to close down existing ones; and a phalanx of natural allies with related campaigns, including animal rights activists, labor unions, religious groups, consumer rights activists, and environmental groups.

"As the circle of interested parties is drawn wider, the alliance ultimately shortens the distance between farmer and consumer," observes Mark Ritchie, president of the Institute for Agriculture and Trade Policy, a research and advocacy group often at the center of these partnerships. This closer proximity may prove critical to the ultimate sustainability of our food supply, since socially and ecologically sound buying habits are not just the passive *result* of changes in the way food is produced, but can actually be the most powerful *drivers of* these changes. The explosion of farmers' markets, community-supported agriculture, and other direct buying arrangements between farmers and consumers points to the growing numbers of nonfarmers who have already shifted their role in the food chain from that of choosing from the tens of thousands of food brands offered by a few dozen companies to bypassing such brands altogether. And, since many of the additives and processing steps that take up the bulk of the food dollar are simply the inevitable consequence of the ever-increasing time commercial food now spends in global transit and storage, this shortening of distance between grower and consumer will not only benefit the culture and ecology of farm communities. It will also give us access to much fresher, more flavorful, and more nutritious food. Luckily, as any food marketer can tell you, these characteristics aren't a hard sell.

Brian Halweil is a staff researcher at the Worldwatch Institute.

When the World's Wells Run Dry

It may seem to defy the logic of a closed planetary system, but the supply of water available for irrigation is indeed diminishing—at an alarming rate

by Sandra Postel

In 1970, farmers in rural Deaf Smith County in the Texas panhandle encountered a small but definite sign that local agriculture was seriously out of balance. An irrigation well that had been drilled in 1936 went dry. After more than 30 years of heavy pumping, the water table had dropped 24 meters. Soon other wells began to dry up too.

Water tables were falling across a wide area of the Texas High Plains, and when energy prices shot up in the 1970s, farmers were forced to close down thousands of wells because they could not longer afford to pump from such depths.

During the last three decades, the depletion of underground water reserves, known as aquifers, has spread from isolated pockets of the agricultural landscape to large portions of the world's irrigated land. Many farmers are now pumping groundwater faster than nature is replenishing it, causing a steady drop in water tables. Just as a bank account dwindles if withdrawals routinely exceed deposits, so will an underground water reserve decline if pumping exceeds recharge. Groundwater overdrafting is now widespread in the crop-producing regions of central and northern China, northwest and southern India, parts of Pakistan, much of the western United States, North Africa, the Middle East, and the Arabian Peninsula.

Many cities are overexploiting groundwater as well. Portions of Bangkok and Mexico City are actually sinking as geologic formations compact after the water is removed. Albuquerque, Phoenix, and Tucson are among the larger U.S. cities that are overdrafting their aquifers.

Globally, however, it is in agriculture where the greatest social risks lie. Irrigated land is disproportionately important to world food production. Some 40 percent of the global harvest comes from the 17 percent of cropland that is irrigated. Because of limited opportunities for expanding rainfed production, we are betting on that share to increase markedly in the decades ahead, in order to feed the world's growing population. As irrigation goes deeper and deeper into hydrologic debt, the possibilities for serious disruption grow ever greater. Should energy prices rise again, for example, farmers in many parts of the world could find it too expensive to irrigate. Groundwater overpumping may now be the single biggest threat to food production.

Our irrigation base is remarkably young: 60 percent of it is less than 50 years old. Yet a number of threats to its continued productivity are already apparent. Along with groundwater depletion, there is the buildup of salts in the soil, the silting up of reservoirs and canals, mounting competition for water between cities and farms and between countries sharing rivers, rapid population growth in regions that are already water-stressed—and on top of all that, the uncertainties of climate change. Any one of these threats could seriously compromise agriculture's productivity. But these stresses are evolving simultaneously—making it increasingly likely that cracks will appear in our agricultural foundation.

Few governments are taking adequate steps to address any of these threats and, hidden below the surface, groundwater depletion often gets the least attention of all. Yet this hydrologic equivalent of deficit financing cannot continue indefinitely. Groundwater withdrawals will eventually come back into balance with replenishment—the only question is whether they do so in a planned and coordinated way that maintains food supplies, or in a chaotic and unexpected way that reduces food production, worsens poverty, and disrupts regional economies.

It is true that there are enormous inefficiencies elsewhere in the agricultural sector—and tackling these could

21. When the World's Wells Run Dry

take some of the pressure off aquifers. A shift in diets, for example, could conserve large amounts of irrigation water. The typical U.S. diet, with its high share of animal products, requires twice as much water to produce as the nutritious but less meat-intensive diets common in some Asian and European nations. If U.S. consumers moved down the food chain, the same volume of water could produce enough food for two people instead of one, leaving more water in rivers and aquifers. But given the rates of groundwater depletion, there is no longer any reasonable alternative to tackling the problem directly. Aquifer management will be an essential part of any strategy for living within the limits imposed by a finite supply of fresh water.

The Groundwater Revolution

During the first century of the modern irrigation age—roughly from 1850 to 1950—efforts to develop water supplies focused mainly on rivers. Government agencies and private investors constructed dams to capture river water and canals to deliver that water to cities and farms. By the middle of this century, engineers had built impressive irrigation schemes in China, India, Pakistan, and the United States, and these nations became the world's top four irrigators. The Indus River system in South Asia, the Yellow and Yangtze Rivers in China, and the Colorado and Sacramento-San Joaquin river systems of the western United States were each irrigating sizable areas by 1950. The global irrigation base then stood at 100 million hectares, up from 40 million in 1900.

Between 1950 and 1995, world irrigated area increased to more than 250 million hectares. Even as the construction of large dams for hydroelectric power, water supply, and flood control picked up pace, a quiet revolution in water use unfolded during this period. Rural electrification, the spread of diesel pumps, and new well-drilling technologies allowed farmers to sink millions of wells into the aquifers beneath their land. For the first time in human history, farmers began to tap groundwater on a large scale.

Aquifers are in many ways an ideal source of water. Farmers can pump groundwater whenever they need it, and that kind of availability typically pays off in higher crop yields. Compare this with the standard scenario for irrigating with river water: river flow is erratic, so a reservoir is usually required to store flood water for use in the dry season. And reservoirs—especially arid-land reservoirs such as Lake Nasser behind Egypt's High Aswan Dam—can lose 10 percent or more of their water to evaporation. In addition, the large canal networks that move water out of reservoirs are often unreliable—they may not deliver enough water when farmers actually need it. Aquifers, on the other hand, have a fairly slow and steady flow that is usually available year-round and they don't lose water to evaporation. Finally, groundwater is generally less expensive to develop than river water. Data from 191 irrigation projects funded by the World Bank show that groundwater schemes cost a third less on average than surface schemes.

Not surprisingly, huge numbers of farmers and investors turned to groundwater as soon as they acquired the means to tap into it. In China, the number of irrigation wells shot up from 110,000 in 1961 to nearly 2.4 million by the mid-1980s. In India, government canal building nearly doubled the area under surface irrigation between 1950 and 1985, but the most impressive growth was in groundwater development: the area irrigated by tubewells ballooned from 100,000 hectares in 1961 to 11.3 million hectares in 1985—a 113-fold rise, most of it privately funded. (A tubewell is a narrow well that is drilled into an aquifer, as opposed to a larger-

Ancient Romans made this water-carrying pipe with cement and crushed rock. Courtesy George E. Bartuska, Winter Park, Florida.

diameter well that is excavated, either by hand or with machinery.) In neighboring Pakistan, groundwater was the fastest-growing form of irrigation from the mid-1960s through the 1980s. A public program of tubewell development failed miserably, but private groundwater investments climbed steeply. The total number of tubewells in that country rose from some 25,000 in 1964 to nearly 360,000 in 1993.

After World War II, the United States experienced a groundwater boom as well. Farmers in California stepped up their pumping of groundwater beneath the rich soils of the Central Valley, which was well on its way to becoming the nation's fruit and vegetable basket. But the greatest aquifer development was in the U.S. Great Plains, a region that straddles the 100th meridian, the nation's transition zone from rain-fed to irrigated agriculture. In a striking bit of good fortune, the drier western portion of the plains is underlain by a vast underground pool called the Ogallala. One of the planet's greatest aquifers, it spans portions of eight states, from South Dakota in the north to Texas in the south. The Ogallala extends for 453,000 square kilometers, and—prior to exploitation—held 3,700 cubic kilometers of water, a volume equal to the annual flow of more than 200 Colorado Rivers. In the years after World War II, a new generation of powerful centrifugal pumps allowed farmers to tap into this water on a large scale, first in northwest Texas and western Kansas, and then gradually farther north into Nebraska. Today, the Ogallala alone waters one-fifth of U.S. irrigated land.

Taking Stock

Like any renewable resource, groundwater can be tapped indefinitely as long as the rate of extraction does not exceed the rate of replenishment. In many regions, however, aquifers get so little natural recharge that they are essentially nonrenewable. These "fossil aquifers" are the remnants of ancient climates that were much wetter than current local conditions. Pumping from fossil aquifers depletes the supply, just as pumping from an oil reserve does. Even where aquifers do get replenished by rainfall, few governments have established rules and regulations to ensure that they are exploited at a sustainable rate. In most places, any farmer who can afford to sink a well and pump water can do so unrestrained. Ownership of land typically implies the right to the water below. The upshot is a classic "tragedy of the commons," in which individuals acting out of self-interest deplete a common resource.

In India, for example, the situation has become so severe that in September 1996 the Supreme Court directed one of the country's premier research centers to examine it. The National Environmental Engineering Research Institute, based in Nagpur, found that "overexploitation of ground water resources is widespread across the country." Water tables in critical agricultural areas are sinking "at an

The Ogallala Aquifer

When Major Stephen Long struck out west, up the South Platte River in 1820, he named the "desolate waste" he encountered west of the 100th meridian the Great American Desert. Attempts to cultivate this arid land led to disasters such as the Dust Bowl. But in the 1950s, new pumps opened up the Ogallala aquifer, one of the world's largest underground reservoirs. Changing the desert into a breadbasket, the aquifer now waters one-fifth of U.S. irrigated land. But overpumping is draining the Ogallala much more quickly than it is recharged. Falling water tables and higher pumping costs have forced many farmers to abandon irrigation: while more than 5.2 million hectares were irrigated by the Ogallala in 1978, a decade later that number had dropped 20 percent, to 4.2 million. Without significant changes, the Ogallala oasis may turn out to be little more than a mirage.

alarming rate," due to rapid proliferation of irrigation wells, which now number at least 6 million, and the failure to regulate pumping adequately. Nine Indian states are now running major water deficits, which in the aggregate total just over 100 billion cubic meters (bcm) a year—and those deficits are growing (see table, "Water Deficits in Key Countries and Regions, Mid-1990's").

The situation is particularly serious in the northern states of Punjab and Haryana, India's principal breadbaskets. Village surveys found that water tables are dropping 0.6 to 0.7 meters per year in parts of Haryana and half a meter per year across large areas of Punjab. In the state of Gujarat, on the northwest coast, 87 out of 96 observation wells showed declining groundwater levels during the 1980s, and aquifers in the Mehsana district are now reportedly depleted. Overpumping in Gujarat has also allowed salt water to invade the aquifers, contaminating drinking water supplies. In the state of Tamil Nadu, in the extreme south, water tables have dropped by as much as 30 meters since the 1970s, and aquifers in the Coimbatore district are now dry.

Farmers usually run into problems before the water disappears entirely. At some point, the pumping costs get out of hand or the well yields drop too low, and they are forced to choose among several options. They can take irrigated land out of production, eliminate a harvest or two, switch to less water-intensive crops, or adopt more-efficient irrigation practices. Apart from shifting out of thirsty nonstaple crops like sugarcane or cotton, improving efficiency is the only option that can sustain food production while lowering water use. Yet in India, investments in efficiency are minuscule relative to the challenge at hand. David Seckler, Director General of the International Water Management Institute in Sri Lanka, estimates that a quarter of India's grain harvest could be in jeopardy from groundwater depletion.

Besides threatening food production, groundwater overpumping is widening the income gap between rich and poor in some areas. As water tables drop, farmers must drill deeper wells and buy more powerful pumps. In parts of Punjab and Haryana, for example, wealthier farmers have installed more expensive, deeper tubewells costing about 125,000 rupees ($2,890). But the poor cannot afford such equipment. So as the shallower wells dry up, some of the small-scale farmers end up renting their land to the wealthier farmers and becoming laborers on the larger farms.

Other countries are facing similar problems. In Pakistan's province of Punjab—the country's leading agricultural region, which is just across the border from the Indian state of the same name—groundwater is being pumped at a rate that exceeds recharge by an estimated 27 percent. In Bangladesh, groundwater use is about half the rate of natural replenishment on an annual basis. But during the dry season, when irrigation is most needed, heavy pumping causes many wells to go dry. On about a third of Bangladesh's irrigated area, water tables routinely drop below the suction level of shallow tubewells during the dry months. Although monsoon rains recharge these aquifers and water tables rise again later in the year, farmers run out of water when they need it most. Again, the greatest hardships befall poor farmers, who cannot afford to deepen their wells or buy bigger pumps.

In China, which is roughly tied with India for the most irrigated land, groundwater conditions are equally unsettling. Northern China is running a chronic water deficit, with groundwater overpumping amounting to some 30 bcm a year. Of the three major river basins in the region, the Hai is always in deficit, the Yellow is almost always in deficit, and the Huai is occasionally. This northern and central plain produces roughly 40 percent of China's grain. Across a wide area, the water table has been dropping 1 to 1.5 meters a year, even as water demands continue to increase.

Modeling work by Dennis Engi of Sandia National Laboratories in New Mexico suggests that the water deficit in the Hai basin could grow by more than half between 1995 and 2025, even assuming that China completes at least part of a controversial plan to divert some Yangtze River water northward. Engi projects a 190 percent deficit increase for the Yellow River basin. Over the time frame of Engi's study, the combined deficit in these two basins could more than double, from 27 bcm to 55 bcm.

As in India, the unsustainable use of groundwater is creating a false sense of the nation's food production potential. The worsening groundwater deficits will eventually force Chinese farmers to either take land out of irrigation, switch to less thirsty crops, or irrigate more efficiently. How they respond will make a big difference to China's grain outlook: that projected 2025 deficit for the Hai and Yellow River basins is roughly equal to the volume of water needed to grow 55 million tons of grain—14 percent of the nation's current annual grain consumption and about a fourth of current global grain exports.

In the United States, farmers are overpumping aquifers in several important crop-producing regions. California is overdrafting groundwater at a rate of 1.6 bcm a year, equal to 15 percent of the state's annual groundwater use. Two-thirds of this depletion occurs in the Central Valley, which supplies about half of the nation's fruits and vegetables. By far the most serious case of depletion, however, is in the region watered by the Ogallala aquifer. Particularly in its southern reaches, the Ogallala gets very little replenishment from rainfall, so almost any pumping diminishes it. Currently the aquifer is being depleted at a rate of some 12 bcm a year. Total depletion to date amounts to some 325 bcm, a volume equal to the annual flow of 18 Colorado Rivers. More than two-thirds of this depletion has occurred in the Texas High Plains.

Driven by falling water tables, higher pumping costs, and historically low crop prices, many farmers who depend on the Ogallala have already abandoned irrigated agriculture. At its peak in 1978, the total area irrigated by the Ogallala in Colorado, Kansas, Nebraska, New Mexico,

Oklahoma, and Texas reached 5.2 million hectares. Less than a decade later, this area had fallen by nearly 20 percent, to 4.2 million hectares. A long-range study of the region, done in the mid-1980s, suggested that more than 40 percent of the peak irrigated area would come out of irrigation by 2020; if this happens, another 1.2 million hectares will either revert to dryland farming or be abandoned over the next two decades.

Desert Fantasies

In North Africa and the Arabian Peninsula, where it rarely rains, a number of countries depend on fossil aquifers. Saudi Arabia, for instance, sits atop several deep aquifers containing some 1,919 cubic kilometers of water—just over half as much as the Ogallala. The Saudis started pumping water on a grand scale after the OPEC oil embargo of the 1970s. Fear of a retaliatory grain embargo prompted the government to launch a major initiative to make the nation self-sufficient in grain by encouraging large-scale wheat production in the desert. The government heavily subsidized land, equipment, and irrigation water. It also bought the wheat at several times the world market price. From a few thousand tons in the mid-1970s, the annual grain harvest grew to a peak of 5 million tons in 1994. Saudi water demand at this time totaled nearly 20 bcm a year, and 85 percent of it was met by mining nonrenewable groundwater. Saudi Arabia not only achieved self-sufficiency in wheat; for a time, it was among the world's wheat exporters.

But this self-sufficiency would not last. Crop production soon crashed when King Fahd's government was forced to rein in expenditures as the nation's revenues declined. Within two years, Saudi grain output fell by 60 percent, to 1.9 million tons in 1996. Today Saudi Arabia is harvesting slightly more grain than in 1984, the year it first became self-sufficient, but because its population has grown from 12 million to more than 20 million since then, the nation has again joined the ranks of the grain importers.

Moreover, the Saudis' massive two-decade experiment with desert agriculture has left the nation much poorer in water. In its peak years of grain production, the nation ran a water deficit of 17 bcm a year, consuming more than 3,000 tons of water for each ton of grain produced in the hot, windy desert. (The standard ratio is 1,000 tons of water per ton of grain.) At that rate, groundwater reserves would have run out by 2040, and possibly sooner. In recent years, the annual depletion rate has dropped closer to the level of the mid-1980s, but the Saudis are still racking up a water deficit on the order of 6 bcm a year.

Africa's northern tier of countries—from Egypt to Morocco—also relies heavily on fossil aquifers, with estimated depletion running at 10 bcm a year. Nearly 40 percent of this depletion occurs in Libya, which is now pursuing a massive water scheme rivaled in size and complexity only by China's diversion of the Yangtze River. Known as the Great Man-Made River Project, the $25 billion scheme pumps water from desert aquifers in the south and transfers it 1,500 kilometers north through some 4,000 kilometers of concrete pipe.

The brainchild of Libyan leader Muammar Qaddafi, the artificial river was christened with great pomp and ceremony in late August 1991. As of early 1998, it was delivering 146 million cubic meters a year to the cities of Tripoli and Benghazi. If all stages are completed, the scheme will eventually transfer up to 2.2 bcm a year, with 80 percent of it destined for agriculture. As in Saudi Arabia, however, the greening of the desert will be short-lived: some water engineers say the wells may dry up in 40 to 60 years.

Some water experts have called the scheme "madness" and a "national fantasy." Foreign engineers involved in the project have even questioned Qaddafi's real motives. Some have pointed out that the pipelines are 4 meters in diameter, big enough to accommodate trucks or troops. Every 85 kilometers or so, engineers are building huge underground storage areas that apparently are more elaborate than needed for holding water. The master pipeline runs through a mountain where Qaddafi is reported to be building a biological and chemical weapons plant. But other engineers have scoffed at the possibility of any military motive, noting, for example, that the pipelines have no air vents.

From the fields of North Africa to those of northern China, the story is essentially the same: many of the world's most important grainlands are consuming groundwater at unsustainable rates. Collectively, annual water depletion in India, China, the United States, North Africa, and the Arabian Peninsula adds up to some 160 bcm a year—equal to the annual flow of two Nile Rivers. (See table, "Water deficits in Key Countries and Regions, Mid-1990's.) Factoring in Australia, Pakistan, and other areas for which this author did not have comparable data would likely raise this figure by an additional 10 to 25 percent.

Water Deficits in Key Countries and Regions, Mid-1990s

Country/Region	Estimated Annual Water Deficit (billion cubic meters per year)
India	104.0
China	30.0
United States	13.6
North Africa	10.0
Saudi Arabia	6.0
Other	unknown
Minimum Global Total	163.6

SOURCE: Global Water Policy Project and Worldwatch Institute.

The vast majority of this overplumped groundwater is used to irrigate grain, the staple of the human diet. Since it takes about 1,000 tons of water to produce one ton of grain (and a cubic meter of water weighs one metric ton), some 180 million tons of grain—roughly 10 percent of the global harvest—is being produced by depleting water supplies. This simple math raises a very unsettling question: If so much of irrigated agriculture is operating under water deficits now, where are farmers going to find the additional water that will be needed to feed the more than 2 billion people projected to join humanity's ranks by 2030?

Texas Ingenuity

The only way to sustain crop production in the face of dwindling water supplies is to use those supplies more efficiently—to get more crop per drop. Few farmers have a better combination of incentive to conserve and opportunity to innovate than those in nortwest Texas. As the Ogallala shrinks, water efficiency is increasingly the ticket to staying in business. And the response of these Texan farmers is grounds for hope: better irrigation technologies and practices can substantially delay the day of reckoning—buying valuable time to make an orderly transition to a more sustainable water economy.

During the 1980s, the steady drop in underground water levels prompted local water officials and researchers to put together a package of technologies and management options that has boosted the region's water productivity. Spearheaded by the High Plains Underground Water Conservation District in Lubbock, which overseas water management in 15 counties of northwest Texas, the effort has involved a major upgrade of the region's irrigation systems.

Many conventional gravity systems, in which water simply flows down parallel furrows, are less than 60 percent efficient: more than 40 percent of the water runs off the field or seeps through the soil without benefiting the crop. Farmers in the High Plains have been equipping their systems with surge valves that raise efficiency to about 80 percent. Just as the name implies, surge irrigation involves sending water down the furrows of a field in a series of pulses rather than in a continuous stream. The initial pulse somewhat seals the soil, letting subsequent surges flow more quickly and uniformly down the field. This evens out the distribution of water, allowing farmers to apply less at the head of their fields while still ensuring that enough water reaches crops at the tail-end. A time-controlled valve alternates the flow of water between rows, and its cycle and flow rates can be adjusted for different soils, furrow lengths, and other conditions. When combined with soil moisture monitoring and proper scheduling of irrigations, surge systems can cut water use by 10 to 40 percent compared with conventional furrow irrigation. Savings in the Texas High Plains have averaged about 25 percent. High Plains farmers have typically recouped their investment in surge equipment—which ranges from $30 to $120 per hectare depending on whether piping is already in place—within two years.

Many farmers in the region are also using more efficient sprinklers. Conventional sprinklers are more efficient than furrow irrigation in most contexts, because they apply water more uniformly. But in dry, windy areas like the U.S. Great Plains, spraying water high into the air can cause large losses from evaporation and wind drift. The High Plains District is encouraging the use of two varieties of low-pressure sprinklers. One type delivers a light spray from nozzles about a meter above the soil surface, and typically registers efficiencies of 80 percent, about the same as surge irrigation (see table, "Efficiencies of Selected Irrigation Methods, Texas High Plains").

A second variety, however, does substantially better. Low-energy precision application (LEPA) sprinklers deliver water in small doses through nozzles positioned just above the soil surface. They nearly eliminate evaporation and wind drift, and can raise efficiency to 95 percent—often cutting water use by 15 to 40 percent over other methods. In the High Plains District, LEPA has also increased corn yields about 10 percent and cotton yields about 15 percent. The water savings plus the yield increases add up to substantial gains in water productivity. Farmers converting to LEPA typically recoup their investment in two to seven years, depending on whether they are upgrading an existing sprinkler or purchasing a new one. Virtually all the sprinklers in the High Plains District are now either the low-pressure spray or LEPA.

More recently, the district has begun experimenting with drip irrigation of cotton. Using a network of perforated plastic tubing installed on or below the surface, drip systems deliver water directly to the roots of plants. Drip irrigation has cut water use by 30 to 70 percent in countries as diverse as India, Israel, Jordan, Spain, and the United States. And because plants grow better with optimal moisture, drip systems often boost yields by 20 to 50 percent. Since drip systems cost on the order of $2,500 per hectare, they have typically been used just for high value crops like fruits and vegetables. But as water itself grows more expensive and as new, lower-cost systems hit developing-country markets, the technology will become more useful. Because cotton is such a thirsty and widely planted crop, using drip systems to irrigate it could save large quantities of water in Texas and elsewhere. Working with local farmers, the district is giving drip a tough test by comparing its performance to that of LEPA—the most water-efficient sprinkler design now on the market. After the first year of trials, drip produced 19 percent more cotton per hectare than the LEPA-irrigated fields.

The Texas High Plains program has also included substantial extension work to help farmers adopt water-saving practices. (Extension programs are outreach efforts by government agricultural agencies and some universities.) For example, extension agents spread the word about furrow diking—one of the most readily accessible water-saving measures. Furrow dikes are small earthen ridges built across furrows at regular intervals down the field. They form small basins that trap both rain and irrigation water, thereby reducing runoff and increasing soil absorption. Furrow dikes are key to obtaining the highest possible irrigation efficiency with LEPA, for example, and to storing as much pre-season rainfall in the soil as possible.

Constructing furrow dikes costs about $10 per hectare. James Jonish, an economist at Texas Tech University, points out that if furrow dikes capture an extra five centimeters of rainfall in the soil, they can boost cotton yields by up to 225 kilograms of lint per hectare, a potential economic gain of $400 per hectare, depending on cotton prices. In contrast, getting those higher yields by pumping an additional five centimeters of groundwater would cost $15 to $22 per hectare and would of course hasten the aquifer's depletion. Overall, the High Plains District program has allowed growers to boost the water productivity of cotton, which accounts for about half the cropland area, by 75 percent over the last two decades. Full irrigation of cotton used to require a well capable of producing at least 10 gallons a minute per acre (four-tenths of a hectare), but the district now considers 2 to 3 gallons a minute sufficient.

Despite these successes, High Plains farmers face an uphill battle. Drought conditions in 1998 forced them to pump more groundwater than usual. Water tables dropped an average of 0.64 meters between early 1998 and early 1999, twice the average annual drop over the last decade. The first half of 1999 was wetter than usual but given the general trend, further improvement is essential. The district is now ramping up a program in which computer systems use real-time weather data and more precise information on crop water needs to adjust irrigation regimes. This approach will, for example, allow a two-and-a-half-day cycle of LEPA irrigation, rather than the usual five- to seven-day cycle. Shorter cycles should make it possible to maintain a nearly ideal moisture environment with even less water than the standard LEPA approach, since the very small volumes of water released can be carefully calibrated to match the crop's immediate demand. Preliminary results with corn and cotton show promising yield increases. Water district assistant manager Ken Carver expects the program to go into widespread use soon after its introduction this year. The potential of this approach is enormous: it offers a way to irrigate corn, wheat and other grains nearly as efficiently as drip systems irrigate fruits, vegetables, and cotton. In areas where groundwater is diminishing, these methods hold out hope that production declines can at least be delayed—and in some areas, perhaps, avoided altogether.

Setting New Rules

No government has made a concerted effort to solve the problem of groundwater overpumping. Indeed, most contribute to the problem by subsidizing groundwater use. Many farmers in India, for example, pay only a flat fee for electricity, which makes the marginal cost of pumping groundwater close to zero. Why invest in more-efficient irrigation technologies if it costs nearly the same to pump 10,000 cubic meters of groundwater as it does to pump 5,000?

Likewise, Texas irrigators get a break on their federal income taxes for depleting the Ogallala aquifer: they receive a "depletion allowance" much as oil companies do for depleting oil reserves. Each year, they measure how much their water table has dropped, calculate the value of that depleted water, and then claim an adjustment on their income tax. This subsidy may partially explain why some farmers use the water saved through efficiency improvements to grow thirstier crops rather than leaving it in the ground. From a social standpoint, it is far more sensible to tax groundwater depletion in order to make

Efficiencies of Selected Irrigation Methods, Texas High Plains

Irrigation Method	Typical Efficiency	Water Application Needed to Add 100 Millimeters to Root Zone	Water Savings Over Conventional Furrow[1]
	(percent)	(millimeters)	(percent)
Conventional Furrow	60	167	—
Furrow with Surge Valve	80	125	25
Low-Pressure Sprinkler	80	125	25
LEPA Sprinkler	90–95	105	37
Drip	95	105	37

[1]Data do not specify what portion of savings result from reduced evaporation versus runoff and seepage.
SOURCE: Based on High Plains Underground Water Conservation District (Lubbock, Texas), *The Cross Section*, various issues.

current users pay more of the real costs of their activities. Such a tax would allow products made with the depleted water—whether beef steaks or cotton shirts—to better reflect their true ecological costs.

Governments have also failed to tackle the task of regulating access to groundwater. To prevent a tragedy of the commons, it's necessary to limit the number of users of the common resource, to reduce the quantity of the resource that each user can take, or to pursue some combination of these two options. This regulating function can be performed by a self-governing communal group—in which rights and responsibilities are determined by the farmers themselves—or by a public agency with authority to impose rules for the social good on private individuals. In reality, however, groundwater conditions are rarely even monitored, much less regulated.

Only recently has the groundwater issue begun to appear on national agendas—and still only in a few countries. Officials in India circulated a "model groundwater bill" in 1992, but none of the Indian states has passed legislation along those lines. Some have made efforts to regulate groundwater use through licensing, credit, or electricity restrictions, or by setting minimum well-spacing requirements. But no serious efforts have been made to control the volume of water extracted. V. Narain, a researcher at the New Delhi-based Tata Energy Research Institute, puts it simply: "groundwater is viewed essentially as a chattel attached to land," and there is "no limit on how much water a landowner may draw."

Indian researchers and policymakers broadly agree that rights to land and water need to be separated. Some have argued for turning de facto private groundwater rights into legal common property rights conferred upon communities in a watershed. But instituting such a reform can be a political high-wire act. Wealthy farmers, who have the ear of politicians, do not want to lose their ability to pump groundwater on their property in any quantity they desire.

The United States has no official national groundwater policy either. As in India, it is up to the states to manage their own aquifers. So far, only Arizona has passed a comprehensive groundwater law that explicitly calls for balancing withdrawal with recharge. Arizona's strategy for meeting this goal by 2025 would take some of the strain off its overpumped groundwater by substituting Colorado River water imported through an expensive, federally-subsidized canal project. But few regions can rely on such an option, which in any case merely replaces one type of excessive water use with another.

An important first step in developing a realistic groundwater policy is for governments to commission credible and unbiased assessments of the long-term rate of recharge for every groundwater basin or aquifer. This would establish the limit of sustainable use. The second step is for all concerned parties—including scientists, farmer and community groups, and government agencies—to devise a plan for balancing pumping with recharge. If current pumping exceeds the sustainable limit, achieving this goal will involve some mix of pumping reductions and artificial recharge—the process of channeling rainfall or surplus river water into the underground aquifer, where this is possible.

Arriving at an equitable way of allocating groundwater rights such that total pumping remains within sustainable levels will not be easy. Legislatures or courts might need to invoke a legal principle that elevates the public interest over private rights. One possibility, for example, is the public trust doctrine, which asserts that governments hold certain rights in trust for the public and can take action to protect those rights from private interests.

Some scholars have recommended use of the public trust doctrine to deal with India's groundwater problem. Recent rulings in the United States show that this legal instrument is potentially very powerful. The California Supreme Court ordered Los Angeles to cut back its rightful diversions of water from tributaries that feed Mono Lake, declaring that the state holds the lake in trust for the people and is obligated to protect it. The applicability of the public trust or similar doctrines may vary somewhat from one legal system to the next, but where a broad interpretation is feasible, there could be sweeping effects since even existing rights can be revoked in order to prevent violation of the public trust.

Once a legal basis for limiting groundwater use is established, the next step is to devise a practical plan for actually making groundwater use sustainable. Mexico is one of the few countries that seem to be tackling this task head on. After enacting a new water law in 1992, Mexico created River Basic Councils, which are intended to be water authorities open to a high degree of public participation. For example, the council for the Lerma–Chapala River basin, an area that contains 12 percent of Mexico's irrigated land, is in the process of setting specific regulations for each aquifer in the region. Technical committees are responsible for devising plans to reduce overpumping. Because these committees are composed of a broad mix of players, including the groundwater users themselves, they lend legitimacy to both the process and the outcome.

Although the details of a workable plan will vary from place to place, it is now possible to draw a rough blueprint for sustainable groundwater use. But nearly everywhere, the first big hurdle is overcoming the out-of-sight, out-of-mind syndrome. When looking at, say, a field of golden wheat, it can be difficult to imagine why harvests like that can't just go on forever. But the future of that crop—and of humanity itself—will depend on how well we manage the water below.

Sandra Postel is director of the Global Water Policy Project in Amherst, Massachusetts, and a senior fellow at the Worldwatch Institute. She is the author of Pillar of Sand: Can the Irrigation Miracle Last? *(W. W. Norton & Company, 1999), from which this article is adapted.*

OCEANS
Are on the Critical List

The primary threats to the planet's seas—overfishing, habitat degradation, pollution, introduction of alien species, and climate change—are largely human-induced.

BY ANNE PLATT MCGINN

OCEANS FUNCTION as a source of food and fuel, a means of trade and commerce, and a base for cities and tourism. Worldwide, people obtain much of their animal protein from fish. Ocean-based deposits meet one-fourth of the world's annual oil and gas needs, and more than half of world trade travels by ship. More important than these economic figures, however, is the fact that humans depend on oceans for life itself. Harboring a greater variety of animal body types (phyla) than terrestrial systems and supplying more than half of the planet's ecological goods and services, the oceans play a commanding role in the Earth's balance of life.

Due to their large physical volume and density, oceans absorb, store, and transport vast quantities of heat, water, and nutrients. The oceans store about 1,000 times more heat than the atmosphere does, for example. Through processes such as evaporation and photosynthesis, marine systems and species help regulate the climate, maintain a livable atmosphere, convert solar energy into food, and break down natural wastes. The value of these "free" services far surpasses that of ocean-based industries. Coral reefs alone, for instance, are estimated to be worth $375,000,000,000 annually by providing fish, medicines, tourism revenues, and coastal protection for more than 100 countries.

Despite the importance of healthy oceans to our economy and well-being, we have pushed the world's oceans perilously close to—and in some cases past—their natural limits. The warning signs are clear. The share of overexploited marine fish species jumped from almost none in 1950 to 35% in 1996, with an additional 25% nearing full exploitation. More than half of the world's coastlines and 60% of the coral reefs are threatened by human activities, including intensive coastal development, pollution, and overfishing.

In January, 1998, as the United Nations was launching the Year of the Ocean, more than 1,600 marine scientists, fishery biologists, conservationists, and oceanographers from across the globe issued a joint statement entitled "Troubled Waters." They agreed that the most pressing threats to ocean health are human-induced, including species overexploitation, habitat degradation, pollution, introduction of alien species, and climate change. The impacts of these five threats are exacerbated by poorly planned commercial activities and coastal population growth.

Yet, many people still consider the oceans as not only inexhaustible, but immune to human interference. Because scientists just recently have begun to piece together how ocean systems work, society has yet to appreciate—much less protect—the wealth of oceans in its entirety. Indeed, current courses of action are rapidly undermining this wealth.

Nearly 1,000,000,000 people, predominantly in Asia, rely on fish for at least 30% of their animal protein supply. Most of these fish come from oceans, but with increasing frequency they are cultured on farms rather than captured in the wild. Aquaculture, based on the traditional Asian practice of raising fish in ponds, constitutes one of the fastest-growing sectors in world food production.

In addition to harvesting food from the sea, people have traditionally relied on oceans as a transportation route. Sea trade currently is dominated by multinational companies that are more influenced by the rise and fall of stock prices than by the tides and trade winds. Modern fishing trawlers, oil tankers, aircraft carriers, and container ships follow a path set by electronic beams, satellites, and computers.

Society derives a substantial portion of energy and fuel from the sea—a trend that was virtually unthinkable a century ago. In an age of falling trade barriers and mounting pressures on land-based resources, new

22. Oceans Are on the Critical List

ocean-based industries such as tidal and thermal energy production promise to become even more vital to the workings of the world economy. Having increased sixfold between 1955 and 1995, the volume of international trade is expected to triple again by 2020, according to the U.S. National Oceanographic and Atmospheric Administration, and 90% of it is expected to move by ocean.

In contrast to familiar fishing grounds and sea passageways, the depths of the ocean were long believed to be a vast wasteland that was inhospitable, if not completely devoid of life. Since the first deployment of submersibles in the 1930s and more advanced underwater acoustics and pressure chambers in the 1960s, scientific and commercial exploration has helped illuminate life in the deep sea and the geological history of the ancient ocean. Mining for sand, gravel, coral, and minerals (including sulfur and, most recently, petroleum) has taken place in shallow waters and continental shelves for decades, although offshore mining is severely restricted in some national waters.

Isolated, but highly concentrated, deep-sea deposits of manganese, gold, nickel, and copper, first discovered in the late 1970s, continue to tempt investors. These valuable nodules have proved technologically difficult and expensive to extract, given the extreme pressures and depths of their location. An international compromise on the deep seabed mining provisions of the Law of the Sea in 1994 has opened the way to some mining in international waters, but it appears unlikely to lead to much as long as mineral prices remain low, demand is largely met from the land, and the cost of underwater operations remains prohibitively high.

Perhaps more valuable than the mineral wealth in oceans are still-undiscovered living resources—new forms of life, potential medicines, and genetic material. For example, in 1997, medical researchers stumbled across a compound in dogfish that stops the spread of cancer by cutting off the blood supply to tumors. The promise of life-saving cures from marine species is gradually becoming a commercial reality for bioprospectors and pharmaceutical companies as anti-inflammatory and cancer drugs have been discovered and other leads are being pursued.

Tinkering with the ocean for the sake of shortsighted commercial development, whether for mineral wealth or medicine, warrants close scrutiny, however. Given how little we know—a mere 1.5% of the deep sea has been explored, let alone adequately inventoried—any development could be potentially irreversible in these unique environments. Although seabed mining is subjected to some degree of international oversight, prospecting for living biological resources is completely unregulated.

During the past 100 years, scientists who work both underwater and among marine fossils found high in mountains have shown that the tree of life has its evolutionary roots in the sea. For about 3,200,000,000 years, all life on Earth was marine. A complex and diverse food web slowly evolved from a fortuitous mix there of single-celled algae, bacteria, and several million trips around the sun. Life remained sea-bound until 245,000,000 years ago, when the atmosphere became oxygen-rich.

Thanks to several billion years' worth of trial and error, the oceans today are home to a variety of species that have no descendants on land. Thirty-two out of 33 animal life forms are represented in marine habitats. (Only insects are missing.) Fifteen of these are exclusively marine phyla, including those of comb jellies, peanut worms, and starfish. Five phyla, including that of sponges, live predominantly in salt water. Although, on an individual basis, marine species count for just nine percent of the 1,800,000 species described for the entire planet, there may be as many as 10,000,000 additional species in the sea that as yet have not been classified.

In addition to hosting a vast array of biological diversity, the marine environment performs such vital functions as oxygen production, nutrient recycling, storm protection, and climate regulation—services that often are taken for granted.

Marine biological activity is concentrated along the world's coastlines (where sunlit surface waters receive nutrients and sediments from land-based runoff, river deltas, and rainfall) and in upwelling systems (where cold, nutrient-rich, deep-water currents run up against continental margins). It provides 25% of the planet's primary biological productivity and an estimated 80–90% of the global commercial fish catch. It is estimated that coastal environments account for 38% of the goods and services provided by the Earth's ecosystems, while open oceans contribute an additional 25%. The value of all marine goods and services is estimated at 21 trillion dollars annually, 70% more than terrestrial systems come to.

Oceans are vital to both the chemical and biological balance of life. The same mechanism that created the present atmosphere—photosynthesis—continues to feed the marine food chain. Phytoplankton—tiny microscopic plants—take carbon dioxide (CO_2) from the atmosphere and convert it into oxygen and simple sugars, a form of carbon that can be consumed by marine animals. Other types of phytoplankton process nitrogen and sulfur, thereby helping the oceans function as a biological pump.

Although most organic carbon is consumed in the marine food web and eventually returned to the atmosphere via respiration, the unused balance rains down to the deep waters that make up the bulk of the ocean, where it is stored temporarily. Over the course of millions of years, these deposits have accumulated to the point that most of the world's organic carbon, approximately 15,000,000 gigatons (a gigaton equals 1,000,000,000 tons), is sequestered in marine sediments, compared with 4,000 gigatons in land-based reserves. On an annual basis, about one-third of the world's carbon emissions—around two gigatons—is taken up by oceans, an amount roughly equal to the uptake by land-based resources. If deforestation continues to diminish the ability of forests to absorb carbon dioxide, oceans are expected to play a more important role in regulating the planet's CO_2 budget in the future as human-induced emissions keep rising.

Perhaps no other example so vividly illustrates the connections between the oceans and the atmosphere than El Niño. Named after the Christ child because it usually appears in December, the El Niño Southern Oscillation takes place when trade winds and ocean surface currents in the eastern and central Pacific Ocean reverse direction. Scientists do not know what triggers the shift, but the aftermath is clear: Warm surface waters essentially pile up in the eastern Pacific and block deep, cold waters from upwelling, while a low pressure system hovers over South America, collecting heat and moisture that would otherwise be distributed at sea. This produces severe weather conditions in many parts of the world—increased precipitation, heavy flooding, drought, fire, and deep freezes—which, in turn, have enormous economic impact. During the 1997-98 El Niño, for instance, Argentina lost more than $3,000,000,000 in agricultural products due to these ocean-climate reactions, and Peru reported a 90% drop in anchovy harvests compared with the previous year.

A sea of problems

As noted earlier, the primary threats to oceans are largely human-induced and synergistic. Fishing, for example, has drastically altered the marine food web and underwater habitat areas. Meanwhile, the ocean's front line of defense—the coastal zone—is crumbling from years of degradation and fragmentation, while its waters have been treated as a waste receptacle for generations. The combination of overexploitation, the loss of buffer areas, and a rising tide of pollution has suffocated marine life and the livelihoods based on it in some areas. Upsetting the marine ecosystem in these ways has, in turn, given the upper hand to invasive species and changes in climate.

Overfishing poses a serious biological threat to ocean health. The resulting reductions in the genetic diversity of the spawning populations make it more difficult for the species to adapt to future environmental changes. The orange roughy, for instance, may have been fished down to the point where future recovery is impossible. Moreover, declines in one species can alter predator-prey relations and leave ecosystems vulnerable to invasive species. The overharvesting of triggerfish and pufferfish for souvenirs on coral reefs in the Caribbean has sapped the health of the entire reef chain. As these fish declined, populations of their prey—sea urchins—exploded, damaging the coral by grazing on the protective layers of algae and hurting the local reef-diving industry.

These trends have enormous social consequences as well. The welfare of more than 200,000,000 people around the world who depend on fishing for their income and food security is severely threatened. As the fish disappear, so do the coastal communities that depend on fishing for their way of life. Subsistence and small-scale fishers, who catch nearly half of the world's fish, suffer the greatest losses as they cannot afford to compete with large-scale vessels or changing technology. Furthermore, the health of more than 1,000,000,000 poor consumers who depend on minimal quantities of fish to constitute their diets is at risk as an ever-growing share of fish—83% by value—continues to be exported to industrial countries each year.

Despite a steadily growing human appetite for fish, large quantities are wasted each year because the fish are undersized or a nonmarketable sex or species, or because a fisher does not have a permit to catch them and must therefore throw them out. The United Nations' Food and Agricultural Organization estimates that discards of fish alone—not counting marine mammals, seabirds, and turtles—total 20,000,000 tons, equivalent to one-fourth of the annual marine catch. Many of these fish do not survive the process of getting entangled in gear, being brought on board, and then tossed back to sea.

Another threat to habitat areas stems from trawling, with nets and chains dragged across vast areas of mud, rocks, gravel, and sand, essentially sweeping everything in the vicinity. By recent estimates, all the ocean's continental shelves are trawled by fishers at least once every two years, with some areas hit several times a season. Considered a major cause of habitat degradation, trawling disturbs bottom-dwelling communities as well as localized species diversity and food supplies.

The conditions that make coastal areas so productive for fish—proximity to nutrient flows and tidal mixing and their place at the crossroads between land and water—also make them vulnerable to human assault. Today, nearly 40% of the world lives within 60 miles of a coastline. As more people move to coastal areas and further stress the seams between land and sea, coastal ecosystems are losing ground.

Human activities on land cause a large portion of offshore contamination. An estimated 44% of marine pollution comes from land-based pathways, flowing down rivers into tidal estuaries, where it bleeds out to sea; an additional 33% is airborne pollution that is carried by winds and deposited far off shore. From nutrient-rich sediments, fertilizers, and human waste to toxic heavy metals and synthetic chemicals, the outfall from human society ends up circulating in the fluid and turbulent seas.

Excessive nutrient loading has left some coastal systems looking visibly sick. Seen from an airplane, the surface waters of Manila Bay in the Philippines resemble green soup due to dense carpets of algae. Nitrogen and phosphorus are necessary for life and, in limited quantities, can help boost plant productivity, but too much of a good thing can be bad. Excessive nutrients build up and create conditions that are conducive to outbreaks of dense algae blooms, also known as "red tides." The blooms block sunlight, absorb dissolved oxygen, and disrupt food-web dynamics. Large portions of the Gulf of Mexico are now considered a biological "dead zone" due to algal blooms.

The frequency and severity of red tides has increased in the past couple of decades. Some experts link the recent outbreaks to increasing loads of nitrogen and phosphorus from nutrient-rich wastewater and agricultural runoff in poorly flushed waters.

Organochlorines, a fairly recent addition to the marine environment, are proving to have pernicious effects. Synthetic organic compounds such as chlordane, DDT, and PCBs are used for everything from electrical wiring to pesticides. Indeed, one reason they are so difficult to control is that they are ubiquitous. The organic form of tin (tributyltin), for example, is used in most of the world's marine paints to keep barnacles, seaweed, and other organisms from clinging to ships. Once the paint is dissolved in the water, it accumulates in mollusks, scallops, and rock crabs, which are consumed by fish and marine mammals.

As part of a larger group of chemicals known collectively as persistent organic pollutants (POPs), these compounds are difficult to control because they do not degrade easily. Highly volatile in warm temperatures, POPs tend to circulate toward colder environments where the conditions are more stable, such as the Arctic Circle. Moreover, they do not dissolve in water, but are lipid-soluble, meaning that they accumulate in the fat tissues of fish that are then consumed by predators at a more concentrated level.

POPs have been implicated in a wide range of animal and human health problems—from suppression of immune systems, leading to higher risk of illness and infection, to disruption of the endocrine system, which is linked to birth defects and infertility. Their continued use in many parts of the world poses a threat to marine life and fish consumers everywhere.

Because marine species are extremely sensitive to fluctuations in temperature, changes in climate and atmospheric conditions pose high risks to them. Recent evidence shows that the thinning ozone layer above Antarctica has allowed more ultraviolet-B radiation to penetrate the waters. This has affected photosynthesis and the growth of phytoplankton and macroalgae. The effects are not limited to the base of the food chain. By striking aquatic species during their most vulnerable stages of life and reducing their food supply at the same time, increases in UV-B could have devastating impacts on world fisheries production.

Because higher temperatures cause water to expand, a warming world may trigger more frequent and damaging storms. Ironically, the coastal barriers, seawalls, jetties, and levees that are designed to protect human settlements from storm surges likely exacerbate the problem of coastal erosion and instability, as they create deeper inshore troughs that boost wave intensity and sustain winds.

Depending on the rate and extent of warming, global sea levels may rise as much as three feet by 2100—up to five times as much as during the last century. Such a rise would flood most of New York City, including the entire subway system and all three major airports. Economic damages and losses could cost the global economy as much as $970,000,000,000 in 2100, according to the Organisation for Economic Co-operation and Development. The human costs would be unimaginable, especially in the low-lying, densely populated river deltas of Bangladesh, China, Egypt, and Nigeria.

These damages could be just the tip of the iceberg. Warmer temperatures would likely accelerate polar ice cap melting and could boost this rising wave by several feet. Just four years after a large portion of Antarctica melted, another large ice sheet fell off into the Southern Sea in February, 1998, rekindling fears that global warming could ignite a massive thaw that would flood coastal areas worldwide. Because oceans play such a vital role in regulating the Earth's climate and maintaining a healthy planet, minor changes in ocean circulation or in its temperature or chemical balance could have repercussions many orders of magnitude larger than the sum of human-induced wounds.

While understanding past climatic fluctuations and predicting future developments are an ongoing challenge for scientists, there is clear and growing evidence of the overuse—indeed abuse—that many marine ecosystems and species are suffering from direct human actions. The situation is probably much worse, for many sources of danger are still unknown or poorly monitored. The need to take preventive and decisive action on behalf of oceans is more important than ever.

Saving the oceans

Scientists' calls for precaution and protective measures are largely ignored by policymakers, who focus on enhancing commerce, trade, and market supply and look to extract as much from the sea as possible, with little regard for the effects on marine species or habitats. Overcoming the interest groups that favor the status quo will require engaging all potential stakeholders and reformulating the governance equation to incorporate the stewardship obligations that come with the privilege of use.

Fortunately for the planet, a new sea ethic is emerging. From tighter dumping regulations to recent international agreements, policymakers have made initial progress toward the goal of cleaning up humans' act. Still, much more is needed in the way of public education to build political support for marine conservation.

To boost ongoing efforts, two key principles are important. First, any dividing up of the waters should be based on equity, fairness, and need as determined by dependence on the resource and the best available scientific knowledge, not simply on economic might and political pressure. In a similar vein, resource users should be responsible for their actions, with decision-making and accountability shared by stakeholders and government officials. Second, given the uncertainty in scientific knowledge and management capabilities, it is necessary to err on the side of caution and take a precautionary approach.

Replanting mangroves and constructing artificial reefs are two concrete steps that help some fish stocks rebound quickly while letting people witness firsthand the results of their labors. Once they see the immediate payoff of their work, they are more likely to stay involved in longer-term protection efforts, such as marine sanctuaries, which involve removing an area from use entirely.

Marine protected areas are an important tool to help marine scientists and resource planners incorporate an ecologically based approach to oceans protection. By limiting accessibility and easing pressures on the resource, these areas allow stocks to rebound and profits to return. Globally, more than 1,300 marine and coastal sites have some form of protection, but most lack effective on-the-ground management.

Meanwhile, efforts to establish marine refuges and parks lag far behind similar efforts on land. The World Heritage Convention, which identifies and protects areas of special significance to mankind, identifies just 31 sites that include either a marine or a coastal component, out of a total of 522. John Waugh, Senior Program Officer of the World Conservation Union-U.S., and others argue that the World Heritage List could be extended to a number of marine hotspots and should include representative areas of the continental shelf the deep sea, and the open ocean. Setting these and other areas aside as off-limits to commercial development can help advance scientific understanding of marine systems and provide refuge for threatened species.

To address the need for better data, coral reef scientists have enlisted the help of recreational scuba divers. Sport divers who volunteer to collect data are given basic training to identify and survey fish and coral species and conduct rudimentary site assessments. The data then are compiled and put into a global inventory that policymakers use to monitor trends and to target intervention. More efforts like these—that engage the help of concerned individuals and volunteers—could help overcome funding and data deficiencies and build greater public awareness of the problems plaguing the world's oceans.

Promoting sustainable ocean use also means shifting demand away from environmentally damaging products and extraction techniques. To this end, market forces, such as charging consumers more for particular fish and introducing industry codes of conduct, can be helpful. In April, 1996, the World Wide Fund for Nature teamed up with one of the world's largest manufacturers of seafood products, Anglo-Dutch Unilever, to create economic incentives for sustainable fishing. Implemented through an independent Marine Stewardship Council, fisheries products that are harvested in a sustainable manner will qualify for an ecolabel. Similar efforts could help convince industries to curb wasteful practices and generate greater consumer awareness of the need to choose products carefully.

Away from public oversight, companies engaged in shipping, oil and gas extraction, deep-sea mining, bioprospecting, and tidal and thermal energy represent a coalition of special interests whose activities help determine the fate of the oceans. It is crucial to get representatives of these industries engaged in implementing a new ocean charter that supports sustainable use. Their practices not only affect the health of oceans, they help decide the pace of a transition toward a more sustainable energy economy, which, in turn, affects the balance between climate and oceans.

Making trade data and industry information publicly available is an important way to build industry credibility and ensure some degree of public oversight. While regulations are an important component of environmental protection, pressure from consumers, watchdog groups, and conscientious business leaders can help develop voluntary codes of action and standard industry practices that can move industrial sectors toward cleaner and greener operations. Economic incentives targeted to particular industries, such as low-interest loans for thermal projects, can aid companies in making a quicker transition to sustainable practices.

The fact that oceans are so central to the global economy and to human and planetary health may be the strongest motivation for protective action. Although the range of assaults and threats to ocean health are broad, the benefits that oceans provide are invaluable and shared by all. These huge bodies of water represent an enormous opportunity to forge a new system of cooperative, international governance based on shared resources and common interests. Achieving these far-reaching goals, however, begins with the technically simple, but politically daunting, task of overcoming several thousand years' worth of ingrained behavior. It requires seeing oceans not as an economic frontier for exploitation, but as a scientific frontier for exploration and a biological frontier for careful use.

For generations, oceans have drawn people to their shores for a glimpse of the horizon, a sense of scale, and awe at nature's might. Today, oceans offer careful observers a different kind of awe—a warning that humans' impacts on the Earth are exceeding natural bounds and in danger of disrupting life. Protection efforts already lag far behind what is needed. How humans choose to react will determine the future of the planet. Oceans are not simply one more system under pressure—they are critical to man's survival. As Carl Safina writes in *The Song for the Blue Ocean*, "we need the oceans more than they need us."

Anne Platt McGinn is a senior researcher, Worldwatch Institute, Washington, D.C.

The Human Impact on Climate

How much of a disruption do we cause? The much-awaited answer could be ours by 2050, but only if nations of the world commit to long-term climate monitoring now

by Thomas R. Karl and Kevin E. Trenberth

"The balance of evidence suggests a discernible human influence on global climate." With these carefully chosen words, the Intergovernmental Panel on Climate Change (jointly supported by the World Meteorological Organization and the United Nations Environmental Program) recognized in 1995 that human beings are far from inconsequential when it comes to the health of the planet. What the panel did not spell out—and what scientists and politicians dispute fiercely—is exactly when, where and how much that influence has and will be felt.

So far the climate changes thought to relate to human endeavors have been relatively modest. But various projections suggest that the degree of change will become dramatic by the middle of the 21st century, exceeding anything seen in nature during the past 10,000 years. Although some regions may benefit for a time, overall the alterations are expected to be disruptive or even severe. If researchers could clarify the extent to which specific activities influence climate, they would be in a much better position to suggest strategies for ameliorating the worst disturbances. Is such quantification possible? We think it is and that it can be achieved by the year 2050—but only if that goal remains an international priority.

Despite uncertainties about details of climate change, our activities clearly affect the atmosphere in several troubling ways. Burning of fossil fuels in power plants and automobiles ejects particles and gases that alter the composition of the atmosphere. Visible pollution from sulfur-rich fuels includes micron-size particles called aerosols, which often cast a milky haze in the sky. These aerosols temporarily cool the atmosphere because they reflect some of the sun's rays back to space, but they stay in the air for only a few days before rain sweeps them to the planet's surface. Certain invisible gases deliver a more lasting impact. Carbon dioxide remains in the atmosphere for a century or more. Worse yet, such greenhouse gases trap some of the solar radiation that the planet would otherwise radiate back to space, creating a "blanket" that insulates and warms the lower atmosphere.

Indisputably, fossil-fuel emissions alone have increased carbon dioxide concentrations in the atmosphere by about 30 percent since the start of the Industrial Revolution in the late 1700s. Oceans and plants help to offset this flux by scrubbing some of the gas out of the air over time, yet carbon dioxide concentrations continue to grow. The inevitable result of pumping the sky full of greenhouse gases is global warming. Indeed, most scientists agree that the earth's mean temperature has risen at least 0.6 degree Celsius (more than one degree Fahrenheit) over the past 120 years, much of it caused by the burning of fossil fuels.

The global warming that results from the greenhouse effect dries the planet by evaporating moisture from oceans, soils and plants. Additional moisture in the atmosphere provides a swollen reservoir of water that is tapped by all precipitating weather systems,

23. Human Impact on Climate

> Climate simulation and prediction will come of age only with an **ongoing record of changes** as they happen.

be they tropical storms, thundershowers, snowstorms or frontal systems. This enhanced water cycle brings on more severe droughts in dry areas and leads to strikingly heavy rain or snowfall in wet regions, which heightens the risk of flooding. Such weather patterns have burdened many parts of the world in recent decades.

Human activities aside from burning fossil fuels can also wreak havoc on the climate system. For instance, the conversion of forests to farmland eliminates trees that would otherwise absorb carbon from the atmosphere and reduce the greenhouse effect. Fewer trees also mean greater rainfall runoff, thereby increasing the risk of floods.

It is one thing to have a sense of the factors climate will have to be able to construct more accurate climate models than have ever been designed before. We will therefore require the technological muscle of supercomputers a million times faster than those in use today. We will also have to continue to disentangle the myriad interactions among the oceans, atmosphere and biosphere to know exactly what variables to feed into the computer models.

Most important, we must be able to demonstrate that our models accurately simulate past and present climate change before we can rely on models to predict the future. To do that, we need long-term records. Climate simulation and prediction will come of age only with an ongoing record of changes as they happen.

Computers and Climate Interactions

For scientists who model climate patterns, everything from the waxing and waning of ice ages to the desertification of central Africa plays out inside the models run on supercomputers. Interactions among the compo-

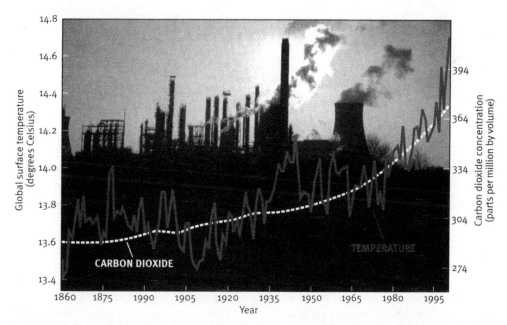

Burning fossil fuels *(photograph)* has increased atmospheric concentrations of carbon dioxide *(white dashes)* and has contributed to a rise in global surface temperatures during the past 140 years *(gray line)*.

Chinch Gryniewicz/*Corbis*; LAURIE GRACE (graph)

that can bring about climate change. It is another to know how the human activity in any given place will affect the local and global climate. To achieve that aim, those of us who are concerned about the human influence on nents of the climate system—the atmosphere, oceans, land, sea ice, freshwater and biosphere—behave according to physical laws represented by dozens of mathematical equations. Modelers instruct the computers to

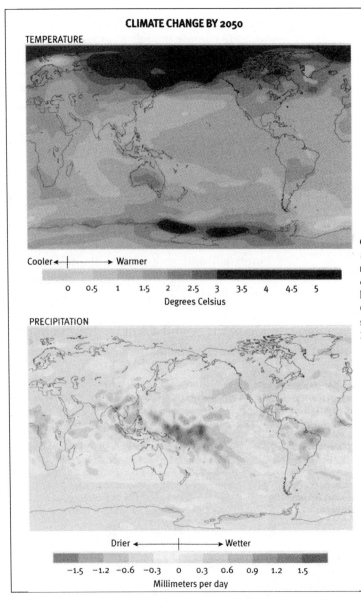

Global warming of up to five degrees Celsius *(top)* could enhance precipitation *(bottom)* in much of the world by the middle of the 21st century. These simulations use 1992 estimates by the Intergovernmental Panel on Climate Change for emissions of greenhouse gases and sulfate aerosols between the years 2000 and 2050.

NOAA/GEOPHYSICAL FLUID DYNAMICS LABORATORY

solve these equations for each box in a three-dimensional grid that covers the globe. Because nature is not constrained by boxes, the chore is not only to incorporate the correct mathematics within each box but also to describe appropriately the transfer of energy and mass into and out of the boxes.

The computers at the world's preeminent climate-modeling facilities can perform between 10 and 50 billion operations per second, but with so many evolving variables, the simulation of a single century can take months. The time it takes to run a simulation, then, limits the resolution (or number of boxes) that can be included within climate models. For typical models designed to mimic the detailed evolution of weather systems, boxes in the three-dimensional grid measure about 250 kilometers (156 miles) square in the horizontal direction and one kilometer in the vertical. Tracking patterns within smaller areas thus proves especially difficult.

Even the most sophisticated of our current global models cannot directly simulate conditions such as cloud cover and the formation of rain. Powerful thunderstorm clouds that can unleash sudden downpours often operate on scales of less than 10 kilometers, and raindrops condense at submillimeter scales. Because each of these events happens in a region smaller than the volume of the smallest grid unit, their characteristics must be inferred by elaborate statistical techniques.

Such small-scale weather phenomena develop randomly. The frequency of these random events can differ extensively from place to place, but most agents that alter climate, such as rising levels of greenhouse gases, af-

Deforestation changes climate in more than one way: Cutting down trees makes the forest less able to scrub carbon dioxide out of the air. Dark-colored forests also absorb more solar energy and keep the region warmer and more moist than do the light-colored areas left when the trees are gone.

fect all areas of the planet much more uniformly. The variability of weather will increasingly mask large-scale climate activity as smaller regions are considered. Lifting that mask thus drains computer time, because it requires running several simulations, each with slightly different starting conditions. The climate features that occur in every simulation constitute the climate "signal," whereas those that are not reproducible are considered weather-related climate "noise."

Conservative estimates indicate that computer-processing speed will have increased by well over a million times by 2050. With that computational power, climate modelers could perform many simulations with different starting conditions and better distinguish climate signals from climate noise. We could also routinely run longer simulations of hundreds of years with less than one-kilometer horizontal resolution and an average of 100-meter vertical resolution over the oceans and atmosphere.

Faster computers help to predict climate change only if the mathematical equations fed into them perfectly describe what happens in nature. For example, if a model atmosphere is simulated to be too cold by four degrees C (not uncommon a decade ago), the simulation will indicate that the atmosphere can hold about 20 percent less water than its actual capacity—a significant error that renders meaningless any subsequent estimates of evaporation and precipitation. Another problem is that we do not yet know how to replicate adequately all the processes that influence climate, such as hiccups in the carbon cycle and modifications in land use. What is more, these changes can initiate feedback cycles that, if ignored, can lead the model astray. Raising temperature, for example, sometimes enhances another variable, such as moisture content of the atmosphere, which in turn amplifies the original perturbation. (In this case, more moisture in the air causes increased warming because water vapor is a powerful greenhouse gas.)

Researchers are only beginning to realize how much some of these positive feedbacks influence the planet's life-giving carbon cycle. The 1991 eruption of Mount Pinatubo in the Philippines, for instance, belched out enough ash and sulfur dioxide to cause a temporary global cooling as those compounds interacted with water droplets in the air to block some of the sun's incoming radiation. This depleted energy can inhibit carbon dioxide uptake in the plants.

Using land in a different way can perturb continental and regional climate systems in ways that are difficult to translate into equations. Clearing forests for farming and ranching brightens the land surface. Croplands are lighter-colored than dark forest and thus reflect more solar radiation, which tends to cool the atmosphere, especially in autumn and summer.

Dearth of Data

Climate simulations can never move out of the realm of good guesses without accurate observations to validate them and to show that the models do indeed reflect reality. In other words, to reduce our uncertainty about the sensitivity of the climate system to human activity, we need to know how the climate has changed in the past. We must be capable of adequately simulating conditions before the Industrial Revolution and especially since that time, when humans have altered irrevocably the composition of the atmosphere.

To understand climate from times prior to the development of weather-tracking satellites and other instruments, we rely on indicators such as air and chemicals trapped in ice cores, the width of tree rings, coral growth, and sediment deposits on the bottoms of oceans and lakes. These snapshots provide us with information that aids in piecing together past conditions. To truly understand the present climate, however, we require more than snapshots of physical, chemical and biological quantities; we also need the equivalent of long-running videotape records of the currently evolving climate. Ongoing measurements of sea ice, snow cover, soil moisture, vegetative cover, and ocean temperature and salinity are just some of the variables involved.

But the present outlook is grim: no U.S. or international institution has the mandate or resources to monitor long-term climate. Scientists currently compile their interpretations of climate change from large networks of satellites and surface sensors such as buoys, ships, observatories, weather stations and airplanes that are being operated for other purposes, such as short-term weather forecasting. As a result, depictions of past climate variability are often equivocal or missing.

The National Oceanic and Atmospheric Administration operates many of these networks, but it does not have the resources to commit to a long-term climate-monitoring program. Even the National Aeronautics and Space Administration's upcoming Earth Observing System, which entails launching several sophisticated satellites to monitor various aspects of global systems, does not include the continuity of a long-term climate observation program in its mission statement.

Whatever the state of climate monitoring may be, another challenge in the next decade will be to ensure that the quantities we do measure actually represent real multidecadal changes in the environment. In other words, what happens if we use a new camera or point it in a different direction? For instance, a satellite typically lasts only four years or so before it is replaced with another in a different orbit. The replacement usually has new instruments and observes the earth at a different time of day. Over a period of years, then, we end up measuring not only climate variability but also the changes introduced by observing the climate in a different way. Unless precautions are taken to quantify the modifications in observing technology and sampling methods before the older technology is replaced, climate records could be rendered useless because it will be impossible to compare the new set of data with its older counterpart.

Future scientists must be able to evaluate their climate simulations with unequivocal data that are properly archived. Unfortunately, the data we have archived from satellites and critical surface sensors are in jeopardy of being lost forever. Long-term surface observations in the U.S. are still being recorded on outdated punched paper tapes or are stored on decaying paper or on old computer hardware. About half the data from our new Doppler radars are lost because the recording system relies on people to deal with the details of data preservation during severe weather events, when warnings and other critical functions are a more immediate concern.

Can We Realize the Vision?

Over the next 50 years we can broadly understand, if we choose to, how human beings are affecting the global, regional and even small-scale aspects of climate. But waiting until then to take action would be foolhardy. Long lifetimes of carbon dioxide and other greenhouse gases in the atmosphere, coupled with the climate's typically slow response to evolving conditions, mean that even if we cut back on harmful human activities today, the planet very likely will still undergo substantial change.

Glaciers melting in the Andes highlands and elsewhere are already confirming the reality of a warming planet. Rising sea level—and drowning coastlines—testify to the projected global warming of perhaps two degrees C or more by the end of the next century. Climate change will in all likelihood capture the most attention when its effects exacerbate other pressures on society. The spread of settlements into coastal regions and low-lying areas vulnerable to flooding is just one of the initial difficulties that we will most likely face. But as long as society can fall back on the uncertainty of human impact on climate, legislative mandates for changing standards of fossil-fuel emissions or forest clear-cutting will be hard fought.

The need to foretell how much we influence our world argues for doing everything we can to develop comprehensive observing and data-archiving systems now. The resulting information could feed models that help make skillful predictions of climate several years in advance. With the right planning we could be in a position to predict, for example, exactly how dams and reservoirs might be better designed to accommodate anticipated floods and to what extent greenhouse gas emissions from new power plants will warm the planet.

Climate change is happening now, and more change is certain. We can act to slow it down, and we can sensibly plan for it, but at present we are doing neither. To anticipate the true shape of future climate, scientists must overcome the obstacles we have outlined above. The need for greater computer power and for a more sophisticated understanding of the nuances of climate interactions should be relatively easy to overcome. The real stumbling block is the long-term commitment to global climate monitoring. How can we get governments to commit resources for decades of surveys, particularly when so many governments change hands with such frequency?

If we really want the power to predict the effects of human activity by 2050—and to begin addressing the disruption of our environment—we must pursue another path. We have a tool to clear such a path: the United Nations Framework Convention on Climate Change, signed by President George Bush in 1992. The convention binds together 179 governments with a commitment to remedy damaging human influence on global climate. The alliance took a step toward stabilizing greenhouse gas emissions by producing the Kyoto Protocol in 1997, but long-term global climate-monitoring systems remain unrealized.

FURTHER INFORMATION

GLOBAL WARMING: IT'S HAPPENING. Kevin E. Trenberth in *naturalSCIENCE*, Vol. 1, Article 9; 1997. Available at naturalscience.com/ns/articles/01-09/ns/_ket.html on the World Wide Web.

ADEQUACY OF CLIMATE OBSERVING SYSTEMS, 1999. Commission on Geosciences, Environment, and Resources. National Academy Press, 1999. Available at www.nap.edu/books/0309063906/html/ on the World Wide Web.

CLIMATE CHANGE AND GREENHOUSE GASES. Tamara S. Ledley et al. in *EOS*, Vol. 80, No. 39, pages 453–458; Sept. 28, 1999. Available at www.agu.org/eos_elec/99148e.html on the World Wide Web.

The United Nations Framework Convention on Climate Change and Kyoto Protocol updates are available at www.unfccc.org/ on the World Wide Web.

THOMAS R. KARL has directed the National Climatic Data Center (NCDC) in Asheville, N.C., since March 1998. The center is part of the National Oceanic and Atmospheric Administration and serves as the world's largest active archive of climate data. Karl, who has worked at the center since 1980, has focused much of his research on climate trends and extreme weather. He also writes reports for the Intergovernmental Panel on Climate Change (IPCC), the official science source for international climate change negotiations.

KEVIN E. TRENBERTH directs the Climate Analysis section at the National Center for Atmospheric Research (NCAR) in Boulder, Colo., where he studies El Niño and climate variability. After several years in the New Zealand Meteorological Service, he became a professor of atmospheric sciences at the University of Illinois in 1977 and moved to NCAR in 1984. Trenberth also co-writes IPCC reports with Karl.

Warming Up

The real evidence for the greenhouse effect.

By Gregg Easterbrook

IT GOT DRY, and that must be the influence of global warming. Wait—then there were downpours, which proved the greenhouse effect. A monster hurricane crossed North Carolina; global warming must be the cause. No, wait—out comes a book about the most deadly hurricane in U.S. history, which hit in 1900, long before there was greenhouse gas buildup. Snowfall has decreased some in recent years, and that must prove global warming. But wait—when cities were snowed in during the winter of 1996–1997, commentators blamed greenhouse disruption. Annual temperatures have been high during the '90s, so the world must be warming. But wait—when winters were frigid during the 1970s, there were congressional hearings on whether an ice age was beginning. It's hot, it's cold, it's dry, it's wet—and don't get us started on El Niño.

As cable TV channels and extended local newscasts increasingly fill airtime by reporting weather details—and even the national networks send their reporters to hurricane areas to scrunch up inside designer slickers and shout, "I think I just felt a raindrop!"—the notion has taken hold that every barometric fluctuation must demonstrate climate change. This anecdotal case for global warming is mostly nonsense, driven by nescience of a basic point, from statistics and probability, that the weather is always weird somewhere. But, even as public attention has fixated on greenhouse claims by anecdote, the less dramatic, technical case for climate change has been gradually strengthening. Artificial warming is far from proven, but today the scientific evidence is stronger than it was five years ago. Most research suggests that a warmer world now seems probable; that altered climate patterns may be in store; that the chance of truly dangerous global warming, though still low, is rising. The worst aspect of attributing every dry spell or cloudburst to the greenhouse effect may be that such reasoning obscures the real concern: that credible signs of climate change are in fact beginning to appear.

One reason there is so much flummery in the global warming debate is that the weather in the Northeast United States, where the opinion-makers live, has a disproportionate effect on whether greenhouse concerns are taken seriously. During the summer of 1999, most of the Northeast experienced a drought; and, between July and September, *The New York Times'* op-ed page ran no fewer than four articles asserting that an artificial greenhouse effect was now proven, in part by dehydration. Then autumn brought drenching rains to the Eastern seaboard. The phase of wetness combined with the phase of dryness to make it an average year for precipitation, the apparent portent having "washed out," in statistician lingo, into normality. This did not prevent commentators from switching to a claim that showers proved global warming. "Go outside: try to understand that the sun beating down, the rain pouring down, the wind blowing by are all human artifacts. We don't live on the planet we were born on," one *Times* op-ed declared. Perhaps one shouldn't speak for all *Times* contributors, but I live on the planet I was born on, and feel it can be said without fear of contradiction that the sun is not a "human artifact."

BECAUSE WEATHER IS short-term and expected to vary, using it to draw conclusions about climate is treacherous. Weather is what happens today; climate is what happens during your lifetime. It is climate, not weather, that society needs to care about. In recent years, whenever a record high temperature is set, it has been common to hear global warming invoked as the reason. But record temperatures, high or low, are statistical "outliers," or nonstandard

24. Warming Up: The Real Evidence for the Greenhouse Effect

effects, that rarely reveal anything. The highest temperature ever recorded in the United States was in the year 1913, long before greenhouse gas accumulation began. Europe's record high occurred in 1881, Australia's in 1889, South America's in 1905, Africa's in 1922.

Hurricane incidence, currently spoken of as disturbingly high, is also inconclusive as an omen of climate change. Distribution of what the National Oceanic and Atmospheric Administration (NOAA) calls "major landfalling hurricanes" is basically equal before and after greenhouse gases, with the 1910s and '20s having about the same number of major Atlantic cyclones as the 1980s and '90s. A devastating hurricane, Andrew, struck in 1992, and an abnormally large hurricane, Floyd, arrived this year. But the strongest U.S. hurricane ever was in 1935. Hurricanes became a media obsession in the mid-'90s, when there were two consecutive above-average years for such storms. But those years stood out mainly because most of the period from 1964 to 1994 was below the historic average; commentators had forgotten how frequently hurricanes can occur.

Tornadoes are currently on a frightening upswing: there were more of them in the fourth quarter of 1998 than during any previous quarter in the nation's history, while January 1999 saw the most U.S. tornadoes ever in a single month. This is a somber concern for those who live in tornado zones, but does it suggest global warming? Greenhouse effect theory calls for warming mainly in high latitudes, which should reduce the poles-to-equator temperature gradients that power storm formation; a greenhouse world might have less violent weather. Charts of tornado incidence in the postwar period show 1998 and 1999 spiking up out of nowhere, not representing any slow-building trend; what's happening may be a harbinger, or it may be random variation.

THIS SUMMER A strain of encephalitis was detected in New York City, and, being close to the media's oxygen, it occasioned one *Times* writer to opine that greenhouse-triggered "outbreaks" of plague-like proportions are "descending on us." If so, our bodies are choosing a funny way to show it: by getting healthier. Though the World Health Organization has cautioned that climate change could cause the spread of some diseases, mainly by extending the range of tropical contagion, no such effect has manifested itself in public health data. As greenhouse gases accumulate, rates for almost all diseases are declining almost everywhere in the world, with the tragic exception of AIDS in Africa. (There is evidence that ocean-temperature changes relate to disease incidence among marine mammals.)

Another popular assertion regarding global warming is that rising insurance claims from weather-related damage show that dire news is afoot. The National Environmental Trust, now running greenhouse effect ads asserting that "the weather has been pretty weird," claims that "severe weather events have cost $272 billion in damages in the '90s." Never mind that the National Climatic Data Center (NCDC) puts the figure at $170 billion; maybe somebody has been inflating claims to the insurance adjusters. The NCDC figure represents an enormous tab and a historic high. But ever-rising insurance claims might be expected in an affluent society where property grows steadily more valuable and a leading construction trend has been building high-end housing in coastal areas, the places most susceptible to storm damage.

The most dubious assertion made about global warming is that it must be happening because computers think so. Several scientific centers around the world now run "general circulation models" (GCMs). These computers, which are important research tools, simulate the atmosphere, and every GCM predicts a warmer future. But the Intergovernmental Panel on Climate Change (IPCC), a U.N.-sponsored scientists' organization that is the leading player in the study of the greenhouse effect and is generally toward the left of the debate, issues regular warnings that GCMs are, at best, approximations. Computational power, even that of advanced supercomputers, is still too modest to recreate a system as complicated as global climate; greenhouse GCMs can't yet take into account basic considerations, such as the role of clouds.

Right now, one GCM, at the Goddard Institute for Space Studies, is predicting that a rapid global temperature rise will occur during the next decade; other GCMs are predicting more gradual effects. Overall, as GCMs have improved, the warming they predict has tended to decline. The IPCC's current "best estimate" is that, if existing trends in greenhouse emissions continue, the world will warm by approximately another 3.5 degrees Fahrenheit by about the year 2100—a significant and possibly dangerous figure, but one-third lower than the organization's estimate of a decade ago.

There is near-unanimous scientific agreement that the world has gotten warmer by about one degree Fahrenheit during the twentieth century. Worldwide, the decade of the '90s was warm, with seven of the past ten years showing above-average global temperatures. Are these findings the ones that prove global warming?

Being a long-term movement, the one-degree twentieth-century rise clearly tells us something. The problem is that we don't know what, because it is not yet known what caused the increase. Greenhouse gases from fossil fuels? Ocean-current shifts? Fluctuation in the output of the sun? Your guess is as good as the next Ph.D.'s "There is a scientific consensus that the greenhouse effect is a serious concern, but not on much else. We are in the early days of understanding the global climate," says Michael Schlesinger, a professor of atmospheric sciences at the University of Illinois. Key puzzling datum: Roughly half of this century's warming occurred *before* 1940, when artificial greenhouse gas levels were not significant.

Only slight increases in mean global temperature are what caused the '90s to be the warmest decade "on record," and slight warming may be natural variation. Moreover, the whole notion of "on record" is a bit misleading. Precise temperature data go back only about 100 years, to the late nineteenth century. At that time, the climate was recovering from the Little Ice Age, which lasted from about 1500 to about 1850. Since the "on record" period begins at a point when the earth was somewhat cool, perhaps a warming trend is no surprise. The doomsday cohort often declares that, during the '90s, temperatures were warmer than at any other time in the past 100,000 years. But no one really knows; the IPCC says that "[d]ata prior to 1400 are too sparse to allow the reliable estimation of global mean temperature."

The IPCC repeatedly cautions that there is no scientific answer for what has caused the mild warming observed in this century. All we can be sure of, the organization says, is that human activity now plays some role in climate. Combustion of fossil fuels is not the sole concern. Deforestation, mainly in the developing world, has become a greenhouse influence. Some human activity is positive in climate terms: high-yield agriculture and managed forestry subtract vast volumes of greenhouse gases from the air by accelerating crop and tree growth. On balance, however, the IPCC suggests, society now influences climate in ways that won't be to our liking. Global climate has consistently been favorable to agriculture during the dramatic twentieth-century population increase. If human action now risks disturbing a climate favorable to humans, that may be all we need know to take greenhouse emissions seriously as a policy concern.

ADDED TO THE general likelihood that human activity now plays a role in climate are two research findings of considerable power. One, by Thomas Karl, head of the NCDC, shows that, over the long term, incidence of heavy rain is increasing. At the turn of the century, about eight percent of U.S. annual precipitation occurred during downpours, and now ten percent does; this is not striking as a percentage, but it is a long-term distinction that may mean more than individual temperature spikes. A decade ago Karl ranked among the most important greenhouse effect skeptics, often pointing out how poorly global warming computer predictions squared with observed climate. In recent years Karl has become a greenhouse moderate, as observational evidence begins to strengthen.

A second important research finding is that the "frost-free" season in the United States is now eleven days longer than it was in the 1950s. This trend, too, is long-term and thus more likely to be significant. Ellen Cooter and Sharon LeDuc, researchers at NOAA, have found that, in recent decades, spring, as reckoned by the night of the last hard freeze, has been arriving one day earlier every three years.

Of course, most people want spring to come earlier, and the longer the frost-free season is, the happier farmers are. The mild warming of this century surely contributed to the postwar global flowering of agriculture, which has staved off widely predicted Malthusian catastrophes. Research shows that the warming that has occurred so far has come mainly in the form of less-cold nighttime lows in winter, rather than hotter daytime highs in summer; less-cold winter nights are a boon, moderating energy demand. The fear is that an artificial greenhouse effect will not follow its current course. In the rainfall and early-spring studies may reside a warning that changing climate patterns may begin to harm agriculture—for instance, by shifting precipitation away from currently tilled regions. Higher summertime daylight maximums may eventually be in store, scorching crops and increasing energy demand. There could be nasty surprises no one currently projects.

The downpour and early-spring findings are not conclusive but are sufficiently worrisome; they seem to be part of the reason Republican presidential candidate George W. Bush recently said, "I believe there is global warming." New patterns of rain and frost might have artificial causes, natural causes, or some combination of the two. Artificial causes are the only ones we can do anything about.

INCREASING SCIENTIFIC INDICATION of climate change has not led to policy action, in part because the global warming debate continues to be conducted as a partisan extravaganza, one in which participants of all stripes are sometimes most interested in name-calling and fund-raising. Anamorphosis rules. Alarmists make a point of citing the IPCC's scary upper-bound projection of a six-degree future warming, rather than the number the organization favors—its "best estimate" of about half that amount. Naysayers harp on the literalist point that, because climate dynamics are poorly understood, it is impossible to be certain how artificial greenhouse gases interact with their natural counterparts. But common sense tells us that billions of tons of compounds added to the air must have *some* effect.

At its most frivolous, the greenhouse debate is dragged down by a sort of "science lite." The George C. Marhsall Institute in Washington publishes anti-greenhouse-effect studies designed to resemble mainstream scientific research but closer in content to editorial writing. One prominent greenhouse Calamity Jane is Michael Oppenheimer of the Environmental Defense Fund (EDF), identified in *New York Times* quotes as a "scientist" despite the fact that he has no academic affiliation and is a full-time employee of an advocacy organization with a fund-raising interest in greenhouse alarm. A decade ago, Oppenheimer wrote that, by now, the greenhouse effect could cause a catastrophic global drought that would destroy the U.S. economy and create dust storms so severe that highways would become impassable. Oppenheimer holds the EDF's Barbra Streisand Global Atmospheric Change Chair, which somehow never turns up in his *Times* identification.

There's a certain amusement value in watching the doomsayers and the naysayers slug it out, but the overwrought views that tend to dominate global warming debate divert attention from the fact that middle-ground compromises are possible. The Pew Center on Global Climate Change, a middle-ground group funded by the Philadelphia philanthropy, has signed up American Electric Power, Boeing, BP Amoco, DuPont, Lockheed Martin, Maytag, Pacific Gas and Electric, Shell, Sunoco, Toyota, Weyerhaeuser, Whirlpool, and other major corporations in a coalition that accepts the need for greenhouse action and advocates moderate policy steps; this list even includes a few oil companies. Some corporate members of the Pew Cen-

24. Warming Up: The Real Evidence for the Greenhouse Effect

ter have already committed to reducing greenhouse emissions in advance of any law to that effect. It is the views of centrists such as those at the Pew Center, not the fandangos on the extremes, that ought to be drawing the attention of the media and the policy world.

IF MODERATE POLICY steps are now justified, what should they be? Figures on the green left have called for such drastic actions as an immediate one-half reduction in global greenhouse emissions, which would be possible only by prohibiting whole categories of present activity—automobile ownership, for instance. Figures on the brown right have suggested that even judicious progress against greenhouse gases would have ruinous economic consequences. Middle-ground responses are possible, however.

The most important centrist point is that greenhouse gas reductions should be sought via energy efficiency. Energy efficiency not only pays for itself in the long run by moderating fuel demand, but research into new forms of clean energy will aid the process of preparing the United States for the post-petroleum economy that must come regardless of whether global temperatures go up or down. Between 1973 and 1986, when oil prices were a national concern, U.S. energy consumption did not increase, while the GNP grew by one-third; this indicates that it is entirely possible to use fossil fuels more efficiently without economic sacrifice. But, in recent years, U.S. energy efficiency gains have stopped. The sport utility vehicle fad is the most obvious reason—this year, overall fuel economy of new vehicles fell to its lowest level in 20 years. Lack of progress on energy efficiency not only leads to avoidable greenhouse gases, it also sustains demand for imported oil, which is not in the U.S. security interest. A renewed commitment to advanced energy technology would both restrain greenhouse gases and be bad news for OPEC, which seems a desirable combination.

The most direct way to encourage clean-energy innovation would be to tax emissions of carbon dioxide and methane, the chief greenhouse gases. Classical economics generally prefers pollution taxes to complicated regulatory schemes in which government imposes specific technical standards—if you're just being taxed, it's up to you to set your own priorities, deciding for yourself, say, whether to switch from an SUV to a family sedan that gets more miles per gallon. Economists, including Martin Feldstein, chairman of the Council of Economic Advisors during the Reagan administration, have said they support greenhouse reduction via revenue-neutral fossil-fuel taxes, meaning carbon charges coupled with dollar-for-dollar reductions in other taxes. In 1997, a group that included six Nobel Prize-winning economists signed a statement saying that moderate reductions in greenhouse gases could be achieved in this fashion without causing economic harm.

AN EXAMPLE OF a successful market-based approach to pollution control comes from the acid rain trading program, enacted in 1990 under President Bush. The program instructed power utilities to reduce acid rain but allowed them to trade emission rights. When the law was passed, rights for a ton of acid rain emissions were expected to sell for about $1,500 each. Instead, today the permits sell for about $90 each, saving billions of dollars annually in control costs, while national acid rain levels are declining more rapidly than expected. When the program was established, the high estimated price of progress was based on control technology available at the time. As soon as they were given a financial incentive to devise anti-pollution ideas, utilities proved highly adept at discovering new, affordable ways to cut emissions. Granting companies the right to trade emissions permits made acid rain reduction a business opportunity rather than a burden. This program creates reason to hope that moderate greenhouse gas reductions could be achieved by a similarly successful initiative—ideally, by a revenue-neutral carbon tax with a trading program.

Resources for the Future (RFF), the centrist Washington think tank that designed and championed the acid rain emissions trading program, has now thrown its weight behind a similar approach to global warming reform. In the RFF proposal, rights to carbon emissions would be traded at $25 a ton, which initially would translate, for consumers, into about six cents per gallon of gasoline. Carbon trading would stay at this level until Congress was satisfied that the system was working, at which point the carbon charge would begin to rise annually. Polls show that a majority of Americans say they would pay somewhat more for gasoline in order to combat greenhouse gases; if other taxes were reduced in the process, the program might be politically realistic. What RFF proposes is what some economists call "shaping" a market—create a situation in which a needed innovation will be rewarded and then see what the market comes up with. "The market is not inspiring greenhouse gas control because there is presently no way anyone can buy the benefits of preventing climate change," notes Kenneth Arrow, a Nobel-winning economist. Moderately priced carbon trading could change that equation.

It might also trigger new energy technology that would create the next great economic boom category. John Browne, CEO of BP Amoco, has called it realistic to think that, by the year 2050, half the world's energy could be supplied by non-polluting, renewable sources. Calculations show that much of the country's energy needs could be supplied by solar-electric conversion, using technology that is now working at the laboratory level and is not terribly far from applied use. Full-sized cars powered by zero-emissions "fuel cells" could realize the equivalent of 100 miles per gallon and are not far off: Chrysler, Ford, Toyota, and Mercedes-Benz are all working on practical fuel-cell cars. Robert Socolow, a professor of engineering at Princeton University, notes that it is even becoming technologically realistic to "decarbonize" fossil fuels, using them with little or no greenhouse emissions.

5 ❖ RESOURCES: Air

CONVERTING TO CLEANER and more efficient energy will entail huge capital costs and take many years. Because of this, "you don't lose much by moving slowly, compared to what you can gain if technology and programs are used wisely," notes Robert Stavins, a resource economist at Harvard University. Global warming is a long-term problem: if it is real, it will develop across decades, while any fix will require decades to reach its full effect. Rushing headlong into sweeping, binding programs, on the other hand, might disrupt prosperity.

The University of Illinois's Schlesinger, who conducts climate research and also writes policy analysis in conjunction with the RAND Corporation, has for years been advocating an "adaptive" greenhouse response—enact mild restrictions, watch to determine whether they inspire innovation, track the science to see if it gets stronger, and then adjust future policies based on whatever is learned. This thinking is right for global warming because, Schlesinger says, "we don't know what energy innovations are practical, and we don't know how much harm global warming will do. It's a policy area where we have no idea what either the costs or the benefits will be." Adaptive response is thus a political median: it assumes that some degree of future warming will be inevitable but also that, over the long run, technology and policy can bring the problem under control and eventually reverse any harm. Schlesinger proposed this gradualist approach in an influential paper in the late '80s. Then-Senator Al Gore denounced the idea as "irresponsible" for not embracing immediate, sweeping rules. Today, adaptive response is the Clinton administration's global warming policy in the form of the proposed Kyoto Protocol, which is currently where the global warming rubber meets the road.

At Kyoto, Japan, in 1997, nations agreed in principle to a treaty whose essence defies easy phrasing: that, by the year 2012, the industrialized world would reduce greenhouse gas emissions to 5.2 percent below the level of 1990. As its share, the United States agreed in principle to cut greenhouse emissions to seven percent below the 1990 level. In 2012, the Kyoto Protocol says, the situation will be reassessed based on new knowledge. This is "adaptive response" thinking all the way, and Gore now firmly favors it.

THE KYOTO PROTOCOL allows industrial nations to obtain part of their cutbacks by reducing emissions in the developing world and then buying the credits: that is, through international carbon trading. Reductions will be much cheaper in developing nations. Chinese coal-fired power plants, for example, run at less than ten percent thermal efficiency, versus almost 40 percent for plants in the United States; retrofitting such plants with Western technology would cut global warming emissions for a lower price per avoided ton than anything an American generating station might do. International trading is also important because, although the industrialized world has so far put the bulk of artificial warming gases into the air, the developing world recently became the leading source of greenhouse emissions. China already burns more coal than the United States does, and, at current rates, within two decades it will surpass America to become the number one source of greenhouse gases.

Many Republicans have denounced Kyoto as giving away economic sovereignty. They rightly point out that the treaty contains a fatal-error clause: If the protocol ever goes into force, it could be amended by a three-quarters vote of the parties, with the amendments binding. This means the developing nations, which outnumber the West by roughly three-quarters, could vote to order the United States to stop using petroleum. Obviously any future attempt by Kyoto parties to dictate the economic policy of the United States would be ignored, but then why sign up for a treaty whose provisions you are planning to disregard? If this clause can't be negotiated out of the protocol, it's back to the drawing board.

Before the protocol was initialed in Japan, the Senate voted 95–0 to reject Kyoto unless it restricts greenhouse gas emissions from the developing world, which it does not; no greenhouse gas limits are imposed on developing nations. President Clinton has not submitted the treaty to the Senate for ratification and says he will not until it imposes "meaningful" restrictions on developing nations; currently, this places the protocol in limbo. If the 2000 presidential race pits Gore against Bush, the treaty may be a prominent issue, since Bush has said he believes global warming exists but that Kyoto would cost American jobs.

Initially cool to the Kyoto concept, developing nations are becoming enthusiastic as they realize potential commercial gains. Sunita Narain, an official of India's Centre for Science and Environment, recently wrote that the treaty "is increasingly being understood not as an environmental agreement but a trading agreement." A theory has arisen among conservatives that Kyoto is a backdoor attempt to start a new foreign-assistance initiative, channeling capital and technology to developing nations via free-market carbon trades. The protocol would indeed improve the fortunes of the developing world, and what's wrong with that? A decline of pollution from the developing world would have both humanitarian value and global benefits.

For its part, when Kyoto was happening, the European Union made numerous greener-than-thou pronouncements about how the United States was the world's font of pollution and enlightened Europe would act decisively to save the world; for this Europe was fondly praised in the U.S. press. Yet no EU member state has ratified Kyoto. The slaps at the United States haven't stopped, though. Last spring, the EU condemned international carbon trading. Europe's economic ministers have calculated that, if there is no trading, this will harm the energy-intensive U.S. economy more than it would European economies—and forget the ostensible goal of global greenhouse gas reduction.

24. Warming Up: The Real Evidence for the Greenhouse Effect

BEYOND THE POLITICAL jockeying are what seem to be two intense objections to Kyoto. The first is that, even if all Western nations smile and ratify, as the West decreases greenhouse emissions, developing-world emissions will assiduously increase. Unrestricted under Kyoto, China, through the year 2012, is expected to add five tons of global warming gases to the air for every one the United States takes out. The second, seemingly killer, objection is that, even if the Kyoto rules realize complete success, total concentration of atmospheric greenhouse gases will diminish by only one percent relative to the anticipated 2012 figure. Ninety-nine percent of the problem will still hang in the air.

Despite these concerns, some version of Kyoto is still desirable, assuming the treaty's faults can be repaired through negotiation. If an international agreement serves only to diminish greenhouse buildup by one percent, that is preferable to allowing global warming gases to continue to accumulate without limitation.

Early technical breakthroughs are usually the hard ones. The nation may find that breakthroughs obtained in trying to meet early, modest greenhouse goals make possible technology that not only serves far-reaching goals but helps move society toward the clean-energy economy that must be achieved regardless of what the global thermostat does. We may also find that involving the West in cleanup of the developing world promotes the human enterprise by reducing global pollution and improving living conditions for several billion people. Basing this involvement on business-to-business trading rather than on government-to-government aid should help accelerate the adoption of market economies in the nations that need them most. Toss in the fact that whoever masters clean energy may have a significant competitive advantage in twenty-first century economics, and moderate greenhouse reform not only seems justified—it looks like an exciting opportunity for a nation whose world leadership is based on being on the cutting edge of technology and market economics.

Unit 6

Unit Selections

25. **Making Things Last: Reinventing Our Material Culture,** Gary Gardner and Payal Sampat
26. **Groundwater Shock: The Polluting of the World's Major Freshwater Stores,** Payal Sampat
27. **POPs Culture,** Anne Platt McGinn
28. **It's a Breath of Fresh Air,** David Whitman

Key Points to Consider

❖ How does an industrial system oriented to high rates of materials consumption contribute more to pollution than one based on sustainable practices? Are there ways to make the transition from material to sustainable industrial societies?

❖ Why is groundwater pollution so difficult to trace and to monitor? What mechanisms might be employed to reduce the contributions of agriculture and industry to the contamination of the world's important freshwater supply?

❖ How can persistent organic pollutants contribute to water pollution and what role does the food chain play in concentrating pollutants? What is the long-term prognosis for POPs in the organic environment?

❖ What are some of the most significant improvements in environmental quality made during the last 30 years in the United States? Do you think the U.S. environment is better or worse than it was 30 years ago?

 **Links** www.dushkin.com/online/

30. **IISDnet**
    http://iisd1.iisd.ca
31. **Persistant Organic Pollutants**
    http://irptc.unep.ch/pops/
32. **School of Labor and Industrial Relations: Hot Links**
    http://www.lir.msu.edu/hotlinks/
33. **Space Research Institute**
    http://arc.iki.rssi.ru/Welcome.html
34. **Worldwatch Institute**
    http://www.worldwatch.org

These sites are annotated on pages 4 and 5.

Pollution: The Hazards of Growth

Of all the massive technological changes that have combined to create our modern industrial society, perhaps none has been as significant for the environment as the chemical revolution. The largest single threat to environmental stability is the proliferation of chemical compounds for a nearly infinite variety of purposes, including the universal use of organic chemicals (fossil fuels) as the prime source of the world's energy systems. The problem is not just that thousands of new chemical compounds are being discovered or created each year, but that their long-term environmental effects are often not known until an environmental disaster involving humans or other living organisms occurs. The problem is exacerbated by the time lag that exists between the recognition of potentially harmful chemical contamination and the cleanup activities that are ultimately required.

A critical part of the process of dealing with chemical pollutants is the identification of toxic and hazardous materials, a problem that is intensified by the myriad ways in which a vast number of such materials, natural and man-made, can enter environmental systems. Governmental legislation and controls are important in correcting the damages produced by toxic and hazardous materials such as DDT or PCBs or CFCs, in limiting fossil fuel burning, or in preventing the spread of living organic hazards such as pests and disease-causing agents. Unfortunately, as evidenced by most of the articles in this unit, we are losing the battle against harmful substances regardless of legislation, and chemical pollution of the environment is probably getting worse rather than better.

The first article in this unit deals with the ultimate causes of the chemical pollution problem: the high rate of consumption in industrial countries. In "Making Things Last: Reinventing Our Material Culture," Gary Gardner and Payal Sampat of the Worldwatch Institute suggest that an overdependence upon material consumption has led to many pollution problems arising from the disposal of solid waste. The solution to these problems and other pollution problems will be found only in the transition to a rational, sustainable materials society.

Emphasis on organic or biological pollution is offered in the second article in this last unit, which focuses on what may be humanity's most important environmental problem: the quality of the global supply of freshwater. In "Groundwater Shock: The Polluting of the World's Major Freshwater Stores," Payal Sampat notes that the vast majority (97 percent) of the world's freshwater supply lies not in the visible surface systems of lakes and streams but in underground aquifers. This precious reserve, used for virtually every purpose from drinking to irrigating crops, is becoming polluted by surface processes related to agricultural, commercial, industrial, domestic, transportation, and other human activities.

Payat notes that while much of the world worries about what is happening in the atmosphere (global warming), what happens below our feet may ultimately be of as much concern.

The section's third selection continues with the theme of chemical pollution and the interrelationship between chemical additives and biological systems. Researcher Anne Platt McGinn investigates the industrial innovation that is producing "POPs" or persistent organic pollutants that are so durable within environmental systems and so toxic that they may be still causing biological damage centuries after their entry into the environment. In "POPs Culture," McGinn describes the release, dispersal, accumulation, and consumption of these pollutants—usually within the world's water systems, both oceanic and terrestrial. The solution is nothing short of a re-engineering of the way we make and use synthetic chemicals and an awareness that chemical manufacturing is as much an ecological process as it is an industrial or economic one.

Finally, the concluding article in the section, "It's a Breath of Fresh Air," offers a breath of optimism. Science reporter David Whitman catalogues the progress that has been made in environmental quality in the United States since President Richard Nixon warned in 1970 that by the year 2000, the United States would be "a country in which we can't drink the water, where we can't breathe the air." Whitman notes that while global issues are far from resolved, the progress made in the United States provides reason for hope. It is worth recalling, says Whitman, "that doom-and-gloom environmental predictions have proved wrong more often than they have proved right."

The pollution problem might appear nearly impossible to solve. Yet as the last article notes, solutions exist: massive cleanup campaigns to remove existing harmful chemicals from the environment and to severely restrict their future use; strict regulation of the production, distribution, use, and disposal of potentially hazardous chemicals; the development of sound biological techniques to replace existing uses of chemicals for such purposes as pest control; the adoption of energy and material resource conservation policies; and more conservative and protective agricultural and construction practices. We now possess the knowledge and the tools to ensure that environmental cleanup is carried through. (It will not be an easy task, and it will be terribly expensive. It will also demand a new way of thinking about humankind's role in the environmental systems upon which all life forms depend.) If we do not complete the task, the support capacity of the environment may be damaged or diminished beyond our capacities to repair it. The consequences would be fatal for all who inhabit this planet.

Making Things Last: Reinventing Our Material Culture.

High rates of consumption threaten the environment. Here are ways we can live well without trashing the world.

By Gary Gardner and Payal Sampat

A bulldozer at work in a giant landfill. Twentieth-century manufacturing has wasted unprecedented amounts of raw materials, but new policies can help create sustainable business practices, according to researchers at the Worldwatch Institute.

An extraterrestrial observer of the earth might conclude that the conversion of raw materials to waste is a major purpose of human economic activity.

In fact, the scale of materials used by Americans, Europeans, Japanese, and other industrial-country citizens in the twentieth century dwarfs that of any previous era. Consumption of metal, glass, wood, cement, and chemicals in industrial countries since 1900 is unprecedented, having grown 18-fold in the United States alone, according to the U.S. Geological Survey.

Modern manufacturing has transformed a global river of materials into a stunning array of new products, from skyscrapers and spacecraft to plastic bags, compact discs, contact lenses, and ball-point pens. The unparalleled waste that characterizes this materially unique century has also wrought extraordinary damage on human and environmental health.

This abuse of the environment is the product of a "frontier" mindset that views materials, and the earth's capacity to absorb wastes, as practically limitless. The frontier perspective may have seemed appropriate in the nineteenth century, when global population had not yet reached 2 billion, but it has led to an increasingly disruptive industrial system that equates progress with materials consumption. A different mindset will be needed in order to prevent industrial economies from further damaging the natural landscape.

Promoting Service Providers

Perhaps the most revolutionary shift toward sustainable materials use is the conversion of manufacturing firms to service-providing firms. Service providers earn their profits not by selling goods, such as wash-

25. Making Things Last: Reinventing Our Material Culture

ing machines or cars, but by providing the services that goods currently deliver—convenient cleaning of clothes, for example, or transportation. Providers could also be responsible for all of the materials and products used to provide their service, maintaining those goods and retrieving them when they wear out. Service firms would thus have a strong incentive to make products that last and can be easily repaired, upgraded, reused, or recycled.

Many service-provider firms would lease their products rather than sell them. The Xerox Corporation, for example, now leases most of its office copy machines as part of a redefined mission to provide document services, rather than to sell photocopiers. The new arrangement gives Xerox a strong incentive to maximize the life of its machines: Between 1992 and 1997, the company doubled its share of remanufactured copiers to 28%, keeping 30,000 tons of waste out of landfills in 1997 alone. Each remanufactured machine meets the same standards, and carries the same warranty, as a newly minted one. In addition, Xerox introduced a product-return program for spent copy and printer cartridges in 1991, and it now recaptures 65% of used cartridges.

Consumers could help save on materials by eliminating goods that spend most of their time idle. For example, using laundry services rather than home washing machines could dramatically cut materials use per wash, because semi-commercial machines are used more intensively than home washers. Home washers are also 10 to 80 times more materials intensive—depending on how they are disposed of—than the machines used in a laundromat. If dismantled and recycled, a home washer uses 10 times as much material per wash as a semi-commercial machine that is disposed of in the same way.

Washing may be a function that consumers would prefer to retain in their homes, but even home washing could be accommodated by a service firm that leases the machines. This option would save less material than the use of a laundromat, but much more than if machines are bought by individuals. In sum, whether service is provided directly (by hiring someone to mow your lawn, for example) or indirectly (by leasing a lawn-mower), replacing infrequently used goods with services can save tons of material.

In some cases, service providers can replace materials with intelligence or labor. As the computer revolution continues to unfold, digital technology—basically embodied intelligence—can be used to breathe new life into products that rapidly become obsolete, such as cameras and televisions. If product capabilities are upgraded through the replacement of a computer chip, then perfectly good casings, lenses, and picture tubes can avoid a premature trip to the landfill. Similarly, labor can be used to extend the useful life of products: Service providers need workers to disassemble, repair, and rebuild their leaseable goods, saving materials and increasing employment at the same time.

Copy machines could be part of a revolutionary shift toward sustainable use in which businesses provide services rather than selling machines. The Xerox Corporation already leases most of its office copiers and maintains old machines longer, keeping them out of landfills.

A Boost in Recycling

The gains from a revolutionized service economy can be augmented by an equally ambitious overhaul of recycling practices. Some products are already designed with recycling in mind. German automakers now bar-code car components to show scrap dealers the mix of materials contained in each piece. And some producers of cars, television sets, and washing machines build their products for easy disassembly at the end of the product's life.

Easy disassembly can bring substantial gains. Xerox's ambitious plan to boost its share of remanufactured machines from the current 28% to 85%, for example, is feasible because of the company's 1997 shift to redesigned, easily disassembled copy machines. Widespread adoption of these "design-for-environment" initiatives could boost recycling rates throughout the economy. And there is much room for improvement: Today, just 17% of durable goods are recycled in the United States.

With the right incentives, much greater materials reductions from recycling are possible. For example, Germany implemented a revolutionary package-waste ordinance in 1993 that holds producers accountable for nearly all of the packaging material they generate. The new law increased the amount of packaging recycled from 12% in 1992 to 86% in 1997. The law also gave producers a strong incentive to cut their use of packaging, which dropped 17% for households and small businesses between 1991 and 1997. The use of secondary packaging—outer containers like the box around a tube of toothpaste—has also declined in Germany. Now several other countries, including Austria, France, and Belgium, have adopted similar recycling legislation.

Making the Most of Materials

As with service firms and recycling, materials efficiency can be imaginatively rethought and powerfully upgraded. If the efficiency of a product were measured not just at the factory gate—in terms of the materials required to produce it—but across its entire life, characteristics such as durability and capacity for reuse would suddenly become more important.

Reducing logging and mining would save gargantuan amounts of energy.

For example, doubling the useful life of a car may involve no improvement in materials efficiency at the factory, but it cuts in half both the resources used and the waste generated per trip over the car's life—a clear increase in total resource efficiency. Recognizing these benefits, many companies are emphasizing the durability of the products they use. Toyota, for example, shifted to entirely reusable shipping containers in 1991, each with a potential lifetime of 20 years. Advances like these, expanded to the entire economy, would sharply reduce container and packaging waste—which account for 30% of inflows to U.S. landfills.

Product life is also extended through the remanufacture, repair, and reuse of spent goods. The environmental impact of beverage consumption in Denmark has fallen considerably since the country switched from aluminum cans to glass containers that can be reused 50 to 100 times. Widespread adoption of these measures would in some ways be a step back to the future. Most grandparents in industrial countries can remember an economy in which milk bottles and other beverage containers were washed and reused, shoes were resoled, clothes were mended, and machines were rebuilt. Some may remember that all but two of the U.S. ships sunk at Pearl Harbor were recovered, overhauled, and recommissioned, in part because of the savings in time and material that this option offered. That such practices seem strange to new generations of consumers is a reflection of how far industrial economies have drifted from the careful use of materials resources.

A logger takes down another tree. The government should end subsidies for the logging and mining industries that make virgin materials seem cheap, the authors say. This policy would encourage greater use of recycled materials.

25. Making Things Last: Reinventing Our Material Culture

People recycle more when disposal is taxed. Waste taxes based on the amount of garbage generated are most effective when combined with curbside recycling, according to the authors.

Materials Efficiency: A Balance Sheet

Product	Efficiency Gains	Factors that Undercut Efficiency Gains
Plastics in Cars	Use of plastics in U.S. cars increased by 26% between 1980 and 1994, replacing steel in many uses and reducing car weight by 6%.	Cars contain 25 chemically incompatible plastics that, unlike steel, cannot be easily recycled. Most plastic in cars winds up in landfills.
Bottles and Cans	Aluminum cans weigh 30% less today than they did 20 years ago.	Cans replaced an environmentally superior product—refillable bottles; 95% of soda containers were refillable in the United States in 1960.
Lead Batteries	A typical automobile battery used 30 pounds of lead in 1974, but only 20 pounds in 1994—with improved performance.	U.S. domestic battery shipments increased by 76% in the same period, more than offsetting the efficiency gains.
Radial Tires	Radial tires are 25% lighter and last twice as long as bias-ply tires.	Radial tires are more difficult to retread. Sales of passenger car retreads fell by 52% in the United States between 1977 and 1997.
Mobile Phones	Weight of mobile phones was reduced 10-fold between 1991 and 1996.	Subscribers to cellular telephone service jumped more than eightfold in the same period, nearly offsetting the gains from making phones lighter. Moreover, the mobile phones did not typically replace older telephones, but were additions to a household's phone inventory.

Source: *Mind Over Matter*

Extending product life offers an array of advantages over the habitual use of virgin materials and the manufacturing of new products. For starters, fewer wastes are generated when products spend more time circulating through an economy. But less apparent gains are at least as important. Reducing logging and mining would save gargantuan amounts of energy: Materials extraction and processing account for an estimated 75% of the energy used by industry in some industrial countries.

Shifting Gears

Overhauling materials practices will require policies that steer economies away from forests, mines, and petroleum stocks as the primary source of materials and away from landfills and incinerators as cheap disposal options. Businesses and consumers need to be encouraged to use far less virgin material and to tap the rich flow of currently wasted resources through product reuse, remanufacturing, or sharing, or through materials recycling.

A key step in this direction is to abandon the government subsidies that make virgin materials seem cheap. Whether in the form of direct payments or as resource giveaways, assistance to mining and logging firms makes virgin materials artificially attractive to manufacturers. The 1872 Mining Law in the United States, for example, continues to give mining firms access to public lands for just $12 per hectare, without requiring payment of royalties or even the cleanup of mining sites. The effect of this virtual giveaway is to encourage virgin materials use at the expense of alternatives such as recycling.

By closing the subsidy spigot for extractive industries, policy makers can earn double dividends. The environmental gains would be substantial because most materials-driven environmental damage occurs at the extractive stage. And the public treasury would be fattened through the elimination of tax breaks or other treasury-draining subsidies, and possibly through payments from the mining and logging operations that remain open. What's more, these benefits would be achieved at little social cost: Mining and logging, for example, provide relatively few jobs. In the United States, metals mining employed 52,000 workers in 1996, just 0.04% of the total U.S. work force that year.

POLLUTION: THE HAZARDS OF GROWTH

Waste generation can also be substantially curtailed, even to the point of near-zero waste in some industries and cities. A handful of firms report achieving near-zero waste levels at some facilities. The city of Canberra, Australia, is pursuing a "No-Waste-by-2010" strategy, while the Netherlands has set a national waste-reduction goal of 70% to 90%. A key way to meet such ambitious targets is to tax waste in all its forms, from smokestack emissions to landfilled solids. Pollution taxes in the Netherlands, for example, were primarily responsible for a 72% to 99% reduction in heavy metals discharges into waterways between 1976 and the mid-1990s. High landfill taxes in Denmark have boosted construction debris reuse from 12% to 82% in eight years—far above the 4% rates of most industrial countries. Such a tax could bring huge materials savings in the United States, where construction-materials use between 2000 and 2020 is projected to exceed total use in the twentieth century.

At the consumer level, a waste tax can take the form of fees that are based on the amount of garbage generated. Cities that have shifted to such a system have seen a substantial reduction in waste generation. "Pay-as-you-throw" programs, in which people are charged by the bag or by volume of trash, illustrate the direct effect of taxes on waste. For example, Dover, New Hampshire, and Crockett, Texas, reduced household waste by about 25% in five years once such programs were introduced. These initiatives are most effective when coupled with curbside recycling programs: As disposal is taxed, people recycle more. Eleven of 17 U.S. communities with record-setting recycling rates use pay-as-you-throw systems.

A modified version of a waste tax is the refundable deposit—essentially a temporary tax that is returned to the payer when the taxed material is brought back. High deposits for refillable glass bottles in Denmark have yielded 98% to 99% return rates.

Recognizing the problems caused by our dependence on materials is a first step in making the leap to a rational, sustainable materials economy. Once we grasp this idea, the opportunities to dematerialize our economies are well within reach. Societies that shed their attachment to things and focus instead on delivering what people actually need might be remembered 100 years from now as the creators of the most durable civilization in history.

About the Authors

Gary Gardner is a senior researcher at the Worldwatch Institute, 1776 Massachusetts Avenue, N.W., Washington, D.C. 20036. Telephone 1-202-452-1999; Web site www.worldwatch.org.

Payal Sampat is a Worldwatch staff researcher who studies human development issues and sustainable materials use.

This article is adapted from their report *Mind Over Matter: Recasting the Role of Materials in Our Lives*. Worldwatch Institute. 1998. 60 pages. Paperback. $5 plus shipping. The report may also be downloaded from the Worldwatch Institute Web site for $5.

Article 26

Groundwater *Shock*

The Polluting of the World's Major Freshwater Stores

Scientists have shown that the world deep beneath our feet is essential to the life above. Ancient myths depicted the Underworld as a place of damnation and death. Now, the spreading contamination of major aquifers threatens to turn the myth into a tragic reality.

by Payal Sampat

The Mississippi River occupies a mythic place in the American imagination, in part because it is so huge. At any given moment, on average, about 2,100 billion liters of water are flowing across the Big Muddy's broad bottom. If you were to dive about 35 feet down and lie on that bottom, you might feel a sense of awe that the whole river was on top of you. But in one very important sense, you'd be completely wrong. At any point in time, only 1 percent of the water in the Mississippi River system is in the part of the river that flows downstream to the Gulf of Mexico. The other 99 percent lies beneath the bottom, locked in massive strata of rock and sand.

This is a distinction of enormous consequence. The availability of clean water has come to be recognized as perhaps the most critical of all human security issues facing the world in the next quarter-century—and what is happening to water buried under the bottoms of rivers, or under our feet, is vastly different from what happens to the "surface" water of rivers, lakes, and streams. New research finds that contrary to popular belief, it is groundwater that is most dangerously threatened. Moreover, the Mississippi is not unique in its ratio of surface to underground water; worldwide, 97 percent of the planet's liquid freshwater is stored in aquifers.

In the early centuries of civilization, surface water was the only source we needed to know about. Human population was less than a tenth of one percent the size it is now; settlements were on river banks; and the water was relatively clean. We still think of surface water as being the main resource. So it's easy to think that the problem of contamination is mainly one of surface water: it is polluted rivers and streams that threaten health in times of flood, and that have made waterborne diseases a major killer of humankind. But in the past century, as population has almost quadrupled and rivers have become more depleted and polluted, our dependence on pumping groundwater has soared—and as it has, we've made a terrible discovery. Contrary to the popular impression that at least the waters from our springs and wells are pure, we're uncovering a pattern of pervasive pollution there too. And in these sources, unlike rivers, the pollution is generally irreversible.

This is largely the work of another hidden factor: the rate of groundwater renewal is very slow in comparison with that of surface water. It's true that some aquifers recharge fairly quickly, but the average recycling time for groundwater is 1,400 years, as opposed to only 20 days for river water. So when we pump out groundwater, we're effectively removing it from aquifers for generations to come. It may evaporate and return to the atmosphere quickly enough, but the resulting rainfall (most of which falls back into the oceans) may take centuries to recharge the aquifers once they've been depleted. And because water in aquifers moves through the Earth with glacial slowness, its pollutants continue to accumulate. Unlike rivers, which flush themselves into the oceans, aquifers become sinks for pollutants, decade after decade—thus further diminishing the amount of clean water they can yield for human use.

Perhaps the largest misconception being exploded by the spreading water crisis is the assumption that the ground we stand on—and what lies beneath it—is solid, unchanging, and inert. Just as the advent of climate change has awakened us to the fact that the air over our heads is an arena of enormous forces in the midst of titanic shifts, the water crisis has revealed that, slow-moving though it may be, groundwater is part of a system of powerful hydrological interactions—between earth, surface water, sky, and sea—that we ignore at our peril. A few years ago, reflecting on how human activity is beginning to affect climate, Columbia University scientist Wallace Broecker warned, "The climate system is an angry beast and we are poking it with sticks." A similar statement might now be made about the system under our feet. If we continue to drill holes into it—expecting it to swallow our waste and yield freshwater in return—we may be toying with an outcome no one could wish.

Valuing Groundwater

For most of human history, groundwater was tapped mainly in arid regions where surface water was in short supply. From Egypt to Iran, ancient Middle Eastern civilizations used periscope-like conduits to funnel spring water from mountain slopes to nearby towns—a technology that allowed settlement to spread out from the major rivers. Over the centuries, as populations and cropland expanded, innovative well-digging techniques evolved in China, India, and Europe. Water became such a valuable resource that some cultures developed elaborate mythologies imbuing underground water and its seekers with special powers. In medieval Europe, people called water witches or dowsers were believed to be able to detect groundwater using a forked stick and mystical insight.

In the second half of the 20th century, the soaring demand for water turned the dowsers' modern-day counterparts into a major industry. Today, major aquifers are tapped on every continent, and groundwater is the primary source of drinking water for more than 1.5 billion people worldwide (see table, Groundwater as a Share of Drinking Water Use, by Region). The aquifer that lies beneath the Huang-Huai-Hai plain in eastern China alone supplies drinking water to nearly 160 million people. Asia as a whole relies on its groundwater for nearly one-third of its drinking water supply. Some of the largest cities in the developing world—Jakarta, Dhaka, Lima, and Mexico City, among them—depend on aquifers for almost all their water. And in rural areas, where centralized water supply systems are undeveloped, groundwater is typically the sole source of water. More than 95 percent of the rural U.S. population depends on groundwater for drinking.

A principal reason for the explosive rise in groundwater use since 1950 has been a dramatic expansion in irrigated agriculture. In India, the leading country in total irrigated area and the world's third largest grain producer, the number of shallow tubewells used to draw groundwater surged from 3,000 in 1960 to 6 million in 1990. While India doubled the amount of its land irrigated by surface water between 1950 and 1985; it increased the area watered by aquifers 113-fold. Today, aquifers supply water to more than half of India's irrigated land. The United States, with the third highest irrigated area in the world, uses groundwater for 43 percent of its irrigated farmland. Worldwide, irrigation is by far the biggest drain on freshwater: it accounts for about 70 percent of the water we draw from rivers and wells each year.

Other industries have been expanding their water use even faster than agriculture—and generating much higher profits in the process. On average, a ton of water used in industry generates roughly $14,000 worth of output—about 70 times as much profit as the same amount of water used to grow grain. Thus, as the world has industrialized, substantial amounts of water have been shifted from farms to more lucrative factories. Industry's share of total consumption has reached 19 percent and is likely to continue rising rapidly. The amount of water available for drinking is thus constrained not only by a limited resource base, but by competition with other, more powerful users.

And as rivers and lakes are stretched to their limits—many of them dammed, dried up, or polluted—we're growing more and more dependent on groundwater for all these uses. In Taiwan, for example, the share of water supplied by groundwater almost doubled from 21 percent in 1983 to over 40 percent in 1991. And Bangladesh, which was once almost entirely river- and stream-dependent, dug over a million wells in the 1970s to substitute for its badly polluted surface-water supply. Today, almost 90 percent of its people use only groundwater for drinking.

Even as our dependence on groundwater increases, the availability of the resource is becoming more limited. On almost every continent, many major aquifers are being drained faster than their natural rate of recharge. Groundwater depletion is most severe in parts of India, China, the United States, North Africa, and the Middle East. Under certain geological conditions, groundwater overdraft can cause aquifer sediments to compact, permanently shrinking the aquifer's storage capacity. This loss can be quite considerable, and irreversible. The amount of water storage capacity lost because of aquifer compaction in California's Central Valley, for example, is equal to more than 40 percent of the combined storage capacity of all human-made reservoirs across the state.

As the competition among factories, farms, and households intensifies, it's easy to overlook the extent to which freshwater is also required for essential ecological services. It is not just rainfall, but groundwater welling up from

26. Groundwater Shock

Groundwater as a Share of Drinking Water Use by Region

Region	Share of Drinking Water from Groundwater (percent)	People Served (millions)
Asia-Pacific	32	1,000 to 1,200
Europe	75	200 to 500
Latin America	29	150
United States	51	135
Australia	15	3
Africa	NA	NA
World		1,500 to 2,000

Sources: UNEP, OECD, FAO, U.S. EPA, Australian EPA.

beneath, that replenishes rivers, lakes, and streams. In a study of 54 streams in different parts of the country, the U.S. Geological Survey (USGS) found that groundwater is the source for more than half the flow, on average. The 492 billion gallons (1.86 cubic kilometers) of water aquifers add to U.S. surface water bodies each day is nearly equal to the daily flow of the Mississippi. Groundwater provides the base contribution for the Mississippi, the Niger, the Yangtze, and many more of the world's great rivers—some of which would otherwise not be flowing year-round. Wetlands, important habitat for birds, fish, and other wildlife, are often largely groundwater-fed, created in places where the water table overflows to the surface on a constant basis. And while providing surface bodies with enough water to keep them stable, aquifers also help prevent them from flooding: when it rains heavily, aquifers beneath rivers soak up the excess water, preventing the surface flow from rising too rapidly and overflowing onto neighboring fields and towns. In tropical Asia, where the hot season can last as long as 9 months, and where monsoon rains can be very intense, this dual hydrological service is of critical value.

Numerous studies have tracked the extent to which our increasing demand on water has made it a resource critical to a degree that even gold and oil have never been. It's the most valuable thing on Earth. Yet, ironically, it's the thing most consistently overlooked, and most widely used as a final resting place for our waste. And, of course, as contamination spreads, the supplies of usable water get tighter still.

Tracking the Hidden Crisis

In 1940, during the Second World War, the U.S. Department of the Army acquired 70 square kilometers of land around Weldon Spring and its neighboring towns near St. Louis, Missouri. Where farmhouses and barns had been, the Army established the world's largest TNT-producing facility. In this sprawling warren of plants, toluene (a component of gasoline) was treated with nitric acid to produce more than a million tons of the explosive compound each day when production was at its peak.

Part of the manufacturing process involved purifying the TNT—washing off unwanted "nitroaromatic" compounds left behind by the chemical reaction between the toluene and nitric acid. Over the years, millions of gallons of this red-colored muck were generated. Some of it was treated at wastewater plants, but much of it ran off from the leaky treatment facilities into ditches and ravines, and soaked into the ground. In 1945, when the Army left the site, soldiers burned down the contaminated buildings but left the red-tinged soil and the rest of the site as they were. For decades, the site remained abandoned and unused.

Then, in 1980, the U.S. Environmental Protection Agency (EPA) launched its "Superfund" program, which required the cleaning up of several sites in the country that were contaminated with hazardous waste. Weldon Spring made it to the list of sites that were the highest priority for cleanup. The Army Corps of Engineers was assigned the task, but what the Corps workers found baffled them. They expected the soil and vegetation around the site to be contaminated with the nitroaromatic wastes that had been discarded there. When they tested the groundwater, however, they found that the chemicals were showing up in people's wells, in towns several miles from the site—a possibility that no one had anticipated, because the original pollution had been completely localized. Geologists determined that there was an enormous plume of contamination in the water below the TNT factory—a plume that over the previous 35 years had flowed through fissures in the limestone rock to other parts of the aquifer.

The Weldon Spring story may sound like an exceptional case of clumsy planning combined with a particularly vulnerable geological structure. But in fact there is nothing exceptional about it all. Across the United States, as well as in parts of Europe, Asia, and Latin America, human activities are sending massive quantities of chemicals and pollutants into groundwater. This isn't entirely new, of course; the subterranean world has always been a receptacle for whatever we need to dispose of—whether our sewage, our garbage, or our dead. But the enormous volumes of waste we now send underground, and the deadly mixes of chemicals involved, have created problems never before imagined.

What Weldon Spring shows is that we can't always anticipate where the pollution is going to turn up in our water, or how long it will be from the time it was deposited until it reappears. Because groundwater typically moves very slowly—at a speed of less than a foot a day, in some cases—damage done to aquifers may not show up for decades. In many parts of the world, we are only just beginning to discover contamination caused by practices of 30 or 40 years ago. Some of the most egregious cases of aquifer contamination now being unearthed date back to Cold War era nuclear testing and weapons-making, for example. And once it gets into groundwater, the

6 ❖ POLLUTION: THE HAZARDS OF GROWTH

pollution usually persists: the enormous volume, inaccessibility, and slow rate at which groundwater moves make aquifers virtually impossible to purify.

As this covert crisis unfolds, we are barely beginning to understand its dimensions. Few countries track the health of their aquifers—their enormous size and remoteness make them extremely expensive to monitor. As the new century begins, even hydrogeologists and health officials have only a hazy impression of the likely extent of groundwater damage in different parts of the world. Nonetheless, given the data we now have, it is possible to sketch a rough map of the regions affected, and the principal threats they face (see map, *Ground Contamination Hotspots* and table, *Some Major Threats to Groundwater*).

The Filter that Failed: Pesticides in Your Water

Pesticides are designed to kill. The first synthetic pesticides were introduced in the 1940s, but it took several decades of increasingly heavy use before it became apparent that these chemicals were injuring non-target organisms—including humans. One reason for the delay was that some groups of pesticides, such as organochlorines, usually have little effect until they bioaccumulate. Their concentration in living tissue increases as they move up the food chain. So eventually, the top predators—birds of prey, for example—may end up carrying a disproportionately high burden of the toxin. But bioaccumulation takes time, and it may take still more time before the effects are discovered. In cases where reproductive systems are affected, the aftermath of this chemical accumulation may not show up for a generation.

Even when the health concerns of some pesticides were recognized in the 1960s, it was easily assumed that the real dangers lay in the dispersal of these chemicals among animals and plants—not deep underground. It was assumed that very little pesticide would leach below the upper layers of soil, and that if it did, it would be degraded before it could get any deeper. Soil, after all, is known to be a natural filter, which purifies water as it trickles through. It was thought that industrial or agricultural chemicals, like such natural contaminants as rock dust, or leaf mold, would be filtered out as the water percolated through the soil.

But over the past 35 years, this seemingly safe assumption has proved mistaken. Cases of extensive pesticide contamination of groundwater have come to light in farming regions of the United States, Western Europe, Latin America, and South Asia. What we now know is that pesticides not only leach into aquifers, but sometimes remain there long after the chemical is no longer used. DDT, for instance, is still found in U.S. waters even though its use was banned 30 years ago. In the San Joaquin Valley of California, the soil fumigant DBCP (dibromochloropropane), which was used intensively in fruit orchards before it was banned in 1977, still lurks in the region's water supplies. Of 4,507 wells sampled by the USGS between 1971 and 1988, nearly a third had DBCP levels that were at least 10 times higher than allowed by the current drinking water standard.

In places where organochlorines are still widely used, the risks continue to mount. After half a century of spraying in the eastern Indian states of West Bengal and Bihar, for example, the Central Pollution Control Board found DDT in groundwater at levels as high as 4,500 micrograms per liter—several thousand times higher than what is considered a safe dose.

The amount of chemical that reaches groundwater depends on the amount used above ground, the geology of the region, and the characteristics of the pesticide itself. In some parts of the midwestern United States, for example, although pesticides are used intensively, the impermeable soils of the region make it difficult for the chemicals to percolate underground. The fissured aquifers of southern Arizona, Florida, Maine, and southern California, on the other hand, are very vulnerable to pollution—and these too are places where pesticides are applied in large quantities.

Pesticides are often found in combination, because most farms use a range of toxins to destroy different kinds of insects, fungi, and plant diseases. The USGS detected two or more pesticides in groundwater at nearly a quarter of the sites sampled in its National Water Quality Assessment between 1993 and 1995. In the Central Columbia Plateau aquifer, which extends over the states of Washington and Idaho, more than two-thirds of water samples contained multiple pesticides. Scientists aren't entirely sure what happens when these chemicals and their various metabolites come together. We don't even have standards for the many hundred *individual* pesticides in use—the EPA has drinking water standards for just 33 of these compounds—to say nothing of the infinite variety of toxic blends now trickling into the groundwater.

While the most direct impacts may be on the water we drink, there is also concern about what occurs when the pesticide-laden water below farmland is pumped back up for irrigation. One apparent consequence is a reduction in crop yields.

In 1990, the now-defunct U.S. Office of Technology Assessment reported that herbicides in shallow groundwater had the effect of "pruning" crop roots, thereby retarding plant growth.

From Green Revolution to Blue Baby: the Slow Creep of Nitrogen

Since the early 1950s, farmers all over the world have stepped up their use of nitrogen fertilizers. Global fertilizer use has grown ninefold in that time. But the larger doses of nutrients often can't be fully utilized by plants. A study conducted over a 140,000 square kilometer region of Northern China, for example, found that crops used on average only 40 percent of the nitrogen that was applied.

An almost identical degree of waste was found in Sri Lanka. Much of the excess fertilizer dissolves in irrigation water, eventually trickling through the soil into underlying aquifers.

Joining the excess chemical fertilizer from farm crops is the organic waste generated by farm animals, and the sewage produced by cities. Livestock waste forms a particularly potent tributary to the stream of excess nutrients flowing into the environment, because of its enormous volume. In the United States, farm animals produce 130 times as much waste as the country's people do—with the result that millions of tons of cow and pig feces are washed into streams and rivers, and some of the nitrogen they carry ends up in groundwater. To this Augean burden can be added the innumerable leaks and overflows from urban sewage systems, the fertilizer runoff from suburban lawns, golf courses, and landscaping, and the nitrates leaking (along with other pollutants) from landfills.

There is very little historical information available about trends in the pollution of aquifers. But several studies show that nitrate concentrations have increased as fertilizer applications and population size have grown. In California's San Joaquin-Tulare Valley, for instance, nitrate levels in groundwater increased 2.5 times between the 1950s and 1980s—a period in which fertilizer inputs grew six-fold. Levels in Danish groundwater have nearly tripled since the 1940s. As with pesticides, the aftermath of this multi-sided assault of excess nutrients has only recently begun to become visible, in part because of the slow speed at which nitrate moves underground.

What happens when nitrates get into drinking water? Consumed in high concentrations—at levels above 10 milligrams (mg) per liter, but usually on the order of 100 mg/liter—they can cause infant methemoglobinemia, or so-called blue-baby syndrome. Because of their low gastric acidity, infant digestive systems convert nitrate to nitrite, which blocks the oxygen-carrying capacity of a baby's blood, causing suffocation and death. Since 1945, about 3,000 cases have been reported worldwide—nearly half of them in Hungary, where private wells have particularly high concentrations of nitrates. Ruminant livestock such as goats, sheep, and cows, are vulnerable to methemoglobinemia in much the same way infants are, because their digestive systems also quickly convert nitrate to nitrite. Nitrates are also implicated in digestive tract cancers, although the epidemiological link is still uncertain.

In cropland, nitrate pollution of groundwater can have a paradoxical effect. Too much nitrate can weaken plants' immune systems, making them more vulnerable to pests and disease. So when nitrate-laden groundwater is used to irrigate crops that are also being fertilized, the net effect may be to reduce, rather than to increase production. This kind of over-fertilizing makes wheat more susceptible to wheat rust, for example, and it makes pear trees more vulnerable to fire blight.

In assembling studies of groundwater from around the world, we have found that nitrate pollution is pervasive—but has become particularly severe in the places where human population—and the demand for high food productivity—is most concentrated. In the northern Chinese counties of Beijing, Tianjin, Hebei, and Shandong, nitrate concentrations in groundwater exceeded 50 mg/liter in more than half of the locations studied. (The World Health Organization [WHO] drinking water guideline is 10 mg/liter.) In some places, the concentration had risen as high as 300 mg/liter. Since then, these levels may have increased, as fertilizer applications have escalated since the tests were carried out in 1995 and will likely increase even more as China's population (and demand for food) swells, and as more farmland is lost to urbanization, industrial development, nutrient depletion, and erosion.

Reports from other regions show similar results. The USGS found that about 15 percent of shallow groundwater sampled below agricultural and urban areas in the United States had nitrate concentrations higher than the 10 mg/liter guideline. In Sri Lanka, 79 percent of wells sampled by the British Geological Survey had nitrate levels that exceeded this guideline. Some 56 percent of wells tested in the Yucatan peninsula in Mexico had levels above 45 mg/liter. And the European Topic Centre on Inland Waters found that in Romania and Moldova, more than 35 percent of the sites sampled had nitrate concentrations higher than 50 mg/liter.

From Tank of Gas to Drinking Glass: the Pervasiveness of Petrochemicals

Drive through any part of the United States, and you'll probably pass more gas stations than schools or churches. As you pull into a station to fill up, it may not occur to you that you're parked over one of the most pervasive threats to ground-water: an underground storage tank (UST) for petroleum. Many of these tanks were installed two or three decades ago and, having been left in place long past their expected lifetimes, have rusted through in places—allowing a steady leakage of gasoline into the ground. Because they're underground, they're expensive to dig up and repair, so the leakage in some cases continues for years.

Petroleum and its associated chemicals—benzene, toluene, and gasoline additives such as MTBE—constitute the most common category of groundwater contaminant found in aquifers in the United States. Many of these chemicals are also known or suspected to be cancer-causing. In 1998, the EPA found that over 100,000 commercially owned petroleum USTs were leaking, of which close to 18,000 are known to have contaminated groundwater. In Texas, 223 of 254 counties report leaky USTs, resulting in a silent disaster that, according to the EPA, "has affected, or has the potential to affect, virtually every major and minor aquifer in the state." Household tanks, which

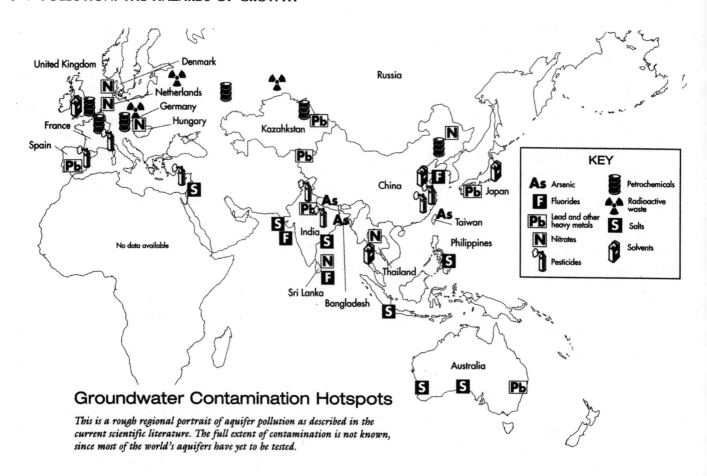

Groundwater Contamination Hotspots

This is a rough regional portrait of aquifer pollution as described in the current scientific literature. The full extent of contamination is not known, since most of the world's aquifers have yet to be tested.

store home heating oil, are a problem as well. Although the household tanks aren't subject to the same regulations and inspections as commercial ones, the EPA says they are "undoubtedly leaking." Outside the United States, the world's ubiquitous petroleum storage tanks are even less monitored, but spot tests suggest that the threat of leakage is omnipresent in the industrialized world. In 1993, petroleum giant Shell reported that a third of its 1,100 gas stations in the United Kingdom were known to have contaminated soil and groundwater. Another example comes from the eastern Kazakh town of Semipalatinsk, where 6,460 tons of kerosene have collected in an aquifer under a military airport, seriously threatening the region's water supplies.

The widespread presence of petrochemicals in groundwater constitutes a kind of global malignancy, the danger of which has grown unobtrusively because there is such a great distance between cause and effect. An underground tank, for example, may take years to rust; it probably won't begin leaking until long after the people who bought it and installed it have left their jobs. Even after it begins to leak, it may take several more years before appreciable concentrations of chemicals appear in the aquifer—and it will likely be years beyond that before any health effects show up in the local population. By then, the trail may be decades old. So it's quite possible that any cancers occurring today as a result of leaking USTs might originate from tanks that were installed half a century ago. At that time, there were gas tanks sufficient to fuel 53 million cars in the world; today there are enough to fuel almost 10 times that number.

From Sediment to Solute: the Emerging Threat of Natural Contaminants

In the early 1990s, several villagers living near India's West Bengal border with Bangladesh began to complain of skin sores that wouldn't go away. A researcher at Calcutta's Jadavpur University, Dipanker Chakraborti, recognized the lesions immediately as early symptoms of chronic arsenic poisoning. In later stages, the disease can lead to gangrene, skin cancer, damage to vital organs, and eventually, death. In the months that followed, Chakraborti began to get letters from doctors and hospitals in Bangladesh, who were seeing streams of patients with similar symptoms. By 1995, it was clear that the country faced a crisis of untold proportions, and that the source of the poisoning was water from tubewells, from which 90 percent of the country gets its drinking water.

Experts estimate that today, arsenic in drinking water could threaten the health of 20 to 60 million Bangladeshis—up to half the country's population—and another 6 to 30 million people in West Bengal. As many as 1 million wells in the region may be contaminated with the heavy metal

26. Groundwater Shock

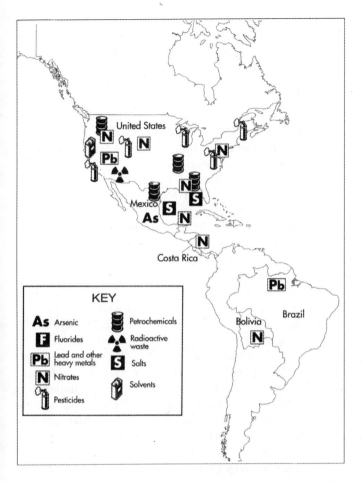

at levels between 5 and 100 times the WHO drinking water guidelines of 0.01 mg/liter.

How did the arsenic get into groundwater? Until the early 1970s, rivers and ponds supplied most of Bangladesh's drinking water. Concerned about the risks of waterborne disease, the WHO and international aid agencies launched a well-drilling program to tap groundwater instead. However, the agencies, not aware that soils of the Ganges aquifers are naturally rich in arsenic, didn't test the sediment before drilling tubewells. Because the effects of chronic arsenic poisoning can take up to 15 years to appear, the epidemic was not addressed until it was well under way.

Scientists are still debating what chemical reactions released the arsenic from the mineral matrix in which it is naturally bound up. Some theories implicate human activities. One hypothesis is that as water was pumped out of the wells, atmospheric oxygen entered the aquifer, oxidizing the iron pyrite sediments, and causing the arsenic to dissolve. An October 1999 article in the scientific journal *Nature* by geologists from the Indian Institute of Technology suggests that phosphates from fertilizer runoff and decaying organic matter may have played a role. The nutrient might have spurred the growth of soil microorganisms, which helped to loosen arsenic from sediments.

Salt is another naturally occurring groundwater pollutant that is introduced by human activity. Normally, water in coastal aquifers empties into the sea. But when too much water is pumped out of these aquifers, the process is reversed: seawater moves inland and enters the aquifer. Because of its high salt content, just 2 percent of seawater mixed with freshwater makes the water unusable for drinking or irrigation. And once salinized, a freshwater aquifer can remain contaminated for a very long time. Brackish aquifers often have to be abandoned because treatment can be very expensive.

In Manila, where water levels have fallen 50 to 80 meters because of overdraft, seawater has flowed as far as 5 kilometers into the Guadalupe aquifer that lies below the city. Saltwater has traveled several kilometers inland into aquifers beneath Jakarta and Madras, and in parts of the U.S. state of Florida. Saltwater intrusion is also a serious problem on islands such as the Maldives and Cyprus, which are very dependent on aquifers for water supply.

Fluoride is another natural contaminant that threatens millions in parts of Asia. Aquifers in the drier regions of western India, northern China, and parts of Thailand and Sri Lanka are naturally rich in fluoride deposits. Fluoride is an essential nutrient for bone and dental health, but when consumed in high concentrations, it can lead to crippling damage to the neck and back, and to a range of dental problems. The WHO estimates that 70 million people in northern China, and 30 million in northwestern India are drinking water with high fluoride levels.

A Chemical Soup

With just over a million residents, Ludhiana is the largest city in Punjab, India's breadbasket state. It is also an important industrial town, known for its textile factories, electroplating industries, and metal foundries. Although the city is entirely dependent on groundwater, its wells are now so polluted with industrial and urban wastes that the water is no longer safe to drink. Samples show high levels of cyanide, cadmium, lead, and pesticides. "Ludhiana City's groundwater is just short of poison," laments a senior official at India's Central Ground Water Board.

Like Ludhiana's residents, more than a third of the planet's people live and work in densely settled cities, which occupy just 2 percent of the Earth's land area. With the labor force thus concentrated, factories and other centers of employment also group together around the same urban areas. Aquifers in these areas are beginning to mirror the increasing density and diversity of the human activity above them. Whereas the pollutants emanating from hog farms or copper mines may be quite predictable, the waste streams flowing into the water under cities contain a witch's brew of contaminants.

Ironically, a major factor in such contamination is that in most places people have learned to dispose of waste—to remove it from sight and smell—so effectively that it is easy to forget that the Earth is a closed ecological system in which nothing permanently disappears. The methods normally used to conceal garbage and other waste—landfills, septic tanks, and sewers—become the major conduits of chemical pollution of groundwater. In the United States, businesses drain almost 2 million kilograms of assorted chemicals into septic systems each year, contaminating the drinking water of 1.3 million people. In many parts of the developing world, factories still dump their liquid effluents onto the ground and wait for it to disappear. In the Bolivian city of Santa Cruz, for example, a shallow aquifer that is the city's main water source has had to soak up the brew of sulfates, nitrates, and chlorides dumped over it. And even protected landfills can be a potent source of aquifer pollution: the EPA found that a quarter of the landfills in the U.S. state of Maine, for example, had contaminated groundwater.

In industrial countries, waste that is too hazardous to landfill is routinely buried in underground tanks. But as these caskets age, like gasoline tanks, they eventually spring leaks. In California's Silicon Valley, where electronics industries store assorted waste solvents in underground tanks, local groundwater authorities found that 85 percent of the tanks they inspected had leaks. Silicon Valley now has more Superfund sites—most of them affecting groundwater—than any other area its size in the country. And 60 percent of the United States' liquid hazardous waste—34 billion liters of solvents, heavy metals, and radioactive materials—is directly injected into the ground. Although the effluents are injected below the deepest source of drinking water, some of these wastes have entered aquifers used for water supplies in parts of Florida, Texas, Ohio, and Oklahoma.

Shenyang, China, and Jaipur, India, are among the scores of cities in the developing world that have had to seek out alternate supplies of water because their groundwater has become unusable. Santa Cruz has also struggled to find clean water. But as it has sunk deeper wells in pursuit of pure supplies, the effluent has traveled deeper into the aquifer to replace the water pumped out of it. In places where alternate supplies aren't easily available, utilities will have to resort to increasingly elaborate filtration set-ups to make the water safe for drinking. In heavily contaminated areas, hundreds of different filters may be necessary. At present, utilities in the U.S. Midwest spend $400 million each year to treat water for just one chemical—atrazine, the most commonly detected pesticide in U.S. groundwater. When chemicals are found in unpredictable mixtures, rather than discretely, providing safe water may become even more expensive.

One Body, Many Wounds

The various incidents of aquifer pollution described may seem isolated. A group of wells in northern China have nitrate problems; another lot in the United Kingdom are laced with benzene. In each place it might seem that the problem is local and can be contained. But put them together, and you begin to see a bigger picture emerging.

Perhaps most worrisome is that we've discovered as much damage as we have, despite the very limited monitoring and testing of underground water. And because of the time-lags involved—and given our high levels of chemical use and waste generation in recent decades—what's still to come may bring even more surprises.

Some of the greatest shocks may be felt in places where chemical use and disposal has climbed in the last few decades, and where the most basic measures to shield groundwater have not been taken. In India, for example, the Central Pollution Control Board (CPCB) surveyed 22 major industrial zones and found that groundwater in every one of them was unfit for drinking. When asked about these findings, CPCB chairman D.K. Biswas remarked, "The result is frightening, and it is my belief that we will get more shocks in the future."

Jack Barbash, an environmental chemist at the U.S. Geological Survey, points out that we may not need to wait for expensive tests to alert us to what to expect in our groundwater. "If you want to know what you're likely to find in aquifers near Shanghai or Calcutta, just look at what's used above ground," he says. "If you've been applying DDT to a field for 20 years, for example, that's one of the chemicals you're likely to find in the underlying groundwater." The full consequences of today's chemical-dependent and waste-producing economies may not become apparent for another generation, but Barbash and other scientists are beginning to get a sense of just how serious those consequences are likely to be if present consumption and disposal practices continue.

Changing Course

Farmers in California's San Joaquin Valley began tapping the area's seemingly boundless groundwater store in the late-nineteenth century. By 1912, the aquifer was so depleted that the water table had fallen by as much as 400 feet in some places. But the farmers continued to tap the resource to keep up with demand for their produce. Over time, the dehydration of the aquifer caused its clay soil to shrink, and the ground began to sink—or as geologists put it, to "subside." In some parts of the valley, the ground has subsided as much as 29 feet—cracking foundations, canals, and aqueducts.

When the San Joaquin farmers could no longer pump enough groundwater to meet their irrigation demands, they began to bring in water from the northern part of the state via the California Aqueduct. The imported water seeped into the compacted aquifer, which was not able to hold all of the incoming flow. The water table then rose to an abnormally high level, dissolving salts and minerals in soils that had not been previously submerged. The salty groundwater, welling up from below, began to poison crop roots. In response, the farmers installed drains under irrigated fields—designed to capture the excess water and divert it to rivers and reservoirs in the valley so that it wouldn't evaporate and leave its salts in the soil.

Some Major Threats to Groundwater

Threat	Sources	Health and Ecosystem Effects at High Concentrations	Principal Regions Affected
Pesticides	Runoff from farms, backyards, golf courses; landfill leaks.	Organochlorines linked to reproductive and endocrine damage in wildlife; organophosphates and carbamates linked to nervous system damage and cancers.	United States, Eastern Europe, China, India.
Nitrates	Fertilizer runoff; manure from livestock operations; septic systems.	Restricts amount of oxygen reaching brain, which can cause death in infants ("blue-baby syndrome"); linked to digestive tract cancers. Causes algal blooms and eutrophication in surface waters.	Midwestern and mid-Atlantic United States, North China Plain, Western Europe, Northern India.
Petro-chemicals	Underground petroleum storage tanks.	Benzene and other petrochemicals can be cancer-causing even at low exposure.	United States, United Kingdom, parts of former Soviet Union.
Chlorinated Solvents	Effluents from metals and plastics degreasing; fabric cleaning, electronics and aircraft manufacture.	Linked to reproductive disorders and some cancers.	Western United States, industrial zones in East Asia.
Arsenic	Naturally occurring; possibly exacerbated by over-pumping aquifers and by phosphorus from fertilizers.	Nervous system and liver damage; skin cancers.	Bangladesh, Eastern India, Nepal, Taiwan.
Other Heavy Metals	Mining waste and tailings; landfills; hazardous waste dumps.	Nervous system and kidney damage; metabolic disruption.	United States, Central America and northeastern South America, Eastern Europe.
Fluoride	Naturally occurring.	Dental problems; crippling spinal and bone damage.	Northern China, Western India; parts of Sri Lanka and Thailand.
Salts	Seawater intrusion; de-icing salt for roads.	Freshwater unusable for drinking or irrigation.	Coastal China and India, Gulf coasts of Mexico and Florida, Australia, Philippines.

Major sources: European Environmental Agency, USGS, British Geological Survey.

But the farmers didn't realize that the rocks and soils of the region contained substantial amounts of the mineral selenium, which is toxic at high doses. Some of the selenium leached into the drainage water, which was routed to the region's wetlands. It wasn't until the mid-1980s that the aftermath of this solution became apparent: ecologists noticed that thousands of waterfowl in the nearby Kesterson Reservoir were dying of selenium poisoning.

Hydrological systems are not easy to outmaneuver, and the San Joaquin farmers' experience serves as a kind of cautionary tale. Each of their stopgap solutions temporarily took care of an immediate obstacle, but led to a longer-term problem more severe than the original one. "Human understanding has lagged one step behind the inflexible realities governing the aquifer system," observes USGS hydrologist Frank Chapelle.

Around the world, human responses to aquifer pollution thus far have essentially reenacted the San Joaquin Valley farmers' well-meaning but inadequate approach. In many places, various authorities and industries have fought back the contamination leak by leak, or chemical by chemical—only to find that the individual fixes simply don't add up. As we line landfills to reduce leakage, for instance, tons of pesticide may be running off nearby farms and into aquifers. As we mend holes in underground gas tanks, acid from mines may be seeping into groundwater. Clearly, it's essential to control the damage we've already inflicted, and to protect communities and ecosystems from the poisoned fallout. But given what we already know—that damage done to aquifers is mostly irreversible, that it can take years before groundwater pollution reveals itself, that chemicals react synergistically, and often in unanticipated ways—its now clear that a patchwork response isn't going to be effective. Given how much damage this pollution inflicts on public health, the environment, and the economy once it gets into the water, it's critical that emphasis be shifted from filtering out toxins to not using them in the first place. Andrew Skinner, who heads the International Association of Hydrogeologists, puts it this way: "Prevention is the only credible strategy."

To do this requires looking not just at individual factories, gas stations, cornfields, and dry cleaning plants, but at the whole social, industrial, and agricultural systems of which these businesses are a part. The ecological untenability of these systems is what's really poisoning the world's water. It is the predominant system of high-input agriculture, for example, that not only shrinks biodiversity with its vast monocultures, but also overwhelms the land—and the underlying water—with its massive applications of agricultural chemicals. It's the system of car-dominated, geographically expanding cities that not only generates unsustainable amounts of climate-disrupting greenhouse gases and acid rain-causing air pollutants, but also overwhelms aquifers and soils with petrochemicals, heavy metals, and sewage. An adequate response will require a thorough overhaul of each of these systems.

Begin with industrial agriculture. Farm runoff is a leading cause of groundwater pollution in many parts of Europe, the United States, China, and India. Lessening its impact calls for adopting practices that sharply reduce this runoff—or, better still, that require far smaller inputs to begin with. In most places, current practices are excessively wasteful. In Colombia, for example, growers spray flowers with as much as 6,000 liters of pesticide per hectare. In Brazil, orchards get almost 10,000 liters per hectare. Experts at the U.N. Food and Agricultural Organization say that with modified application techniques, these chemicals could be applied at one-tenth those amounts and still be effective. But while using more efficient pesticide applications would constitute a major improvement, there is also the possibility of reorienting agriculture to use very little synthetic pesticide at all. Recent studies suggest that farms can maintain high yields while using little or no synthetic input. One decade-long investigation by the Rodale Institute in Pennsylvania, for example, compared traditional manure and legume-based cropping systems which used no synthetic fertilizer or pesticides, with a conventional, high-intensity system. All three fields were planted with maize and soybeans. The researchers found that the traditional systems retained more soil organic matter and nitrogen—indicators of soil fertility—and leached 60 percent less nitrate than the conventional system. Although organic fertilizer (like its synthetic counterpart) is typically a potent source of nitrate, the rotations of diverse legumes and grasses helped fix and retain nitrogen in the soil. Yields for the maize and soybean crops differed by less than 1 percent between the three cropping systems over the 10-year period.

In industrial settings, building "closed-loop" production and consumption systems can help slash the quantities of waste that factories and cities send to landfills, sewers, and dumps—thus protecting aquifers from leaking pollutants. In places as far-ranging as Tennessee, Fiji, Namibia, and Denmark, environmentally conscious investors have begun to build "industrial symbiosis" parks in which the unusable wastes from one firm become the input for another. An industrial park in Kalundborg, Denmark diverts more than 1.3 million tons of effluent from landfills and septic systems each year, while preventing some 135,000 tons of carbon and sulfur from leaking into the atmosphere. Households, too, can become a part of this systemic change by reusing and repairing products. In a campaign organized by the Global Action Plan for the Earth, an international nongovernmental organization, thoughtful consumption habits have enabled some 60,000 households in the United States and Europe to reduce their waste by 42 percent and their water use by 25 percent.

As it becomes clearer to decisionmakers that the most serious threats to human security are no longer those of military attack but of pervasive environmental and social decline, experts worry about the difficulty of mustering sufficient political will to bring about the kinds of systemic—and therefore revolutionary—changes in human life necessary to turn the tide in time. In confronting the now heavily documented assaults of climate change and biodiversity loss, leaders seem on one hand paralyzed by how bleak the big picture appears to be—and on the other hand too easily drawn into denial or delay by the seeming lack of immediate consequences of such delay. But protecting aquifers may provide a more immediate incentive for change, if only because it simply may not be possible to live with contaminated groundwater for as long as we could make do with a gradually more irritable climate or polluted air or impoverished wildlife. Although we've damaged portions of some aquifers to the point of no return, scientists believe that a large part of the resource still remains pure—for the moment. That's not likely to remain the case if we continue to depend on simply stepping up the present reactive tactics of cleaning up more of the chemical spills, replacing more of the leaking gasoline tanks, placing more plastic liners under landfills, or issuing more fines to careless hog farms and copper mines. To save the water in time requires the same fundamental restructuring of the global economy as does the stabilizing of the climate and biosphere as a whole—the rapid transition from a resource-depleting, oil- and coal-fueled, high-input industrial and agricultural economy to one that is based on renewable energy, compact cities, and a very light human footprint. We've been slow to come to grips with this, but it may be our thirst that finally makes us act.

"Heaven is Under Our Feet"

Throughout human history, people have feared that the skies would be the source of great destruction. During the Cold War, industrial nations feared nuclear attack from above, and spent vast amounts of their wealth to avert it. Now some of that fear has shifted to the threats of atmospheric climate change: of increasing ultraviolet radiation through the ozone hole, and the rising intensity of global warming-driven hurricanes and typhoons. Yet, all the while, as the worldwide pollution of aquifers now reveals, we've been slowly poisoning ourselves from beneath. What lies under terra firma may, in fact, be of as much concern as what happens in the firmament above.

The ancient Greeks created an elaborate mythology about the Underworld, or Hades, which they described as a dismal, lifeless place completely lacking the abundant fertility of the world above. Science and human experience have taught us differently. Hydrologists now know that healthy aquifers are essential to the life above ground—that they play a vital role not just in providing water to drink, but in replenishing rivers and wetlands and, through their ultimate effects on rainfall and climate, in nurturing the life of the land and air as well. But ironically, our neglectful actions now threaten to make the Greek myth a reality after all. To avert that threat now will require taking to heart what the hydrologists have found. As Henry David Thoreau observed a century-and-a-half ago, "Heaven is under our feet, as well as over our heads."

Payal Sampat is a staff researcher at the Worldwatch Institute.

A Few Key Sources

Francis H. Chapelle, *The Hidden Sea: Ground Water, Springs, and Wells* (Tucson, AZ: Geoscience Press, Inc., 1997).

U.N. Environment Programme, *Groundwater: A Threatened Resource* (Nairobi: 1996).

European Environmental Agency, *Groundwater Quality and Quantity in Europe* (Copenhagen: 1999).

U.S. Geological Survey, *The Quality of Our Nation's Waters—Nutrients and Pesticides* (Reston, VA: 1999).

British Geological Survey et al., *Characterisation and Assessment of Groundwater Quality Concerns in Asia-Pacific Region* (Oxfordshire, UK: 1996).

POPs Culture

If there's one form of industrial innovation that we can definitely do without, it's the kind that is continually producing new Persistent Organic Pollutants—toxins so potent and durable that current emissions may still be causing cancer and birth defects 1,000 years from now.

by Anne Platt McGinn

Between 1962 and 1970, U.S. soldiers and their South Vietnamese allies sprayed nearly 12 million gallons of herbicide over vast tracts of Southeast Asian forest and more than half of South Vietnam's arable land. The program was designed to eliminate any cover that might conceal North Vietnamese Army units or Viet Cong guerrillas. The crews on the planes that did the spraying devised a slogan for themselves—a variation on a famous "Smokey the Bear" public service message back in the United States. They said, "only you can prevent forests."

The herbicide came in orange-striped drums, so it was called "agent orange." It was a mixture of two chemicals: 2,4,5-T and 2,4-D, both of them commonly used herbicides at the time. As with complex synthetic chemicals in general, these herbicides contained trace amounts of various unwanted substances that arose as byproducts of production. Among the byproducts were some of the chemicals called dioxins. A 1985 report by the U.S. Environmental Protection Agency called dioxins "the most potent carcinogen ever tested in laboratory animals." More recent laboratory work has linked dioxins with birth defects, spontaneous abortion, and injury to the immune system. When those two herbicides were sold in the United States, they typically contained dioxin concentrations of about 0.05 parts per million. But agent orange had dioxin concentrations up to 1,000 times as high.

At the time, the spraying of agent orange seemed a relatively minor part of the conflict. The dioxins, however, will linger in Vietnam's soil long after the war has vanished from living memory. Yet no one is really sure how much damage has been done. Medical doctors in Vietnam do not, by and large, have the resources to carry out longterm public health studies, but some doctors report that in sprayed areas, certain birth defects have become more common: anencephaly (absence of all or part of the brain), spina bifida (a malformation of the vertebral column), and hydrocephaly (overproduction of cerebrospinal fluid, causing a "swelling" of the skull). Immune deficiency diseases and learning disabilities may also be higher in sprayed areas. And if the human damage is uncertain, the broader ecological impact is a complete mystery.

In part because it is so vague, the agent orange legacy illustrates some of the worst aspects of dealing with dangerous synthetic chemicals like dioxins. For purposes of environmental analysis, dioxins are grouped in a loose class of potent toxins known as POPs, short for "persistent organic pollutants." The full definition of a POP, however, is somewhat more complex than the acronym implies. In addition to being persistent (that is, not liable to break down rapidly), organic (having a carbon-based molecular structure), and polluting (in the sense of being significantly toxic), POPs have two other properties. They are fat soluble and therefore liable to accumulate in living tissue; and they occur in the environment in forms that allow them to travel great distances.

If you put all five of these properties together, you can begin to see the potential for "agent orange scenarios" in many places. We know that POPs are very dangerous, but we can never be sure exactly who will be injured by them. In the 1970s, for example, a group of children developed leukemia (a usually fatal blood disorder involving uncontrolled production of white blood cells) in Woburn, a small town in Massachusetts. The leukemia had apparently been caused by solvents in the tap water. But why did the disease emerge only in certain children and not in many others who also presumably ingested the solvents? It often takes sophisticated statistical analysis to find any connection at all between contamination and injury—that's one of the reasons it's so difficult to assess the public health risks from POPs. But of course statistics can't capture the *experience* of contamination: such a threat can seem like an evil lottery.

The apparent randomness of the threat is exacerbated by the fact that injury is often delayed or indirect. Extremely toxic chemicals can bide their time, then poison their victims in ways that are very hard to see. Benzene, for example, is a common solvent. It's an ingredient in some paints, degreasing products, gasoline, and various other shop and industrial compounds. If you're heavily exposed to it, you stand a heightened chance of developing cancer—and so may any children that you have after your exposure. That's true even if you're a man, since fetal exposure isn't the only way benzene may poison children: it can reach right into your chromosomes and injure the genes your child will inherit. Benzene may do its damage without ever touching the child directly at all.

POPs are potent ecological poisons as well. And just as in the human body, their ecological effects often exhibit a kind of weird indirection. In the United States in the 1960s, for example, biologists began to find strong field evidence that the pesticide DDT (dichlorodiphenyltrichloroethane) and similar chemicals were dangerous. But the evidence didn't come from the organisms that had absorbed the pesticides directly. It came from birds of prey—eagles and falcons—who were suffering widespread reproductive failure. Too few eggs and egg shells so thin they cracked soon after laying: these were the results of a type of indirect poisoning known as bioaccumulation. The fat solubility of the pesticides allowed them to concentrate in the tissues of their hosts as they moved up the food chain, from insects to rodents to raptors. Even today, the North American Great Lakes basin is showing the effects of certain POPs, like DDT, which have not been used in the region for decades. Eagle populations are still depressed; tumors continue to appear in fish, birds, and mammals.

But there is one way in which the agent orange scenario deviates from the norm. Most POPs owe their presence in the environment not to the horrible exigencies of war, but to ordinary industrial processes—plastic and pesticide manufacturing, leaky transformers, waste incineration, and so forth. POPs are an inevitable byproduct of business as usual. By design and by accident, we are continually introducing new chemicals into the environment without any clear notion of what they will eventually do—or whether we may one day find ourselves in a desperate scramble to remove them. And among the tens of thousands of chemicals that have been in circulation for decades, relatively few have been studied for their health and environmental effects. Consequently, no one knows exactly how many POPs there are, but it's likely that many thousands of chemicals could qualify for the term.

And beyond their number is the question of their effect: while POPs are toxic by definition, their longterm health and environmental impacts are still largely unknown. Even more complex than evaluating individual POPs is the looming need to understand what kinds of synergistic interactions could be triggered by overlapping exposure—to multiple POPs or to POPs combined with other chemicals. Multiple contamination is the rule, rather than the exception, but virtually nothing is known about it. What we do know is that most of the world's living things are now steeping in a diffuse bath of POPs. And that almost certainly includes you. No matter where you live, you're likely to be contaminated by trace amounts of POPs. They're in your food and water; they may be in the air you breathe; they're probably on your skin from time to time—if, for instance, you handle paints, solvents, or fuels.

Currently, 140 nations are negotiating a treaty to phase out 12 specific POPs (see table, "Production and Use of the Dirty Dozen POPs"). This so-called "dirty dozen" includes nine pesticides, one group of industrial compounds known as polychlorinated biphenyls (PCBs), and two types of industrial byproducts, the dioxins and their chemical cousins, the furans. The treaty is called the "International Legally Binding Instrument for Implementing International Action on Certain Persistent Organic Pollutants" and as its name suggests, it is a laudable but rather timid effort. Its supporters hope that it will eventually serve as a mechanism to phase out dozens of other POPs. But at least in its present form, it doesn't address the fundamental problem. If we want to reduce the risks from the vast and growing number of synthetic chemicals that are being released into the environment, we will have to rethink some of our basic notions of industrial development.

Every Twenty Seven Seconds

There are now over 20 million synthetic chemicals, and that number is increasing by more than 1 million a year. As a rough global average, a new chemical is synthesized every 27 seconds of the day. Very few of these substances ever go into commercial production—something like 99.5 percent remain academic curiosities, or rapidly forgotten attempts to produce a new pesticide, or solvent, or whatever. But every year another 1,000 or so new compounds enter the chemical economy, either as ingredients in finished products, or as "intermediates"—chemicals used to make other chemicals. The total number of synthetics in commerce is probably now somewhere between 50,000 and 100,000. But the total number of synthetics *in the*

POLLUTION: THE HAZARDS OF GROWTH

RELEASE

A Monsanto Chemical works factory in Alabama, circa 1947, made a quantity of PCBs and shipped them in fluid form to a GE factory in Massachusetts, which loaded the fluid into electric transformers as insulation. Transformers were shipped out and installed on thousands of utility poles and buildings.

DISPERSAL

Over the decades, transformers deteriorated or were destroyed—some by lightning, others by demolition. The PCBs leaked into the ground and were dispersed by runoff into streams or slow seepage into aquifers. Some lodged in soil that baked in the sun, turned to dust, and blew away, eventually settling throughout the global environment.

ACCUMULATION

PCB-containing dust that settled in lakes or rivers became a nutrient for algae. Water fleas ate the algae. Small shrimp ate the fleas—each shrimp eating many fleas and bioaccumulating the PCBs that lodged in its fat. Small fish called smelt ate the shrimp, and trout ate many of the smelt—each stage increasing the concentration of the toxin.

and CONSUMPTION

A woman cooks a trout, which has bioaccumulated the PCBs from hundreds of shrimp and thousands of fleas. The PCBs are then added to other POPs she has consumed in cow's milk, beef, and other foods. The last step is the baby, whose first food is her mother's milk.

environment is probably far greater than that, because of the byproducts (like dioxins) unintentionally generated during production, and because of the breakdown products that result from the decay of commercial substances.

Chemical innovation on this scale creates an enormous biological risk, despite the fact that many synthetic chemicals are probably harmless, and many naturally-occurring chemicals are extremely dangerous. To understand the risk, it's useful to have a general sense for what usually happens with natural toxins. Most really potent natural toxins break down far more readily than POPs. Powerful natural toxins also tend to be geographically isolated—they aren't usually dispersed throughout the environment. And while it's true that there are some natural forms of "mass poisoning," such events are generally episodic rather than continual—think of "red tide" algal blooms along ocean shorelines, for example. Finally, apart from such mass poisonings, any really powerful poison produced by a living thing is likely to be "trophically isolated"—that is, it will tend to affect only organisms that play certain ecological roles. To be poisoned by a toxic frog, for example, you almost have to be a frog predator. Don't mess with the frog and you'll be fine. The toxic frog paradigm does not, however, apply to our current chemical economy, which is causing broadscale, chronic exposure to powerful toxins at virtually every ecological level.

Not all manufactured chemicals are organic (that is, carbon-containing); inorganic chemicals play key industrial roles as well. Sulfuric acid (H_2SO_4), for example, is a key feedstock for much chemical production, especially fertilizer. But most commercially important inorganics, like sulfuric acid, aren't synthetic in the sense of being completely artificial—they occur in nature. And synthetic or not, only around 100,000 inorganic chemicals are known. Contrast that with the many millions of organic compounds now known—most of them wholly artificial—and you can begin to get an idea of the stupifying variety in molecular structure that carbon permits.

Large-scale industrial production of organic chemicals was well underway by the middle of the 19th century. Refineries in both Europe and the United States were using coal to produce kerosene—or "coal oil," as it was then called. In 1859, western Pennsylvania became the site of the world's first oil well. As other oil fields opened in the United States, Europe, and east Asia, those coal refineries became oil refineries, and industry acquired a vast and extremely versatile supply of lubricants and fuels. Synthesis of completely novel compounds began in European laboratories at about the same time. DDT, for example, dates from 1874, when it was synthesized by a German chemistry student, although its pesticidal properties were not appreciated until the 1930s. The first plastics were synthesized from cellulose (the primary constituent of wood) in the 1890s. By the end of the century, organic chemistry had revolutionized a major industry—the production of dyes.

The key to that development was the realization that synthetics could be produced in abundance directly from oil, instead of from living plant products. With a cheap source of raw material at hand, synthetics offered an answer to war-time shortages of often much more expensive natural products. Vinyl, for instance, was developed in the 1920s as a rubber substitute; during World War II, it helped ease the demand for this essential plant product—tires still had to be made of rubber, but vinyl worked well as a wire insulator.

In the years following the war, synthetics flooded one manufacturing process after another, since they were often much cheaper than such traditional materials as rubber, wood, metal, glass, and plant fiber. In some cases the synthetic displaced a traditional material outright, but arguably just as important has been the interest in combining old and new—the metal that has a specialty coating to make it more durable, the flooring laminate composed of resin and wood fiber, and so forth. In ways large and

Production and Use of the "Dirty Dozen" POPs

Material	Date of Introduction	Cumulative World Production (tons)
Aldrin (insecticide)	1949	240,000
Chlordane (insecticide)	1945	70,000
DDT (insecticide)	1942	2.8–3 million
Dieldrin (insecticide)	1948	240,000
Endrin (insecticide and rodenticide)	1951	(3,119 tons in 1977)
Heptachlor (insecticide)	1948	(900 tons used in 1974 in the U.S.)
Hexachlorobenzene (fungicide and byproduct of pesticide production)	1945	1–2 million
Mirex™ (insecticide and flame retardant)	1959	no data
Toxaphene™ (insecticide)	1948	1.33 million
PCBs (liquid insulators in transformers, hydraulic fluids; ingredients in some paints, adhesives, and resins. No longer generally produced in industrialized countries.)	1929	1–2 million
Dioxins (byproducts of organochlorine production and incineration, and of wood pulp bleaching)	1920s	(10.5 tons International Toxic Equivalency of dioxins and furans combined in 1995)
Furans (same as with dioxins)	1920s	

SOURCE: Anne Platt McGinn, "Phasing Out Persistent Organic Pollutants," in Lester R. Brown et al., *State of the World 2000* (New York: W.W. Norton & Company, 2000), page 223, note 13.

---

small, synthetics have transformed our built environments—and not simply by replacing things that were made before out of some other material, but by allowing for the creation of products that probably wouldn't otherwise have existed, at least on a mass scale. Plastic, for instance, is as fundamental in electronics manufacturing as microchips. Today, synthetic organic chemicals flow through just about every pipe in the chemical economy (see box, "What Does the Chemical Industry Produce?").

Not surprisingly, the volume of synthetic organic chemical production has moved continually upwards ever since large-scale manufacturing began in the 1930s. Global production escalated from near zero in 1930 to an estimated 300 million tons by the late 1980s. In the United States alone, production has soared from about 150,000 tons in 1935 to nearly 150 million tons by 1995—almost a thousandfold increase. Cinema fans may recall the one word of advice given to the confused young man played by Dustin Hoffman in the 1968 film, "The Graduate": "Plastics." The trend was as clear then as it is now: U.S. production of plastics has increased 6-fold since 1960.

The chemical structure of synthetic organics varies enormously, of course, but when it comes to assessing the potential of any particular chemical to cause trouble, either in the human body or in the environment, one question is of overriding importance: does it contain chlorine? Chlorine is highly reactive—that is, it combines very readily with certain other elements and it tends to bind to them very tightly. (The big exception to this rule involves a looser association called an ionic bond. For example, sodium chloride, or table salt, is the product of an ionic bond between a chlorine and a sodium atom. Such a bond is weak enough to allow the two atoms to separate from each other in solution.) Carbon is one of the elements that chlorine will bond to, although in nature such combinations, known as organochlorines, are rarely abundant. (There are a few exceptions, such as salt marsh emissions of methyl chloride.) But chemists have found that by adding chlorine to carbon-based compounds, an even greater molecular variety becomes possible. Chlorine's ability to snap firmly into place—and to anchor all sorts of chemical structures—has made it, in the words of W. Joseph Stearns,

6 ❖ POLLUTION: THE HAZARDS OF GROWTH

Director of Chlorine Issues for the Dow Chemical Company, "the single most important ingredient in modern [industrial] chemistry."

Take a sophisticated chemical sector, like that of the United States, and consider the importance of chlorine in it. Chlorine is used to make thousands of chemicals—solvents, pesticides, pharmaceuticals, bleaches, and so on. Around 11,000 organochlorines are in production. The biggest readily identifiable category of these is plastic. Of the more than 10 million tons of chlorine that the U.S. industry consumes each year, about one-third goes to produce 14 different types of plastic. The most common of those is polyvinyl chloride (PVC), which is light, strong, and easy to mold. PVC is used to make plastic wrap, shoe soles, automobile components, siding, pipes, and medical products, among other things. In less than a decade, from 1988 to 1996 (the most recent year for which figures were available), global production of PVC expanded by more than 70 percent, from 12.8 million tons to 22 million tons. In the use of products like PVC, you can see how thoroughly we've enveloped ourselves in organochlorines.

Although many organochlorines are not known to be dangerous, a substantial number of them do create major risks. In large measure, those risks are the result of three common characteristics. Organochlorines are very stable—that's obviously part of their manufacturing appeal, but it also means that they don't go away. They tend to be fat soluble, which means that they can bioaccumulate. And many of them have substantial chronic toxicity—that is, while exposure over the short term may not be dangerous, long-term exposure frequently is. (The reasons for toxicity vary. Some organochlorines can "mimic" naturally-occurring chemicals such as hormones, thereby upsetting the body's chemical processes; some weaken the immune system; some affect organ development, some promote cancer, and so on.) Stability, fat solubility, and chronic toxicity: does that begin to sound like a POP? Chlorine certainly isn't *required* to make a POP. Among the non-chlorinated POPs are various organometals (used, for example, in marine paints) and organobromines (used as pesticides and as liquid insulators in electrical equipment). But most known POPs—including all of the "dirty dozen"—are organochlorines.

Organochlorine pesticides are the class of products that has produced what are probably the most notorious POPs (see the table for some examples). It's hardly surprising that pesticides are a major ingredient in our stew of dangerous chemicals—after all, pesticides are designed to be toxic and they are produced in enormous quantities. Since 1945, global production of pesticides has increased an estimated 26-fold, from 0.1 million tons to 2.7 million tons, although growth has slowed in the last 15 years, as health and environmental concerns have inspired an increasing number of bans, primarily in industrialized countries. These restrictions have reduced the total quantity of pesticides used in the industrialized countries, but the *toxicity* of particular pesticides has continued to grow. Current pesticide formulations are 10 to 100 times as toxic as they were in 1975.

Today, pesticide manufacturers usually want their products to have a high acute toxicity and low chronic toxicity. They're looking for compounds that will kill quickly but that don't haunt the field indefinitely, so organochlorines, with their substantial chronic toxicities, no longer have the universal appeal they once did. Newer pesticides are less likely to contain chlorine. That's obviously good, but not good enough, for two reasons: non-organochlorine pesticides also sometimes turn out to be POPs, and nearly all the old products are still with us anyway. They persist in the environment and most are still used in developing countries.

A more obscure array of POPs involves a family of organochlorines that have been used as liquid insulators in electrical equipment, as hydraulic fluids, and as trace additives to plastics, paints, even carbonless copy paper. These are the polychlorinated biphenyls, or PCBs. For decades, the extreme stability, low flammability, and low conductivity of POPs made them the standard liquid insulation in transformers—and since transformers are a near-ubiquitous part of every electrical grid, PCB contamination is now a standard form of landscape poisoning. In industrialized countries, PCBs were manufactured mostly between the 1920s and the late 1970s; they are still manufactured in Russia and are still in use in many developing countries. Scientists estimate that up to 70 percent of all PCBs ever manufactured are still in use or in the environment, often in landfills where they are gradually seeping into water tables. The United Nations Environment Programme (UNEP) recently published guidelines for helping officials in developing countries identify PCBs. But given their multiple uses and more than 90 trade names, simply finding them is going to be a mind-boggling task—let alone cleaning them up.

But the overwhelming majority of POPs are not intentionally produced—they're by-products, like dioxins and furans, two classes of POPs that result primarily from organochlorine production, the bleaching of wood pulp, and the incineration of municipal waste. A 1995 UNEP emissions inventory of 15 countries traced some 7,000 kilograms of dioxin and furan releases to incinerator emissions—that's 69 percent of the total releases of those substances in these countries. (Seven thousand kilograms may not sound like all that much—but bear in mind that these are extremely toxic substances usually produced in trace quantities.) There are 210 known dioxins and furans. And among the byproducts of organochlorine production and use, it's almost certain that many more thousands of POPs remain to be discovered.

Do we really need it?

Over the past three decades or so, attempts to regulate the chemical industry in the industrialized world have grown to a phenomenal degree. In the United States, for example, the effort now involves four federal agencies on a regular basis, and at least seven major pieces of federal legislation, which address pesticides, pollution, and attempt to promote cleaner industries. Any new synthetic produced in Europe or the United States is now subject to some degree of toxicity testing before it can be injected into commerce.

But despite this gargantuan bureaucratic effort, the current regulatory approach is no match for the threat. In the first place, most of the toxicity testing is done by the companies themselves—a practice that invites obvious conflicts of interest. Nor do current efforts offer a realistic possibility of dealing with the testing backlog. Tens of thousands of chemicals entered commerce in the decades before testing was required—and we still have no clear notion of the risks most of them pose. Fewer than 20 percent of the chemicals in commerce have been adequately evaluated for toxicity, according to a 1984 National Academy of Sciences report. (It's perhaps a reflection of the magnitude of the problem that this 16-year-old report should still be widely cited.)

And then there is our uncertainty over what we ought to be testing *for*. The toxicology of synthetic organics is in a near constant state of flux, and the difficulty of establishing a causal link between exposure and injury opens the science up to all sorts of tendentious reinterpretation. Anyone familiar with the smoking and health debates will recognize this problem. Take dioxins, for example. Chloracne—the severe skin deformity that is the hallmark of dioxin poisoning—was identified more than a century ago, in 1899. In 1998, the UN World Health Organization (WHO) reduced its standard for tolerable daily intake of dioxin-like substances from 10 picograms per kilogram of body weight per day to 1–4 picograms. So a person who weighs 68 kilograms (about 150 pounds) shouldn't be exposed to more than 4 trillionths of a gram per day. For infants, the safe levels are even more minuscule. Yet just a couple of years ago, a consultant to the Chemical Manufacturers Association announced that "dioxin has not been shown to pose any health threat to the general public."

Even where obfuscation is not an issue, advances in toxicology tend to create a second testing backlog, since thousands of previously-screened chemicals may need to be re-evaluated. In 1996, for example, the United States launched a major pesticide re-evaluation program, in the light of new research on how these chemicals can affect children, whose high metabolism and rapid rate of physical development make them more vulnerable to certain kinds of toxins. Thus far, screening has been completed on less than a quarter of U.S. pesticide "registrations." (The United States regulates pesticides by designating specific uses permitted for each chemical; each such use is known as a registration.)

The shifting horizon of toxicology can call into question even widely accepted synthetics. The plasticizers known as phthalates, for example, are believed to be among the most common industrial compounds in the environment. Yet recent laboratory research in animals has linked phthalates to damage to the liver, kidney, and testicles, as well as to miscarriage, birth defects, and reduced fertility. Incineration of phthalates produces dioxins. Phthalates occur in everything from construction materials to children's teething rings. And among the 1,000 new chemicals that will enter the economy this year, who knows how many more such discoveries will eventually be made?

In its current form, the chemical sector is clearly at odds with our collective obligation to maintain human and environmental health. What is needed is fundamental reform—a change that goes far deeper than conventional regulation. That reform could start with a very simple but revolutionary idea: it's wise to avoid unnecessary risk. This is the kernel of one of the environmental movement's core concepts: the precautionary principle. The principle states that when any action is contemplated that could affect the environment, those who advocate the action should show that the risks are either negligible, or that they are decisively outweighed by the benefits.

The principle reverses the usual burden of proof. In most environmental controversies today, that burden effectively rests with those who argue against an action: they must usually persuade the public or policy makers that the benefits are outweighed by the risks. But of course, we rarely understand the risks until after the fact—and maybe not even then. That's the problem the principle is meant to address; it's a kind of insurance policy against our own ignorance.

In terms of our chemical use, a reasonable application of the precautionary principle would require us to assume that in certain chemical classes—organochlorines, for example—any new compound is dangerous. The next step would be to ask: do we really need it? This kind of inquiry would tend to foster a different kind of inventiveness, both within the chemical industry and within society as a whole. The emphasis would tend to shift from inventing new chemicals, to inventing new uses for chemicals thought to be reasonably safe, and to inventing new procedures that may not be dependent on chemicals at all. Fewer new chemicals would come into commerce; a growing number of established ones would come out.

There is already a strong precedent for this kind of chemical "stand down" in the impending ban of chlorofluorocarbons (CFCs), the now-notorious class of chemicals once almost universally used as refrigerants and spray-can propellents. CFCs were found to be weakening the stratospheric ozone layer, which shields the Earth's surface from harmful ultraviolet radiation. Under the Montreal Protocol of 1987, CFCs are being phased out in favor of other compounds that are less harmful to the ozone layer. In many parts of the chemical economy, you can see the potential for similar developments. Consider three examples.

6 ❖ POLLUTION: THE HAZARDS OF GROWTH

> ## What Does the Chemical Industry Produce?
>
> ### Some Major Product Categories
>
> **Tars and primary petroleum derivatives** (used to make asphalt, fuels, lubricants, and many of the products listed below)
>
> **Plastics** (used in—you name it)
>
> **Resins** (used, for example, in adhesives, protective coatings, and paints)
>
> **Intermediates** (chemicals used to produce other chemicals)
>
> **Solvents** (liquids used to keep other materials in solution, as for example, in paints and cleaning compounds)
>
> **Surfactants** ("surface-active agents" used in products like detergents to promote an interaction between the product and the material to which it is applied)
>
> **Elastomers** (synthetic rubbers such as neoprene)
>
> **Rubber-processing chemicals**
>
> **Plasticizers** (used in plastics to confer flexibility)
>
> **Pesticides**
>
> **Pharmaceuticals**
>
> **Flavors and perfumes** (manufacturers commonly rely on synthetics to make their products taste and smell the way they want)
>
> **Dyes and pigments** (everything from the paint on your car, to the color of your clothes, to the food you eat)

Pesticides: the phase-out may already have begun

Pesticides are the mainstay of monoculture farming. They are the mechanism that allows for vast expanses of pure corn, cotton, or soybeans—a highly unnatural condition that is very vulnerable to infestation. But pesticides are also expensive and dangerous, and these liabilities underlie the growing boom in organic agriculture. In the industrialized countries, organic production (which uses no synthetic pesticides) is the strongest market within the agricultural sector. In the United States, the organic market has been growing at a rate of 20 percent per year since 1989. Some 35 percent of U.S. consumers look for the organic label at least part of the time. In Europe, one-third of the continent's farmland is expected to be in organic production by the end of this decade. Organic and other forms of low-pesticide farming usually involve more careful stewardship of the soil and more diverse plantings, which tend to have fewer pest problems than conventional monocultures. Even though the yield of a particular crop may be lower than in conventional agriculture, an organic farm can do just as well in terms of total productivity (that is, in terms of all the crops coming off a unit of land) and in terms of financial return—and that's before you factor in the environmental benefits.

A thornier set of pesticide problems involves public health. DDT may be eliminated by the new treaty as an agricultural pesticide, but it's still key to malaria control in many parts of the tropics. Malaria kills 2.7 million people every year—a death toll greater than that of AIDS. In much of sub-Saharan Africa and tropical Asia, control of the mosquitoes that carry the disease is a life-and-death issue, and that has frequently involved the broadscale spraying of DDT. But even here, more careful targetting of the mosquitoes would permit enormous reductions in pesticide use—and might even improve malaria control. Researchers in Africa, for example, have demonstrated that bednets soaked in alternative, less-toxic insecticides can reduce malaria transmission by 30 to 60 percent and childhood mortality by up to 30 percent. And bednets are relatively cheap: a net plus a year's supply of insecticide costs about $11. In 1993, WHO dropped its blanket spraying recommendation for DDT, in favor of targeted spraying of the insecticide indoors only.

PVC: taking the POPs out of the products

The incineration of solid waste is a primary generator of dioxins and furans. While better incineration procedures can greatly reduce this kind of contamination, the single most effective way to lower dioxin output is to get as much chlorine as possible out of the waste stream. PVC is the source of an estimated 80 percent of the chlorine that flows into municipal waste incinerators and nearly all the chlorine in medical waste incinerators (these are among the most important second-rank dioxin emitters, after the municipal incinerators). A top priority for the new chemical economy should therefore be the elimination of PVC, which is 45 percent chlorine by weight, in favor of low-chlorine or chlorine-free materials. Initially, the substitutes are liable to be more expensive than PVC, but even incipient demand could rapidly generate an economy of scale. The market prospects have already led the

Exxon Corporation, one of the world's largest PVC producers, to begin planning a shift from PVC to chlorine-free polyolefin plastics.

Bleaching and benzene: removing POPs from industrial processes

POPs often haunt industrial processes to a far greater degree than they contaminate the products themselves. Thus, for example, paper is not ordinarily a source of organochlorine contamination while it's being used. But paper production certainly is, and paper disposal can be as well, because the huge volume of paper converging on an incinerator may allow trace contaminants to concentrate. Both forms of contamination are caused by the use of chlorine bleaches to whiten woodpulp. Bleaching can produce up to 35 tons of organochlorines per day per mill. Yet this type of pollution is now almost wholly unnecessary—and the paper you're looking at right now proves it. (WORLD WATCH is printed on paper that is bleached without any chlorine or chlorine-based compounds, although unfortunately, that is not yet true of our cover stock.) Thus far, only 6 percent of global bleached pulp production is "totally chlorine free," but that includes more than a quarter of Scandinavian production, so the economic viability of the process is not in question. Some 54 percent of global bleached pulp production is now "elementally chlorine free"—meaning that a chlorine bleach was used, but at least it wasn't raw chlorine. (Our cover stock falls in this category.) It's true that converting a mill to chlorine-free production is expensive, but the picture is very different when you start from scratch. It's actually cheaper to build a chlorine-free mill than a conventional one.

At least some of the more dangerous "intermediates" within the industry are probably susceptible to replacement as well. Benzene, for example, is a major feedstock chemical in the production of a wide range of materials—for example, dyes, film developing agents, solvents, and nylon. For some applications, however, it may be possible to replace benzene with the simple blood sugar, glucose. That may sound like a bizarre substitution, but it's the ring structure of both molecules that allows for a degree of interchangeability. Glucose is cheaper to make than benzene (6 versus 13 cents per pound) and for all practical purposes, it's harmless. As a feedstock, however, the processes for handling glucose are more expensive than the better-established processes for benzene, but these costs don't take into account emissions control costs for benzene. In any case, the costs would presumably decline if the use of glucose as a feedstock became more common. Such adjustments deep within the industrial machinery may seem rather obscure, but they could be major news: the possibility of substituting an innocuous substance for an extremely dangerous one suggests that there may be all sorts of hidden opportunities for re-engineering the chemical sector.

If such re-engineering is to succeed, it will have to proceed from a much broader understanding of what we're doing when we make and use synthetic chemicals. Whether we intend it this way or not, chemical manufacturing is as much an ecological process as it is an economic or industrial one. Any industry executive knows that a chemical plant has to make some sort of economic sense. The POPs legacy is telling us that it had better make environmental sense as well.

Anne Platt McGinn is a senior researcher at the Worldwatch Institute.

It's a breath of fresh air

Thirty years after Earth Day, America is getting its environmental act together

By David Whitman

Shortly after the first Earth Day in 1970, President Richard M. Nixon warned that without far-reaching reforms the United States would, by the year 2000, be "a country in which we can't drink the water, where we can't breathe the air, and in which our children... will not be able to [enjoy] the beautiful open spaces... [of] the American landscape."

Today, as the world prepares to celebrate Earth Day 2000 on April 22, the nation's air is cleaner, its water purer. There is more protected open space in national parks, wildlife refuges, and wilderness areas. Yet surveys indicate that only 14 percent to 36 percent of Americans believe the environment has improved a "great deal" since 1970. And according to a 1999 Roper poll, 56 percent worry that the next 10 years will be "the last decade when humans will have a chance to save the earth from an environmental catastrophe."

Not all the news is good, of course. Urban sprawl, loss of habitat, spoilage of wetlands, and global warming are all serious problems getting worse. And beyond America's borders, the picture is even bleaker. Across the former Soviet Union, environmental contamination is rampant. In Asia, in even some of the most cosmopolitan cities, the air is literally unfit to breathe.

America's record of environmental progress, by contrast, shows that despite the grave problems that persist, there is reason for hope. At the time of the first Earth Day, America was a place where oil-drenched rivers caught fire, loggers lopped down great swaths of national forests, recycling was rare, motorists routinely littered, and fabled icons like the bald eagle were headed toward extinction in the lower 48 states. The Environmental Protection Agency did not exist, and industry and government casually dumped millions of tons of hazardous wastes. In the nation's cities, killer smogs blanketed downtowns, lead emissions addled children's minds, and many municipalities treated urban waterways like open sewers. "In 1970, people were concerned that industrial society was going to choke itself," says George Frampton, chair of the White House Council on Environmental Quality. "The environment is in a lot better condition in the U.S. today."

A look back might start with Thanksgiving Day 1966, when millions of New Yorkers awoke to a surprise far more pungent than the whiff of singed turkey. A smog as thick as a London fog had descended on the city, filled with noxious fumes that left thousands of people with respiratory illnesses literally gasping for air. The smog obliterated chunks of Manhattan's skyline, prompting the shutdown of all 11 of the city's incinerators. Health authorities advised motorists to use their cars only when absolutely necessary; landlords were told to lower thermostats to 60 degrees to curb pollutants.

Several days later, the smog dissipated, but not before an estimated 168 people died as a result—the same number killed in the 1995 Oklahoma City bombing. A mayoral task force appointed to assess the crisis warned that unless polluters were curtailed, New York could become "uninhabitable within a decade."

Three years later, *Life* magazine would report that scientists had amassed "solid experimental... evidence" indicating that by the 1980s urban dwellers "will have to wear gas masks to survive air pollution." Nowhere was the problem more severe than in Los Angeles, which had been intermittently choked by smog since World War II. Smog attacks dense enough to be dubbed "daylight dim-outs" had prompted local officials to call for public prayer to end the scourge, and the city council had once donned gas masks when fumes infiltrated its chambers.

Under federal standards issued in 1971, the most serious smog emergency requires a "Stage 3" alert, during which health authorities advise everyone in affected areas to remain indoors and minimize physical activity. The last such alert in this country was in 1974, in San Bernardino County, Calif. Then-governor Ronald Reagan was no tree hugger. But he urged residents of the car-crazy Los Angeles basin to avoid driving if possible until the smog abated.

As late as 1979, Los Angeles suffered through 120 days of Stage 1 smog alerts, the level at which the federal government deems the air very unhealthy and advises everyone to avoid rigorous outdoor exercise. Last year, for the first time since record-keeping began in the mid-'50s, Los Angeles did not record one ozone reading high enough to trigger a smog alert. Nationwide, emissions of all but one of the six major air pollutants tracked by the EPA since the 1970 Clean Air Act was enacted have declined.

Fire on the water. One May 1967 night in Kansas City, hundreds of passersby and motorists stopped to gawk at the Kaw River: A giant swath of viscous oil, some 9,000 square feet in all, was aflame, floating down the dirty waterway. The Kaw wasn't the only river to catch fire before the first Earth Day. Even more notorious was Cleveland's Cuyahoga, where drifting oil, picnic tables, and debris generated a five-story-high conflagration in 1969 that inspired a Randy Newman song ("Burn On, Big River").

Thirty years ago, many cities still discharged raw sewage directly into rivers and the oceans. Factories dumped millions of tons of poisonous byproducts into streams, the Great Lakes, and the sea. "People feared that the Great Lakes were becoming vast cesspools of death," says Denis Hayes, national coordinator of the first Earth Day and chair of Earth Day Network 2000. In 1970 alone, the federal government recalled a million cans of tuna for possible mercury contamination. In the nation's capital, fecal coliform counts in the Potomac River rose to 4,000 times the safe level. At the same time, Manhattan's worldly West Side was discharging a mind-numbing 300 million gallons of raw sewage *a day* into the Hudson River—nearly 30 times what the Exxon Valdez released in its infamous, one-time 1989 oil spill. And long before George Bush made Boston Harbor a campaign issue in 1988, the Standells crooned about the harbor's squalid state in their 1966 rock hit "Dirty Water." ("Well I love that dirty water, Oh, Boston, you're my home....")

Today, the nation's waterways are no longer a waste bucket, mostly thanks to the 1972 Clean Water Act, which prompted a huge drop in uncontrolled "point source" discharges from municipal and industrial polluters. Now two thirds of the nation's waters are safe for fishing and swimming, compared with only a third back then. Where the Kaw once burned, kayakers and swimmers can frolic. And the once foul banks of the Cuyahoga today host a slew of boutiques and bistros.

Deep forest. Last fall, President Clinton announced a ban on logging and road building on more than 40 million acres of national forests—an area larger than Georgia. Even before that, the nation's forests had grown modestly since the 1970s, and they now contain more timber and fewer small trees. "We have more forested land today in the U.S. than at the turn of the century," says U.S. Forest Service Chief Mike Dombeck.

By contrast, in 1970, the national forests, as Wyoming Sen. Gale McGee put it, had been "depleted to the point that would shame Paul Bunyan." At the Forest Service's behest, contract loggers then relied heavily on clear-cutting—leveling sections of a forest, setting fire to the debris, and "scarifying" the area on occasion by scraping it bare with bulldozers. After an air tour of some clear-cuts, a shaken McGee likened the devastation to a B-52 bombing run, a "shocking desecration that has to be seen to be believed."

All told, loggers in 1970 clear-cut 564,000 acres in national forests. Controversy over clear-cutting peaked in Montana's Bitterroot National Forest, where logging eyesores were headline news in 1970. Between 1967 and 1970, 11,211 acres were clear-cut in the Bitterroot alone, compared with just 8 acres last year. When Gifford Pinchot Jr., the son of Theodore Roosevelt's legendary Forest Service chief, toured the Bitterroot clear-cuts, he declared: "If my father had seen this, he would have cried."

In some cases, the country has taken an ecological step forward only to lurch backward. Nuclear waste sites, for instance, are now more carefully regulated

---

Environmental health indicators

Many ecological trends have improved in the nation—without forestalling economic growth

THE LAND

Protected wilderness
1970: 10 million acres
1997: 104 million acres

National Park System
1970: 30 million acres
1997: 83 million acres

THE AIR

Sulfur dioxide emissions
1970: 31.2 million tons
1997: 20.4 million tons

Lead emissions
1970: 221 million tons
1997: 4 million tons

THE WATER

Oil-polluting incidents reported in and around U.S. waters
1970: 15.2 million gallons
1998: 885,000 gallons

THE FOOD SUPPLY

Conventional pesticide usage
1970: 760 million pounds
1997: 975 million pounds

Percent of U.S. food samples found to contain pesticide residues
1978: 47%
1998: 35%

WASTE DISPOSAL

Municipal solid waste discarded
1970: 121 million tons
1997: 217 million tons

Amount recycled
1970: 8 million tons
1997: 49 million tons

GLOBAL WARMING

Atmospheric carbon dioxide concentrations
1970: 325 parts per million
1998: 367 parts per million

SPENDING ON POLLUTION

Private sector
1972: $32 billion
1994: $72.9 billion

than in earlier decades, when workers at Paducah's Gaseous Diffusion Plant brushed green uranium dust off their lunches. And with the end of the Cold War, there is less new nuclear waste being produced. But because radioactivity is both long lasting and cumulative, the nation's inventory of nuclear waste and the capacity for additional contamination have grown dramatically.

The biodiversity of animals and plants also is in decline, with a few notable exceptions. Several animals that were threatened or near extinction in the lower 48 states at the time of the 1973 Endangered Species Act, including peregrine falcons, bald eagles, gray wolves, and California condors, have had their ranks grow in recent decades. But on the whole, the nation's biodiversity has diminished, as sprawl and alien species have wreaked havoc with native habitat. According to the most recent federal assessment, one third of threatened or endangered species listed by the government are in decline, roughly a quarter are stable, and less than a tenth are improving in status. Dwindling salmon runs in the Pacific Northwest are perhaps the most visible loss since 1970.

In many developing nations, the environment is worsening. Tropical forests have shrunk rapidly, accelerating the irreversible extinctions of rare animals and plants that are heavily concentrated in biological "hot spots" there. Water and air pollution are still deadly plagues: More than 3 million Third World children die each year from preventable waterborne diseases. And the world's oceans are being depleted. From 1975 to 1995, the global marine fish catch nearly doubled.

Eve of destruction? At the time of the first Earth Day, President Nixon, the pope, and Nobel laureates such as Gunnar Myrdal all worried that the Earth was on the verge of environmental calamity. Pope Paul VI in 1970 cautioned that "rivers, lakes, and even the oceans are already polluted to the point where there is reason to fear a real 'biological death' in the near future." Since that time, a series of looming ecodisasters, such as a widening ozone hole and acid-rain die-offs, either have turned out to be exaggerated or been curbed by antipollution regulation. "To some extent the environmental movement has been organized around the self-unfulfilling prophecy, but our critics forget that people and government intervened to change the trend line," says Earth Day Chair Hayes.

The intervention Hayes and fellow Earth Day 2000 participants are pushing now: a shift to cleaner energy sources and reductions in greenhouse gases that contribute to climate change. Numerous scientists have warned that global warming could one day wipe out several island states, displace millions of people in shoreline communities, and create epic droughts and food shortages. But its ultimate impact may depend just as much on nature as on policy changes. Human-induced emissions of carbon dioxide, for example, are dwarfed by the planet's natural carbon cycle, which stores vast amounts of this greenhouse gas in deep oceans and forests. Several decades hence, utilities and industry may be able to bury most man-made carbon dioxide emissions in the deep sea, allowing them to reduce atmospheric concentrations of carbon dioxide.

It is too early to say whether such carbon sequestration strategies are safe and can really arrest global warming. But as the world celebrates Earth Day 2000, it may be worth recalling that doom-and-gloom environmental predictions have proved wrong more often than they have proved right.

With Laura Tangley, Nancy L. Bentrup, and Frank McCoy

ON FINDING OUT MORE

There is probably more printed information on environmental issues, regulations, and concerns than on any other major topic. So much is available from such a wide and diverse group of sources, that the first effort at finding information seems an intimidating and even impossible task. Attempting to ferret out what agencies are responsible for what concerns, what organizations to contact for specific environmental information, and who is in charge of what becomes increasingly more difficult.

To list all of the governmental agencies private and public organizations, and journals devoted primarily to environmental issues is, of course, beyond the scope of this current volume. However, we feel that a short primer on environmental information retrieval should be included in order to serve as a springboard for further involvement; for it is through informed involvement that issues, such as those presented, will eventually be corrected.

I. SELECTED OFFICES WITHIN FEDERAL AGENCIES AND FEDERAL-STATE AGENCIES FOR ENVIRONMENTAL INFORMATION RETRIEVAL

Appalachian Regional Commission
1666 Connecticut Avenue, NW, Suite 700, Washington, DC 20009-1068 (202) 884-7799
http://www.arc.gov

Council on Environmental Quality
722 Jackson Place, N.W., Washington, DC 20503 (202) 395-5750
http://www.whitehouse.gov/CEQ/

Delaware River Basin Commission
25 State Police Dr.
P.O. Box 7360, West Trenton, NJ 08628-0360 (609) 883-9500 http://www.state.nj.us/drbc/drbc.htm

Department of Agriculture
14th and Independence Avenue, SW, Washington, DC 20250 (202) 720-2791 http://www.usda.gov

Department of the Army (Corps of Engineers)
20 Massachusetts Ave., NW, Washington, DC 20314-1000 (202) 761-0660 http://www.usace.army.mil

Department of Commerce
14th and Constitution Ave. NW, Washington, DC 20230 (202) 482-2000 http://www.doc.gov

Department of Defense
Public Affairs, 1400 Defense Pentagon, Room 1E757, Washington, DC 20301-1400 (703) 697-5737
http://www.defenselink.mil/index.html

Department of Health and Human Services
200 Independence Avenue, SW, Washington, DC 20201 1 (877) 696-6775 http://www.os.dhhs.gov

Department of the Interior
1849 C Street, NW, Washington, DC 20240-0001 (202) 208-3100 http://www.doi.gov
- Bureau of Indian Affairs (202) 208-3711
- Bureau of Land Management (202) 452-5125
- National Park Service (202) 208-6843
- United States Fish and Wildlife Service (202) 208-4131

Department of State, Bureau of Oceans and International Environmental and Scientific Affairs
2201 C Street, NW, Washington, DC 20520 (202) 647-2492
http://www.state.gov/www/global/oes/

Department of the Treasury, U.S. Customs Service
1300 Pennsylvania Avenue, NW, Washington, DC 20229 (202) 927-1000 http://www.customs.ustreas.gov

Environmental Protection Agency (EPA)
401 M Street, SW, Washington, DC 20460 (202) 260-2090
- *Region 1,* One Congress Street, Suite 1100, Boston, MA 02203-0001 (617) 565-3420 (888) 372-7341 http://www.epa.gov/region01/ (Connecticut, Maine, Massachusetts, New Hampshire, Rhode Island, Vermont)
- *Region 2,* 290 Broadway, New York, NY 10007-1866 (212) 637-3000 http://www.epa.gov/region02/epd/in-news.htm (New Jersey, New York, Puerto Rico, Virgin Islands)
- *Region 3,* 1650 Arch Street, Philadelphia, PA 19103-2029 (215) 814-2900 (800) 438-2474
http://www.epa.gov/region3/news.htm (Delaware, District of Columbia, Maryland, Pennsylvania, Virginia, West Virginia)
- *Region 4,* Atlanta Federal Center, 61 Forsyth Street, SW, Atlanta, GA 30303-3104 (404) 562-9900
http://www.epa.gov/earth/ (Alabama, Florida, Georgia, Kentucky, Mississippi, North Carolina, South Carolina, Tennessee)
- *Region 5,* 77 West Jackson Blvd., Chicago, IL 60604-3507 (312) 353-2000 http://www.epa.gov/region5/ (Illinois, Indiana, Michigan, Minnesota, Ohio, Wisconsin)
- *Region 6,* 1445 Ross Avenue, Suite 1200, Dallas, TX 75202 (214) 665-2200 (800) 887-6063
http://www.epa.gov/earthlr6/ (Arkansas, Louisiana, New Mexico, Oklahoma, Texas)
- *Region 7,* 901 N. 5th Street, Kansas City, KS 66101 (913) 551-7003 (800) 223-0425
http://www.epa.gov/region07/index.html (Iowa, Kansas, Missouri, Nebraska)
- *Region 8,* 999 18th Street, Suite 300, Denver, CO 80202-2466 (303) 312-6312 (800) 227-8917
http://www.epa.gov/region08/ (Colorado, Montana, North Dakota, South Dakota, Utah, Wyoming)
- *Region 9,* 75 Hawthorne Street, San Francisco, CA 94105 (415) 744-1500 http://www.epa.gov/region09/ (Arizona, California, Hawaii, Nevada, American Samoa, Guam, Trust Territories of Pacific Islands, Wake Island)
- *Region 10,* 1200 Sixth Avenue, Seattle, WA 98101 (206) 553-1200 (800) 424-4372
http://www.epa.gov/r10earth/index.htm (Alaska, Idaho, Oregon, Washington)

Federal Energy Regulatory Commission
888 First Street, NE, Washington, DC 20426 (202) 208-2990 http://www.ferc.fed.us

Interstate Commission on Potomac River Basin
6110 Executive Boulevard, Suite 300, Rockville, MD 20852-3903 (301) 984-1908
http://www.potomacriver.org

Nuclear Regulatory Commission
U.S. Nuclear Regulatory Commission, Washington, DC, 20555
(301) 415-8200 http://www.nrc.gov

Susquehanna River Basin Commission
1721 North Front Street, Harrisburg, PA 17102 (717) 238-0423 http://www.srbc.net

Tennessee Valley Authority
400 West Summit Hill Drive, Knoxville, TN 37902 (865) 632-2101 http://www.tva.gov

II. SELECTED STATE, TERRITORIAL, AND CITIZENS' ORGANIZATIONS FOR ENVIRONMENTAL INFORMATION RETRIEVAL

A. Government Agencies

Alabama:
Department of Conservation and Natural Resources, 64 North Union St., Montgomery, AL 36130-1450 (334) 242-3486
http://www.dcnr.state.al.us

Alaska:
Department of Environmental Conservation, 410 Willoughby Avenue, Suite 303, Juneau, AL 99801-1795 (907) 465-5065
http://www.state.ak.us/local/akpages/ENV.CONSERV/home.htm

Arizona:
Department of Water Resources, 500 North 3rd Street, Phoenix, AZ 85004-3226 (602) 417-2400 (800) 352-8400
http://www.water.az.us

Natural Resources Division, 1616 West Adams Street, Phoenix, AZ 85007 (602) 542-4625

Arkansas:
Department of Environmental Quality, 8001 National Drive, Little Rock, AR 72209 (501) 682-0744
http://www.adeq.state.ar.us

California:
Conservation Department Resources Agency, 801 K Street, MS24-01, Sacramento, CA 95814 (916) 322-1080
http://www.consrv.ca.gov

Environmental Protection Agency, 1001 I Street, Sacramento, CA 95814 (916) 445-3846
http://www.calepa.ca.gov

Colorado:
Department of Natural Resources, 1313 Sherman Street, Room 718, Denver, CO 80203-2239 (303) 866-3311
http://www.dnr.state.co.us

Connecticut:
Department of Environmental Protection, State Office Building, 79 Elm Street, Hartford, CT 06106-5127 (860) 424-3000
http://www.dep.state.ct.us

Delaware:
Natural Resources and Environmental Control Department, 89 Kings Highway, Dover, DE 19901 (302) 739-4403
http://www.dnrec.state.de.us

District of Columbia:
Environmental Regulation Administration, 2100 Martin Luther King Jr. Avenue, SE, Suite 203, Washington, DC 20020 (202) 404-1167

Florida:
Department of Environmental Protection, 3900 Commonwealth Blvd., M.S.49, Tallahassee, FL 32399-3000 (850) 921-1222
http://www.dep.state.fl.us

Georgia:
Department of Natural Resources, 205 Butler Street, SE, Suite 1152, East Tower, Atlanta, GA 30334-4100 (404) 657-5947
http://www.ganet.org/dnr/

Guam:
Environmental Protection Agency, IT&E Harmon Plaza, Complex Unit D-107, 130 Rojas St., Harmon, Guam 96911 (617) 646-9402

Hawaii:
Department of Land and Natural Resources, Kalanimoku Bldg., 1151 Punchbowl St, Honolulu, HI 96813 (808) 587-0400
http://www.state.hi.us/dlnr/

Idaho:
Department of Lands, 954 W. Jefferson St., Boise, ID 83720-0050 (208) 334-0200
http://www2.state.id.us/lands/index.htm

Department of Water Resources, 1301 North Orchard Street, Boise, ID 83706 (208) 327-7900
http://www.idwr.state.id.us

Illinois:
Department of Natural Resources, Lincoln Tower Plaza, 524 South 2nd Street, Springfield, IL 62701 (217) 785-0075
http://dnr.state.il.us

Indiana:
Department of Natural Resources, 402 W. Washington St., Indianapolis, IN 46204-2212 (317) 232-4020
http://www.state.in.us/dnr/

Iowa:
Department of Natural Resources, 502 E. 9th St., Wallace State Office Bldg., Des Moines, IA 50319 (515) 281-5918
http://www.state.ia.us/dnr/

Kansas:
Department of Health and Environment, Capital Tower Bldg., Topeka, KS 66603 (785) 296-0461
http://www.state.ks.us/public/kdhe/

Kentucky:
Natural Resources and Environmental Protection Cabinet Capital Plaza Tower, Frankfort, KY 40601 (502) 564-5525
http://www.nr.state.ky.us/nrhome.htm

Louisiana:
Department of Environmental Quality, 7290 Bluebonnet Blvd., Baton Rouge, LA 70810 (225) 765-0741
http://www.deq.state.la.us

Department of Natural Resources, 625 North 4th Street, P.O. Box 94396, Baton Rouge, LA 70804-9396 (225) 342-2707
http://www.dnr.state.la.us/

Maine:
Department of Environmental Protection, 17 State House Station, Augusta, ME 04333-0017 (207) 287-7688 (800) 452-1942
http://www.state.me.us/dep/

Maryland:
Department of Natural Resources, 580 Taylor Avenue, Tawes State Office Bldg., Annapolis, MD 21401 (410) 260-8100 (877) 620-8367
http://www.dnr.state.md.us

Massachusetts:
Department of Environmental Management, 251 Causeway St., Suite 600, Boston, MA 02114 (617) 973-8700
http://www.state.ma.us/dem/dem.htm

Michigan:
Department of Natural Resources, Mason Bldg, 6th Floor, Box 30028, Lansing, MI 48909 (517) 373-2329
http://www.dnr.state.mi.us

Minnesota:
Department of Natural Resources, 500 Lafayette Road, St. Paul, MN 55155-4040 (651) 296-6157
http://www.dnr.state.mn.us

Mississippi:
Department of Environmental Quality, P.O. Box 20305, Jackson, MS 39289 (601) 961-5171
http://www.deq.state.ms.us/

Missouri:
Department of Natural Resources, P.O. Box 176, Jefferson City, MO 65102 (800) 334-6946
http://www.dnr.state.mo.us/

Montana:
Department of Natural Resources and Conservation, 1625 11th Avenue, P.O. Box 201601, Helena, MT 59620-1601 (406) 444-2074
http://www.dnrc.state.mt.us

Nebraska:
Department of Environmental Quality, P.O. Box 98922, Lincoln, NE 68509 (402) 471-2186
http://www.deq.state.ne.us

Nevada:
Department of Conservation and Natural Resources, 123 W. Nye Lane, Room 230 Carson City, NV 89706 (775) 687-4360
http://www.state.nv.us/cnr/

New Hampshire:
Department of Environmental Services, 6 Hazen Drive, P.O. Box 95, Concord, NH 03302-0095 (603) 271-3503
http://www.des.state.nh.us

Department of Resources and Economic Development, P.O. Box 1856, Concord, NH 03302-1856 (603) 271-2411
http://www.dred.state.nh.us

New Jersey:
Department of Environmental Protection, 401 E. State Street, P.O. Box 402, Trenton, NJ 08625-0402 (609)

292-2885
http://www.state.nj.us/dep/

New Mexico:
Environmental Department, 1190 Saint Francis Drive, Santa Fe, NM 87505 (505) 827-2855 (800) 219-6157
http://www.nmenv.state.nm.us

New York:
Department of Environmental Conservation, 50 Wolf Road, Albany, NY 12233 (518) 485-8940
http://www.dec.state.ny.us

North Carolina:
Department of Environment and Natural Resources, 1601 Mail Service Center, Raleigh, NC 27699 (919) 733-4984
http://www.ehnr.state.nc.us/

North Dakota:
Game & Fish Department, 100 North Bismarck Expressway, Bismarck, ND 58501 (701) 328-6300
http://www.state.nd.us/gnf/

Ohio:
Department of Natural Resources, 1952 Belcher Drive, Building C-1, Columbus, OH 43224 (614) 265-6565
http://www.dnr.state.oh.us

Environmental Protection Agency, P.O. Box 1049, Columbus, OH 43216-1049 (614) 644-3020
http://www.epa.state.oh.us

Oklahoma:
Conservation Commission, 2800 North Lincoln Boulevard, Suite 160, Oklahoma City, OK 73105-4210 (405) 521-2384
http://www.okcc.state.ok.us/

Department of Environmental Quality, P.O. Box 1677, Oklahoma City, OK 73101 (405) 702-1000
http://www.deq.state.ok.us

Oregon:
Department of Environmental Quality, 811 S.W. 6th Avenue, Portland, OR 97204-1390 (503) 229-5696
http://www.deq.state.or.us

Pennsylvania:
Department of Environmental Resources, 400 Market Street, Harrisburg, PA 17105 (717) 787-2814
http://www.dep.state.pa.us

Puerto Rico:
Department of Natural Resources, P.O. Box 5887, San Juan, PR 00906 (787) 723-3090

Rhode Island:
Department of Environmental Management, 235 Promenade Street, Providence, RI 02908 (401) 222-2771
http://www.state.ri.us/dem/

South Carolina:
Department of Health and Environmental Control, 2600 Bull Street, Columbia, SC 29201 (803) 898-3432
http://www.scdhec.net

Department of Natural Resources, 1000 Assembly Street, Columbia, SC 29201 (803) 734-3888
http://water.dnr.state.sc.us

South Dakota:
Department of Environment and Natural Resources, Joe Foss Bldg., 523 East Capitol, Pierre, SD 57501 (605) 773-3151
http://www.state.sd.us/denr/denr.html

Tennessee:
Department of Environment and Conservation, 401 Church St., 21st Floor, Nashville, TN 37243 (888) 891-8332
http://www.state.tn.us/environment/

Texas:
Natural Resources Conservation Commission, P.O. Box 13087, Austin, TX 78711 (512) 239-1000
http://www.tnrcc.state.tx.us

Utah:
Department of Natural Resources, 1594 West North Temple, Suite 3710, Salt Lake City, UT 84114 (801) 538-7200
http://www.nr.state.ut.us

Vermont:
Agency of Natural Resources, 103 South Main Street, Waterbury, VT 05671-0301 (802) 241-3600
http://www.anr.state.vt.us

Virgin Islands:
Department of Planning & Natural Resources, 396-1 Annas Retreat, Foster Bldg., Charlotte Amalie, U.S. Virgin Islands 00802 (340) 774-3320
http://www.gov.vi/pnr/

Virginia:
Secretary of Natural Resources, P.O. Box 1475, Richmond, VA 23212 (804) 786-0044
http://snr.vipnet.org

Washington:
Department of Ecology, P.O. Box 47600, Olympia, WA 98504-7600 (360) 407-6000
http://www.ecy.wa.gov/

Department of Natural Resources, P.O. Box 47001, Olympia, WA 98504-7001 (360) 902-1004
http://www.wa.gov/dnr/

West Virginia:
Division of Natural Resources, State Capitol Bld 3, 1900 Kanawha Blvd., Charleston, WV 25305 (304) 558-2754
http://www.dnr.state.wv.us

Wisconsin:
Department of Natural Resources, Box 7921, Madison, WI 53707 (608) 266-2621
http://www.dnr.state.wi.us

Wyoming:
Department of Environmental Quality, 122 West 25th Street, Herschler Bldg., Cheyenne, WY 82002 (307) 777-7758
http://deq.state.wy.us

B. Citizens' Organizations

Advancement of Earth & Environmental Sciences, International Association for, Northeastern Illinois University, Geography and Environmental Studies Department 5500 North St. Louis Avenue, Chicago, Illinois 60625 (312) 794-2628

Air Pollution Control Association 1 Gateway Center, 3rd Floor, Pittsburgh, PA 15222 (412) 232-3444

American Association for the Advancement of Science, 1200 New York Avenue, NW, Washington, DC 20005 (202) 326-6400
http://www.aaas.org/

American Chemical Society, 1155 16th Street, NW, Washington, DC 20036 (800) 227-5558
http://www.acs.org/

American Farm Bureau Federation, 225 Touhy Avenue, Park Ridge, IL 60068 (847) 685-8600
http://www.fb.com

American Fisheries Society, 5410 Grosvenor Lane, Suite 110, Bethesda, MD 20814-2199 (301) 897-8616
http://www.fisheries.org

American Forest & Paper Association, 1111 19th Street, NW, Suite 800, Washington, DC 20036 (202) 463-2700
http://www.afandpa.org

American Forests, P.O. Box 2000, Washington, DC 20013 (202) 955-4500
http://www.americanforests.org

American Institute of Biological Sciences, 1444 I Street NW, Washington, DC 20005 (202) 628-1500
http://www.aibs.org

American Museum of Natural History, Central Park West at 79th Street, New York, NY 10024-5192 (212) 769-5100
http://www.amnh.org

American Petroleum Institute, 1220 L Street, NW, Washington, DC 20005-4070 (202) 682-8000
http://www.api.org

American Rivers, 1025 Vermont Avenue NW, Suite 720, Washington, DC 20005 (202) 347-7550
http://www.amrivers.org

Association for Conservation Information, P.O. Box 12559, Charleston, SC 29412 (803) 762-5032

Boone and Crockett Club, 250 Station Dr., Missoula, MT 59801 (406) 542-1888
http://www.boone-crockett.org

Center for Marine Conservation, 1725 DeSales Street, NW, Suite 600, Washington, DC 20036 (202) 429-5609
http://www.cmc-ocean.org

Citizens for a Better Environment, 3255 Hennepin Avenue South, Suite 70, Minneapolis, MN 55408 (612) 824-8637
http://www.cbemw.org

Coastal Conservation Association, 4801 Woodway, Suite 220W, Houston, TX 77056 (713) 626-4222
http://www.ccatexas.org

Conservation Foundation, 1250 24th Street, NW, Suite 400, Washington, DC 20037 (202) 293-4800

Conservation Fund, 1800 North Kent Street, Suite 1120, Arlington, VA 22209-2156 (703) 525-6300
http://www.conservationfund.org

Conservation International, 2501 M Street, NW, Suite 200, Washington, DC 20037 (202) 429-5660 (800) 429-5660
http://www.conservation.org

Defenders of Wildlife, 1101 14th Street, NW, #1400, Washington, DC, 20005 (202) 682-9400
http://www.defenders.org

Ducks Unlimited, Inc., One Waterfowl Way, Memphis, TN (901) 758-3825 (800) 45DUCKS
http://www.ducks.org

Earthwatch Institute International, 3 Clock Tower Place, Suite 100, Box 75 Maynard, MA 01754 (800) 776-0188
http://www.earthwatch.org

Environmental Action Foundation, Inc., 6930 Carroll Ave., Suite 600, Takoma Park, MD 20912 (301) 891-1100

Food and Agriculture Organization of the United Nations (FAO), Via delle Terme di Caracalla, 1-00100, Rome, Italy phone tt39.06.570-51 http://www.fao.org

Friends of the Earth, 26-28 Underwood St., London, N1 7JQ 020-7490-1555
http://foe.co.uk

Greenpeace U.S.A., 702 H Street, NW, Washington, DC 20001 (800) 326-0959
http://greenpeaceusa.org

International Association of Fish and Wildlife Agencies, 444 North Capitol Street, NW, Suite 544, Washington, DC 20001 (202) 624-7890
http://www.teaming.com/iafwa.htm

International Fund for Agricultural Development (IFAD), 107, Via del Serafico, Rome, Italy 00142 (3906) 54591
http://www.ifad.org

Keep America Beautiful, Inc., 1010 Washington Blvd., Stamford, CT 06901 (203) 323-8987
http://www.kab.org

National Association of Conservation Districts, 509 Capitol Court, NE, Washington, DC 20002 (202) 547-6223
http://www.nacdnet.org

National Audubon Society, 700 Broadway, New York, NY 10003 (212) 979-3000
http://www.audubon.org

National Environmental Health Association, 720 South Colorado Boulevard, 970 South Tower, Denver, CO 80246 (303) 756-9090
http://www.neha.org

National Fisheries Institute, 1901 N. Fort Myer Dr., Suite 700, Arlington, VA 22209 (703) 524-8880
http://www.nfi.org/main.html

National Geographic Society, 1145 17th Street, NW, Washington, DC 20036 (800) 647-5463
http://www.nationalgeographic.com

National Parks and Conservation Association, 1300 19th St., NW, Washington, DC 20036 (202) 223-6722 (800) 628-7275
http://www.npca.org/flash.html

National Wildlife Federation, 8925 Leesburg Pike, Vienna, VA 22184 (703) 790-4000
http://www.nwf.org

Natural Resources Council of America, 801 Pennsylvania Avenue, SE, No. 410, Washington, DC 20003 (202) 333-0411

Nature Conservancy, 4245 North Fairfax Dr., Arlington, VA 22203 (800) 628-6860
http://www.tnc.org

Population Association of America, 8630 Fenton St., Suite 722, Silver Spring, MD 20910 (301) 565-6710
http://www.popassoc.org/

Rainforest Alliance, 65 Bleecker Street, New York, NY 10012 (212) 677-1900
http://www.rainforest-alliance.org

Save-the-Redwoods League, 114 Sansome Street, Room 1200, San Francisco, CA 94104 (415) 362-2352
http://www.savetheredwoods.org

Sierra Club, 85 Second Street, 2nd Floor, San Francisco, CA 94105 (415) 977-5500
http://www.sierraclub.org

Smithsonian Institution, SI Building, Room 153, Washington, DC 20560 (202) 357-2700
http://www.si.edu/

Society of American Foresters, 5400 Grosvenor Lane, Bethesda, MD 20814-2198 (301) 897-8720
http://www.safnet.org

Sport Fishing Institute, 1010 Massachusetts Avenue, NW, Suite 320, Washington, DC 20001 (202) 898-0770

United Nations Educational, Scientific, and Cultural Organization (UNESCO), UNESCO House, 7, Place de Fontenoy, 75352 Paris 07 SP France, (331) 45 68 10 00
http://www.unesco.org

United Nations Environment Programme/Industry & Environment Centre, Tour Mirabeau 39-43, quai André Citroën 75739 Paris, cedex 15, France (331) 44 37 14 50
http://www. unepie.org

Wilderness Society, 1615 M St., NW, Washington, DC 20036-2596 1 (800) THE-WILD
http://www.wilderness.org

World Wildlife Fund, 1250 24th Street, NW, Washington, DC 20037 (800) CALL-WWF
http://www.wwf.org

Zero Population Growth, Inc., 1400 16th Street, NW, Suite 320, Washington, DC 20036 (202) 332-2200
http://www.zpg.org

III. CANADIAN AGENCIES AND CITIZENS' ORGANIZATIONS

A: Government Agencies

Alberta:

Alberta Environmental Protection, Information Centre Main Floor, 9920-108 St., Edmonton, AB T5K 2M6 (780) 944-0313 http://www.gov.ab.ca/env

British Columbia:

Ministry of Environment, Lands, and Parks, P.O. Box 9360 STN PROV GOVT, Victoria, BC V8W 9M2 (250) 387-9422 http://www.gov.bc.ca/ep/

Manitoba:

Manitoba Environment, 333 Legislative Bldg., Winnipeg, MB R3C 0V8 (204) 945-3730
http://www.gov.mb.ca/environ/index.html

Manitoba Natural Resources, Box 24, 200 Saulteaux Crescent, Winnipeg, MB R3J 3W3 (800) 214-6497
http://www.gov.mb.ca/natres/index.html

New Brunswick:

Department of Environment, P.O. Box 6000, Fredericton, NB E3B 5H1 (506) 453-3827
http://www.nb.ca/eig-egl/index.htm

Department of Natural Resources and Energy, P.O. Box

6000, Fredericton, NB E3B 5H1 (506) 453-2614
http://www.gov.nb.ca/0078/index.htm

Newfoundland and Labrador:
Department of Environment and Labour, 4th Floor, West Block, Confederation Bldg., PO Box 8700, St. John's, NF A1B 4J6 (709) 729-2664
http://www.gov.nf.ca/env/Labour/OHS/default.asp

Northwest Territories:
Department of Resources, Wildlife, and Economic Development, #600 Scotia Centre, Bldg. Box 21, 5102-50 Avenue, Yellowknife, NT X1A 3S8 (867) 669-2366
http://www.rwed.gov.nt.ca

Nova Scotia:
Department of Natural Resources, P.O. Box 698, Halifax, NS Canada B3J 2T9 (902) 424-5935
http://www.gov.ns.ca/natr/

Department of the Environment, P.O. Box 697, Halifax, NS Canada B3J 3B7 (902) 424-5300
http://www.gov.ns.ca/enla/

Ontario:
Ministry of Natural Resources, 300 Water Street, P.O. Box 7000, Peterborough, ON K9J 8M5 (705) 755-2000 (416) 314-2000
http://www.mnr.gov.on.ca/mnr/

Prince Edward Island:
Department of Fisheries, Agriculture, and Environment, Jones Bldg., 11 Kent Street, 4th Floor, P.O. Box 2000, Charlottetown, PEI C1A 7N8 (902) 368-5000
http://www.gov.pe.ca/fae/index.php3

Quebec:
Ministère de l'Environnement et de la Faune, Edifice Marie-Guyart, 675, boulevard René-Lévesque, Est, Québec, PC G1R 5V7 (418) 521-3830 (800) 561-1616
http://www.menv.gouv.qc.ca

Ministère des Ressources Naturelles, #B-302, 5700, 4e Avenue Ouest, Charlesbourg, Quebec G1H 6R1 (418) 627-8600
http://www.mrn.gouv.qc.ca

Saskatchewan:
Saskatchewan Environment and Resource Management, 3211 Albert Street, Regina, SK S4S 5W6 (306) 787-2700
http://www.serm.gov.sk.ca/

Yukon Territory:
Government of Yukon, Box 2703, Whitehorse, YT Y1A 2C6 (867) 667-5811 http://www.gov.yk.ca

Department of Renewable Resources, Box 2703, Whitehorse, YT Y1A 2C6 (867) 667-5652 http://www.renres.gov.yk.ca/

B. Citizens' Groups

Alberta Wilderness Association
Box 6398, Station D, Calgary, AB T2P 2E1 (403) 283-2025 http://albertawilderness.ab.ca/

BC Environmental Network (BCEN) 610-207 Hastings St., Vancouver, BC V6B 1H7 (604) 879-2279
http://www.bcen.bc.ca

Canadian EarthCare Society
1476 Water Street, Kelowna, BC V1Y 8P2 (604) 861-4788
http://www.earthcare.org

Ducks Unlimited Canada
P.O. Box 1150, Stonewall, MB Canada R0C 2Z0 (204) 467-3000 (800) 665-3825 http://www.ducks.ca

Federation of Ontario Naturalists
355 Lesmill Road, Don Mills, ON M3B 2W8 (416) 444-8419
http://www.ontarionature.org

L'Association des Entrepreneurs de Service en Environnement du Quebec (AESEQ)
911 Jean-Talon, Est 220, Montreal, PQ H2R 1V5 (514) 270-7110

New Brunswick Environment Industry Association
P.O. Box 637, Stn. A, Fredericton, Canada NB E3B 5B3 (506) 455-0212
http://www.nbeia.nb.ca/index.html

Prince Edward Island Environmental Network (PEIEN)
126 Richmond Street, Charlottetown, PEI C1A 1H9 (902) 566-4170 http://www.isn.net/~network/index.html

Yukon Conservation Society (YCS)
P.O. Box 4163, Whitehorse, YT Y1A 3T3 (403) 668-6637

IV. SELECTED JOURNALS AND PERIODICALS OF ENVIRONMENTAL INTEREST

American Forests
American Forests
P.O. Box 2000, Washington, DC 20013 (202) 955-4500 (800) 368-5748
http://www.americanforests.org/

American Scientist
Scientific Research Society, P.O. Box 13975, Research Triangle Park, NC 27709-3975 (919) 549-0097
http://www.amsci.org/amsci/amsci.html

Audubon
National Audubon Society, 700 Broadway, New York, NY 10003 (212) 979-3000
http://magazine.audubon.org

BioScience
American Institute of Biological Sciences, 1444 I St. NW, Suite 200, Washington, DC 20005 (202) 628-1500
http://www.aibs.org

California Environmental Directory
California Institute of Public Affairs, P.O. Box 189040, Sacramento, CA 95818 (916) 442-2472
http://www.cipahq.org/

The Canadian Field-Naturalist
Box 35069 Westgate, Ottawa, ON, Canada K1Z 1A2 (613) 722-3050
http://www.achilles.net/ofnc/cfn.htm

Conservation Directory
National Wildlife Federation, 8925 Leesburg Pike, Vienna, VA 22184 (703) 790-4000
http://www.nwf.org/printandfilm/publications/consdir/index.html

Earth First! Journal
P.O. Box 1415, Eugene, OR 97440-1415 (541) 344-8004
http://www.earthfirstjournal.org/frontcover.cfm

E: The Environmental Magazine
Earth Action Network, P.O. Box 5098, Westport, CT 06881 (203) 854-5559 http://www.emagazine.com

Environment
Heldref Publications, 1319 18th Street, NW, Washington, DC 20036-1802 (202) 296-6267
http://www.heldref.org

Environment Reporter
Bureau of National Affairs, Inc. 1231 25th Street, NW, Washington, DC 20037 (800) 372-1033
http://www.bna.com/prodcatalog/desc/ER.html

Environmental Action Magazine
Environment Action, Inc. 6930 Carroll Ave., Suite 600, Takoma Park, MD 20912-4414 (301) 891-1106

Environmental Science and Technology
American Chemical Society Publications Support Services, 1155 16th Street, NW, Washington, DC 20036 (202) 872-4554 (800) 227-5558
http://pubs.acs.org/journals/esthag/

Focus (bimonthly newsletter)
World Wildlife Fund, 1250 24th Street, NW, Washington, DC 20037 (202) 293-4800
http://www.wwf.org

The Futurist
World Future Society, 7910 Woodmont Avenue, Suite 450, Bethesda, MD 20814 (301) 656-8274 (800) 989-8274
http://www.wfs.org/wfs/

Greenpeace Magazine
  Greenpeace USA, 1436 U Street, NW, Washington, DC 20009 (202) 462-1177 (800) 326-0959
  http://www.greenpeaceusa.org

Journal of Soil and Water Conservation
  Soil and Water Conservation Society, 7515 Northeast Ankeny Road, Ankeny, IA 50021-9764 (515) 289-2331
  http://www.swcs.org

Journal of Wildlife Management
  The Wildlife Society, Tall Timbers Research Station, 13093 Henry Beadel Dr., Tallahassee, FL 32312 (850) 893-4153
  http://www.wildlife.org/journal.html

Mother Earth News
  Mother Earth News, P.O. Box 56302, Boulder, CO 80322-6302 (800) 234-3368
  http://www.MotherEarthNews.Com

National Wildlife
  National Wildlife Federation, 8925 Leesburg Pike, Vienna, VA 22184 (703) 790-4510
  http://www.nwf.org/natlwild/

Natural Resources Journal
  University of New Mexico, School of Law, 1117 Stanford, NE, Albuquerque, NM 87131 (505) 277-4820
  http://www.unm.edu/~natresj/NRJ/NRJ.html

Nature
  Macmillan Publishers Ltd., Porter South, Grinan Street, London N1 9XW, England (44) 0171 833-4000
  http://www.nature.com

Nature Canada
  Canadian Nature Federation, One Nicholas Street, Suite 606, Ottawa, ON, Canada K1N 787 (613) 562-3447
  http://www.cnf.ca/naturecanada/

Nature Conservancy Magazine
  4245 North Fairfax Dr., Arlington, VA 22209-2003 (703) 841-5300 (800) 267-4088
  http://www.tnc.org/home2.html

Pollution Abstracts
  Cambridge Scientific Abstracts, 7200 Wisconsin Avenue, Suite 601, Bethesda, MD 20814-4823 (301) 961-6700 (800) 843-7751
  http://www.csa.com

Science
  American Association for the Advancement of Science, 1200 New York Avenue NW, Washington, DC 20005 (202) 326-6550
  http://www.sciencemag.org

Sierra Magazine
  Sierra Club, P.O. Box 52968, Boulder, CO 80328 (800) 765-7904
  http://www.sierraclub.org/sierra/

Smithsonian
  900 Jefferson Drive, Washington, DC 20560 (202) 786-2900
  http://smithsonianmag.com

Technology Review
  201 Vassar Street, W59-200 Cambridge, MA 02139 (617) 253-8250
  http://www.techreview.com

U.S. News and World Report
  2400 N Street, NW, Washington, DC 20037-1196 (202) 955-2000
  http://www.usnews.com/usnews/home.htm

The World & I
  New World Communications, 3600 New York Avenue, NE, Washington, DC 20002 (800) 822-2822 (202) 635-4000
  http://www.worldandi.com

Glossary

This glossary of environmental terms is included to provide you with a convenient and ready reference as you encounter general terms in your study of environment that are unfamiliar or require a review. It is not intended to be comprehensive, but taken together with the many definitions included in the articles themselves, it should prove to be quite useful.

Abiotic Without life; any system characterized by a lack of living organisms.
Absorption Incorporation of a substance into a solid or liquid body.
Acid Any compound capable of reacting with a base to form a salt; a substance containing a high hydrogen ion concentration (low pH).
Acid Rain Precipitation containing a high concentration of acid.
Adaptation Adjustment of an organism to the conditions of its environment, enabling reproduction and survival.
Additive A substance added to another in order to impart or improve desirable properties or suppress undesirable ones.
Adsorption Surface retention of solid, liquid, or gas molecules, atoms, or ions by a solid or liquid.
Aerobic Environmental conditions where oxygen is present; aerobic organisms require oxygen in order to survive.
Aerosols Tiny mineral particles in the atmosphere onto which water droplets, crystals, and other chemical compounds may adhere.
Air Quality Standard A prescribed level of a pollutant in the air that should not be exceeded.
Alcohol Fuels The processing of sugary or starchy products (such as sugar cane, corn, or potatoes) into fuel.
Allergens Substances that activate the immune system and cause an allergic response.
Alpha Particle A positively charged particle given off from the nucleus of some radioactive substances; it is identical to a helium atom that has lost its electrons.
Ammonia A colorless gas comprised of one atom of nitrogen and three atoms of hydrogen; liquefied ammonia is used as a fertilizer.
Anthropocentric Considering humans to be the central or most important part of the universe.
Aquaculture Propagation and/or rearing of any aquatic organism in artificial "wetlands" and/or ponds.
Aquifers Porous, water-saturated layers of sand, gravel, or bedrock that can yield significant amounts of water economically.
Atom The smallest particle of an element, composed of electrons moving around an inner core (nucleus) of protons and neutrons. Atoms of elements combine to form molecules and chemical compounds.
Atomic Reactor A structure fueled by radioactive materials that generates energy usually in the form of electricity; reactors are also utilized for medical and biological research.
Autotrophs Organisms capable of using chemical elements in the synthesis of larger compounds; green plants are autotrophs.

Background Radiation The normal radioactivity present; coming principally from outer space and naturally occurring radioactive substances on Earth.
Bacteria One-celled microscopic organisms found in the air, water, and soil. Bacteria cause many diseases of plants and animals; they also are beneficial in agriculture, decay of dead matter, and food and chemical industries.
Benthos Organisms living on the bottom of bodies of water.
Biocentrism Belief that all creatures have rights and values and that humans are not superior to other species.
Biochemical Oxygen Demand (BOD) The oxygen utilized in meeting the metabolic needs of aquatic organisms.
Biodegradable Capable of being reduced to simple compounds through the action of biological processes.
Biodiversity Biological diversity in an environment as indicated by numbers of different species of plants and animals.
Biogeochemical Cycles The cyclical series of transformations of an element through the organisms in a community and their physical environment.

Biological Control The suppression of reproduction of a pest organism utilizing other organisms rather than chemical means.
Biomass The weight of all living tissue in a sample.
Biome A major climax community type covering a specific area on Earth.
Biosphere The overall ecosystem of Earth. It consists of parts of the atmosphere (troposphere), hydrosphere (surface and ground water), and lithosphere (soil, surface rocks, ocean sediments, and other bodies of water).
Biota The flora and fauna in a given region.
Biotic Biological; relating to living elements of an ecosystem.
Biotic Potential Maximum possible growth rate of living systems under ideal conditions.
Birthrate Number of live births in one year per 1,000 midyear population.
Breeder Reactor A nuclear reactor in which the production of fissionable material occurs.

Cancer Invasive, out-of-control cell growth that results in malignant tumors.
Carbon Cycle Process by which carbon is incorporated into living systems, released to the atmosphere, and returned to living organisms.
Carbon Monoxide (CO) A gas, poisonous to most living systems, formed when incomplete combustion of fuel occurs.
Carcinogens Substances capable of producing cancer.
Carrying Capacity The population that an area will support without deteriorating.
Chlorinated Hydrocarbon Insecticide Synthetic organic poisons containing hydrogen, carbon, and chlorine. Because they are fat-soluble, they tend to be recycled through food chains, eventually affecting nontarget systems. Damage is normally done to the organism's nervous system. Examples include DDT, Aldrin, Deildrin, and Chlordane.
Chlorofluorocarbons (CFCs) Any of several simple gaseous compounds that contain carbon, chlorine, fluorine, and sometimes hydrogen; they are suspected of being a major cause of stratospheric ozone depletion.
Circle of Poisons Importation of food contaminated with pesticides banned for use in this country but made here and sold abroad.
Clear-Cutting The practice of removing all trees in a specific area.
Climate Description of the long-term pattern of weather in any particular area.
Climax Community Terminal state of ecological succession in an area; the redwoods are a climax community.
Coal Gasification Process of converting coal to gas; the resultant gas, if used for fuel, sharply reduces sulfur oxide emissions and particulates that result from coal burning.
Commensalism Symbiotic relationship between two different species in which one benefits while the other is neither harmed nor benefited.
Community Ecology Study of interactions of all organisms existing in a specific region.
Competitive Exclusion Resulting from competition; one species forced out of part of an available habitat by a more efficient species.
Conservation The planned management of a natural resource to prevent overexploitation, destruction, or neglect.
Conventional Pollutants Seven substances (sulfur dioxide, carbon monoxide, particulates, hydrocarbons, nitrogen oxides, photochemical oxidants, and lead) that make up the largest volume of air quality degradation, as identified by the Clean Air Act.
Core Dense, intensely hot molten metal mass, thousands of kilometers in diameter, at Earth's center.
Cornucopian Theory The belief that nature is limitless in its abundance and that perpetual growth is both possible and essential.
Corridor Connecting strip of natural habitat that allows migration of organisms from one place to another.
Critical Factor The environmental factor closest to a tolerance limit for a species at a specific time.
Cultural Eutrophication Increase in biological productivity and ecosystem succession resulting from human activities.
Crankcase Smog Devices (PCV System) A system, used principally in automobiles, designed to prevent discharge of combustion emissions into the external environment.

221

Death Rate Number of deaths in one year per 1,000 mid-year population.

Decomposer Any organism that causes the decay of organic matter; bacteria and fungi are two examples.

Deforestation The action or process of clearing forests without adequate replanting.

Degradation (of water resource) Deterioration in water quality caused by contamination or pollution that makes water unsuitable for many purposes.

Demography The statistical study of principally human populations.

Desert An arid biome characterized by little rainfall, high daily temperatures, and low diversity of animal and plant life.

Desertification Converting arid or semiarid lands into deserts by inappropriate farming practices or overgrazing.

Detergent A synthetic soap-like material that emulsifies fats and oils and holds dirt in suspension; some detergents have caused pollution problems because of certain chemicals used in their formulation.

Detrivores Organisms that consume organic litter, debris, and dung.

Diversity Number of species present in a community (species richness), as well as the relative abundance of each species.

DNA (Deoxyribonucleic Acid) One of two principal nucleic acids, the other being RNA (Ribonucleic Acid). DNA contains information used for the control of a living cell. Specific segments of DNA are now recognized as genes, those agents controlling evolutionary and hereditary processes.

Dominant Species Any species of plant or animal that is particularly abundant or controls a major portion of the energy flow in a community.

Drip Irrigation Pipe or perforated tubing used to deliver water a drop at a time directly to soil around each plant. Conserves water and reduces soil waterlogging and salinization.

Ecological Density The number of a singular species in a geographical area, including the highest concentration points within the defined boundaries.

Ecological Succession Process in which organisms occupy a site and gradually change environmental conditions so that other species can replace the original inhabitants.

Ecology Study of the interrelationships between organisms and their environments.

Ecosystem The organisms of a specific area, together with their functionally related environments; considered as a definitive unit.

Ecotourism Wildlife tourism that could damage ecosystems and disrupt species if strict guidelines governing tours to sensitive areas are not enforced.

Edge Effects Change in ecological factors at the boundary between two ecosystems. Some organisms flourish here; others are harmed.

Effluent A liquid discharged as waste.

El Niño Climatic change marked by shifting of a large warm water pool from the western Pacific Ocean toward the East.

Electron Small, negatively charged particle; normally found in orbit around the nucleus of an atom.

Eminent Domain Superior dominion exerted by a governmental state over all property within its boundaries that authorizes it to appropriate all or any part thereof to a necessary public use, with reasonable compensation being made.

Endangered Species Species considered to be in imminent danger of extinction.

Endemic Species Plants or animals that belong or are native to a particular ecosystem.

Environment Physical and biological aspects of a specific area.

Environmental Impact Statement (EIS) A study of the probable environmental impact of a development project before federal funding is provided (required by the National Environmental Policy Act of 1968).

Environmental Protection Agency (EPA) Federal agency responsible for control of air and water pollution, radiation and pesticide problems, ecological research, and solid waste disposal.

Erosion Progressive destruction or impairment of a geographical area; wind and water are the principal agents involved.

Estuary Water passage where an ocean tide meets a river current.

Eutrophic Well nourished; refers to aquatic areas rich in dissolved nutrients.

Evolution A change in the gene frequency within a population, sometimes involving a visible change in the population's characteristics.

Exhaustible Resources Earth's geologic endowment of minerals, nonmineral resources, fossil fuels, and other materials present in fixed amounts.

Extinction Irrevocable elimination of species due to either normal processes of the natural world or through changing environmental conditions.

Fallow Cropland that is plowed but not replanted and is left idle in order to restore productivity mainly through water accumulation, weed control, and buildup of soil nutrients.

Fauna The animal life of a specified area.

Feral Refers to animals or plants that have reverted to a non-cultivated or wild state.

Fission The splitting of an atom into smaller parts.

Floodplain Level land that may be submerged by floodwaters; a plain built up by stream deposition.

Flora The plant life of an area.

Flyway Geographic migration route for birds that includes the breeding and wintering areas that it connects.

Food Additive Substance added to food usually to improve color, flavor, or shelf life.

Food Chain The sequence of organisms in a community, each of which uses the lower source as its energy supply. Green plants are the ultimate basis for the entire sequence.

Fossil Fuels Coal, oil, natural gas, and/or lignite; those fuels derived from former living systems; usually called nonrenewable fuels.

Fuel Cell Manufactured chemical systems capable of producing electrical energy; they usually derive their capabilities via complex reactions involving the sun as the driving energy source.

Fusion The formation of a heavier atomic complex brought about by the addition of atomic nuclei; during the process there is an attendant release of energy.

Gaia Hypothesis Theory that Earth's biosphere is a living system whose complex interactions between its living organisms and nonliving processes regulate environmental conditions over millions of years so that life continues.

Gamma Ray A ray given off by the nucleus of some radioactive elements. A form of energy similar to X rays.

Gene Unit of heredity; segment of DNA nucleus of the cell containing information for the synthesis of a specific protein.

Gene Banks Storage of seed varieties for future breeding experiments.

Genetic Diversity Infinite variation of possible genetic combinations among individuals; what enables a species to adapt to ecological change.

Germ Plasm Genetic material that may be preserved for future use (plant seeds, animal eggs, sperm, and embryos).

Green Revolution The great increase in production of food grains (as in rice and wheat) due to the introduction of high-yielding varieties, to the use of pesticides, and to better management techniques.

Greenhouse Effect The effect noticed in greenhouses when shortwave solar radiation penetrates glass, is converted to longer wavelengths, and is blocked from escaping by the windows. It results in a temperature increase. Earth's atmosphere acts in a similar manner.

Gross National Product (GNP) The total value of the goods and services produced by the residents of a nation during a specified period (such as a year).

Groundwater Water found in porous rock and soil below the soil moisture zone and, generally, below the root zone of plants. Groundwater that saturates rock is separated from an unsaturated zone by the water table.

Habitat The natural environment of a plant or animal.

Hazardous Waste Waste that poses a risk to human or ecological health and thus requires special disposal techniques.

Herbicide Any substance used to kill plants.

Heterotroph Organism that cannot synthesize its own food and must feed on organic compounds produced by other organisms.

Hydrocarbons Organic compounds containing hydrogen, oxygen, and carbon. Commonly found in petroleum, natural gas, and coal.

Hydrogen Lightest-known gas; major element found in all living systems.

Hydrogen Sulfide Compound of hydrogen and sulfur; a toxic air contaminant that smells like rotten eggs.

Hydropower Electrical energy produced by flowing or falling water.

Infiltration Process of water percolation into soil and pores and hollows of permeable rocks.

Intangible Resources Open space, beauty, serenity, genius, information, diversity, and satisfaction are a few of these abstract commodities.

Integrated Pest Management (IPM) Designed to avoid economic loss from pests, this program's methods of pest control strive to minimize the use of environmentally hazardous, synthetic chemicals.

Invasive Refers to those species that have moved into an area and reproduced so aggressively that they have replaced some of the native species.

Ion An atom or group of atoms, possessing a charge; brought about by the loss or gain of electrons.

Ionizing Radiation Energy in the form of rays or particles that have the capacity to dislodge electrons and/or other atomic particles from matter that is irradiated.

Irradiation Exposure to any form of radiation.

Isotopes Two or more forms of an element having the same number of protons in the nucleus of each atom but different numbers of neutrons.

Keystone Species Species that are essential to the functioning of many other organisms in an ecosystem.

Kilowatt Unit of power equal to 1,000 watts.

Leaching Dissolving out of soluble materials by water percolating through soil.

Limnologist Individual who studies the physical, chemical, and biological conditions of aquatic systems.

Malnutrition Faulty or inadequate nutrition.

Malthusian Theory The theory that populations tend to increase by geometric progression (1, 2, 4, 8, 16, etc.) while food supplies increase by arithmetic means (1, 2, 3, 4, 5, etc.).

Metabolism The chemical processes in living tissue through which energy is provided for continuation of the system.

Methane Often called marsh gas (CH_4); an odorless, flammable gas that is the major constituent of natural gas. In nature it develops from decomposing organic matter.

Migration Periodic departure and return of organisms to and from a population area.

Monoculture Cultivation of a single crop, such as wheat or corn, to the exclusion of other land uses.

Mutation Change in genetic material (gene) that determines species characteristics; can be caused by a number of agents, including radiation and chemicals, called mutagens.

Natural Selection The agent of evolutionary change by which organisms possessing advantageous adaptations leave more offspring than those lacking such adaptations.

Niche The unique occupation or way of life of a plant or animal species; where it lives and what it does in the community.

Nitrate A salt of nitric acid. Nitrates are the major source of nitrogen for higher plants. Sodium nitrate and potassium nitrate are used as fertilizers.

Nitrite Highly toxic compound; salt of nitrous acid.

Nitrogen Oxides Common air pollutants. Formed by the combination of nitrogen and oxygen; often the products of petroleum combustion in automobiles.

Nonrenewable Resource Any natural resource that cannot be replaced, regenerated, or brought back to its original state once it has been extracted, for example, coal or crude oil.

Nutrient Any nutritive substance that an organism must take in from its environment because it cannot produce it as fast as it needs it or, more likely, at all.

Oil Shale Rock impregnated with oil. Regarded as a potential source of future petroleum products.

Oligotrophic Most often refers to those lakes with a low concentration of organic matter. Usually contain considerable oxygen; Lakes Tahoe and Baikal are examples.

Organic Matter Plant, animal, or microorganism matter, either living or dead.

Organophosphates A large group of nonpersistent synthetic poisons used in the pesticide industry; include parathion and malathion.

Ozone Molecule of oxygen containing three oxygen atoms; shields much of Earth from ultraviolet radiation.

Particulate Existing in the form of small separate particles, various atmospheric pollutants are industrially produced particulates.

Peroxyacyl Nitrate (PAN) Compound making up part of photochemical smog and the major plant toxicant of smog-type injury; levels as low as 0.01 ppm can injure sensitive plants. Also causes eye irritation in people.

Pesticide Any material used to kill rats, mice, bacteria, fungi, or other pests of humans.

Pesticide Treadmill A situation in which the cost of using pesticides increases while the effectiveness decreases (because pest species develop genetic resistance to the pesticides).

Petrochemicals Chemicals derived from petroleum bases.

pH Scale used to designate the degree of acidity or alkalinity; ranges from 1 to 14; a neutral solution has a pH of 7; low pHs are acid in nature, while pHs above 7 are alkaline.

Phosphate A phosphorous compound; used in medicine and as fertilizers.

Photochemical Smog Type of air pollution; results from sunlight acting with hydrocarbons and oxides of nitrogen in the atmosphere.

Photosynthesis Formation of carbohydrates from carbon dioxide and hydrogen in plants exposed to sunlight; involves a release of oxygen through the decomposition of water.

Photovoltaic Cells An energy-conversion device that captures solar energy and directly converts it to electrical current.

Physical Half-Life Time required for half of the atoms of a radioactive substance present at some beginning to become disintegrated and transformed.

Pioneer Species Hardy species that are the first to colonize a site in the beginning stage of ecological succession.

Plankton Microscopic organisms that occupy the upper water layers in both freshwater and marine ecosystems.

Plutonium Highly toxic, heavy, radioactive, manmade, metallic element. Possesses a very long physical half-life.

Pollution The process of contaminating air, water, or soil with materials that reduce the quality of the medium.

Polychlorinated Biphenyls (PCBs) Poisonous compounds similar in chemical structure to DDT. PCBs are found in a wide variety of products ranging from lubricants, waxes, asphalt, and transformers to inks and insecticides. Known to cause liver, spleen, kidney, and heart damage.

Population All members of a particular species occupying a specific area.

Predator Any organism that consumes all or part of another system; usually responsible for death of the prey.

Primary Production The energy accumulated and stored by plants through photosynthesis.

Rad (Radiation Absorbed Dose) Measurement unit relative to the amount of radiation absorbed by a particular target, biotic or abiotic.

Radioactive Waste Any radioactive by-product of nuclear reactors or nuclear processes.

Radioactivity The emission of electrons, protons (atomic nuclei), and/or rays from elements capable of emitting radiation.

Rain Forest Forest with high humidity, small temperature range, and abundant precipitation; can be tropical or temperate.

Recycle To reuse; usually involves manufactured items, such as aluminum cans, being restructured after use and utilized again.

Red Tide Population explosion or bloom of minute single-celled marine organisms (dinoflagellates), which can accumulate in protected bays and poison other marine life.
Renewable Resources Resources normally replaced or replenished by natural processes; not depleted by moderate use.
Riparian Water Right Legal right of an owner of land bordering a natural lake or stream to remove water from that aquatic system.

Salinization An accumulation of salts in the soil that could eventually make the soil too salty for the growth of plants.
Sanitary Landfill Land waste disposal site in which solid waste is spread, compacted, and covered.
Scrubber Antipollution system that uses liquid sprays in removing particulate pollutants from an airstream.
Sediment Soil particles moved from land into aquatic systems as a result of human activities or natural events, such as material deposited by water or wind.
Seepage Movement of water through soil.
Selection The process, either natural or artificial, of selecting or removing the best or less desirable members of a population.
Selective Breeding Process of selecting and breeding organisms containing traits considered most desirable.
Selective Harvesting Process of taking specific individuals from a population; the removal of trees in a specific age class would be an example.
Sewage Any waste material coming from domestic and industrial origins.
Smog A mixture of smoke and air; now applies to any type of air pollution.
Soil Erosion Detachment and movement of soil by the action of wind and moving water.
Solid Waste Unwanted solid materials usually resulting from industrial processes.
Species A population of morphologically similar organisms, capable of interbreeding and producing viable offspring.
Species Diversity A ratio between the number of species in a community and the number of individuals in each species. Generally, the greater the species diversity composing a community, the more stable is the community.
Strip Mining Mining in which Earth's surface is removed in order to obtain subsurface materials.
Strontium-90 Radioactive isotope of strontium; it results from nuclear explosions and is dangerous, especially for vertebrates, because it is taken up in the construction of bone.
Succession Change in the structure and function of an ecosystem; replacement of one system with another through time.
Sulfur Dioxide (SO_2) Gas produced by burning coal and as a by-product of smelting and other industrial processes. Very toxic to plants.

Sulfur Oxides (SO_x) Oxides of sulfur produced by the burning of oils and coal that contain small amounts of sulfur. Common air pollutants.
Sulfuric Acid (H_2SO_4) Very corrosive acid produced from sulfur dioxide and found as a component of acid rain.
Sustainability Ability of an ecosystem to maintain ecological processes, functions, biodiversity, and productivity over time.
Sustainable Agriculture Agriculture that maintains the integrity of soil and water resources so that it can continue indefinitely.

Technology Applied science; the application of knowledge for practical use.
Tetraethyl Lead Major source of lead found in living tissue; it is produced to reduce engine knock in automobiles.
Thermal Inversion A layer of dense, cool air that is trapped under a layer of less dense warm air (prevents upward-flowing air currents from developing).
Thermal Pollution Unwanted heat, the result of ejection of heat from various sources into the environment.
Thermocline The layer of water in a body of water that separates an upper warm layer from a deeper, colder zone.
Threshold Effect The situation in which no effect is noticed, physiologically or psychologically, until a certain level or concentration is reached.
Tolerance Limit The point at which resistance to a poison or drug breaks down.
Total Fertility Rate (TFR) An estimate of the average number of children that would be born alive to a woman during her reproductive years.
Toxic Poisonous; capable of producing harm to a living system.
Tragedy of the Commons Degradation or depletion of a resource to which people have free and unmanaged access.
Trophic Relating to nutrition; often expressed in trophic pyramids in which organisms feeding on other systems are said to be at a higher trophic level; an example would be carnivores feeding on herbivores, which, in turn, feed on vegetation.
Turbidity Usually refers to the amount of sediment suspended in an aquatic system.

Uranium 235 An isotope of uranium that when bombarded with neutrons undergoes fission, resulting in radiation and energy. Used in atomic reactors for electrical generation.

Zero Population Growth The condition of a population in which birthrates equal death rates; it results in no growth of the population.

Index

A

"acid rain," 11, 17, 18, 19, 73, 75
aerosols, 174
affluence, as cause of environmental degradation, 44
Africa, 11; postpartum maternal deaths in, 52
Agenda 21, 119
"Agent Orange," 204
agrarian services, 156–160
agricultural biodiversity, 159
agricultural reform, 12
agriculture, and the environment, 53–54
AIDS, 11
air pollution, 147; coal use and, 73, 75
alternative energy, 79–82
Amazonia, 10
anemia, caused by iron deficiency, 52
animals: mass extinction of, 123; as providers of human essentials, 126
aquaculture, 118
aquifers, 193–203; groundwater scarcity and, 162–169
Árnason, Bragi, 91, 93, 95
Armillaria, 17
Arrow, Kenneth, 183
arsenic poisoning, 198, 199, 200
Asia: fluoride pollution in, 199; postpartum maternal deaths in, 52
Asian tiger mosquito, 117–118
Aspergillus sydowii, 21
Atomic Age, 79

B

backyard wildlife, 135
Bangladesh, arsenic poisoning in, 198
Becqurel, Edmund, 79
benzene, 205, 211
beta-carotene, 57
Benyus, Janini, 28–29
bicycles, 33–34
biodiversity, 108–115, 124, 126, 214
bioinvasions, 18, 19, 114, 116–122, 124
biological diversity, 171
biological pollution, 116
Biomimicry (Benyus), 28–29
biophilia, 133–134
Biosecurity Act of 1993, 119
Biosphere II project, 24
birders, 131–132
birth defects, due to dioxins, 204
birth rates, population growth and, 45
Black Sea, 14, 15, 21
black-band disease, 20
"bleaching" action, in coral, 20
blue-baby syndrome, 196–197
blue bunting, 130
boats, methanol used in, 93
Bongaarts, John, 45–46
biotechnology, 54, 56–58
Brazil, farmers of, 150–151
breastfeeding campaigns, 63
British Petroleum, 35
Broecker, Wallace, 194
Brooks, Thomas M., 110–111
brown tree snake, 114
Brownell, Kelly, 64, 67, 69
Bruce, Judith, 45
buses, hydrogen fuel cells used in, 93
Bush, George W., 82, 101, 182, 183

C

California, deregulation plan in, 97–102
California Fuel Cell Partnership, 94
California Power Exchange, 100
California Rice Industry Association, 28
Call of Distant Mammoths, The (Ward), 128
campylobacter, 159
carbon dioxide, 54, 75, 76, 174, 178
Carter, Jimmy, 79, 83
chemical contamination, 9
China, 31–32, 37, 38; coal uses in, 72; groundwater scarcity and, 163–164, 165; overeating in, 64; wind energy in, 104–105
chlorinated solvents, as groundwater pollutant, 201
chlorine, 207–208
chlorofluorocarbons (CFCs), 209
cholera, 14, 19
citizen science, 134–135
civilization, in the twenty-first century, 29
Clean Air Act, 12
Clean Water Act of 1972, 127, 213
climate change, 9, 18–19; as threat to Earth's oceans, 170; use of coal and, 75–77
climate, human impact on, 174–179
Clinton, Bill, 121, 213
coal, decline in use of, 72–78
Commoner, Barry, 44
common-pool resources, tragedy of the commons and, 142–149
Competitive Enterprise Institute (CEI), 12
complexity theory, 15
computers and climate interactions, 175–177
ConAgra, 154
conflict resolution, the commons and, 147
conglomerates, 154, 155–156
consumption, of the environment, 44–51
contraception, 45
Convention on International Trade in Endangered Species of Wild Fauna and Flora (CITES), 123–124, 125, 127
coral reefs, 19–22, 170
corporate power, 24–29
corrosion, 9, 11
"Credit with Education" program, in Ghana, 67

D

DaimlerChrysler, 92, 93, 94
Danube River, 14, 15
DDT pesticide, 126, 196, 205, 206, 210
De Alessi, Michael, 12
"dead zone," 172
death rates, population growth and, 45
decarbonization, 47
DeFazio, Peter A., 99
deforestation, 9, 10, 12, 182
degradation, of the environment, 44–51
dematerialization, 47
"demographic transition," 42, 45
dengue fever, 118
Denmark, wind energy in, 104
Devonian extinction, 109, 110, 128
Diadema antillarum, 20
dinoflagellates, 14
dioxins, 204
discontinuity, of environmental change, 9
Diversity of Life, The (Wilson), 123
diversity, of monoculture technologies, 11–12
dodo, 113

Doubly Green Revolution, 54–56, 58; policies for, 56
Dow Louisiana Operations, 25
drip irrigation, groundwater scarcity and, 167
Duffy, Mike, 153
Durning, Alan, 48, 112, 115

E

Earth Day 2000, 44, 85, 212–214
Earth Report 2000, 12
Earth Summit in Rio de Janeiro, 119
earthquakes, in Iceland, 88
ecological poisons, 205
ecological stability, 158
ecology: application of, 58; consumption and, 46
economics, consumption and, 46
ecotourism, 135, 136–137
Ecotourism and Sustainable Development: Who Owns Paradise? (Honey), 136
Ehrenfeld, David, 15
Ehrlich, Paul, 44, 81
El Niño, 19–20, 33–39
electrolysis process, 91, 94, 95
Elton, Charles, 113
encephalitis, 118
endangered species, 152
Endangered Species Act, 126, 127
"energy crisis," 79, 102
energy regulation bill, 99
energy transformation, 46–47
Enron, 35, 97–102
Environmental Protection Agency (EPA), 80, 195, 196, 197, 198, 200
Ethiopia, malnutrition in, 61
evaporation, 170
exotic species. *See* bioinvasions

F

farmers, 150–161; participation of, 58–59
fatal error clause, 184
Federal Energy Regulatory Commission (FERC), 99, 100, 101
Feldstein, Martin, 183
fertility rates, population growth and, 42, 43
Finland, nutrition in, 66
fire cycles, 18, 19
Fischer, Hanns, 27
fishing, 143–146
fluoride, as natural groundwater pollutant, 199, 201
food, for all, in the twenty-first century, 52–60
food security, 55
forests, 16–19, 21; fires in, 9–10
fossil aquifers, 164, 166
fossil-fuel emission, 174
frost-free season, 182
fuel cell, 91, 93
furrow diking, groundwater scarcity and, 167–168

G

"gangster capitalism," 29
gas, prices of, in the United States, 81
gender bias, and cultural prejudices, 62
genes, of commercial crops, 54
genetic diversity, 126, 172
genetic engineering, 56, 57
genetically modified (GM) crops, 54
genetically modified organisms (GMO), 160

225

geothermal energy, 89, 91
Germany, 36, 75; wind energy in, 104
global invasion database, 121
global pollution, 54
global transition, 45
global warming, 174–175, 180, 181
Goldschmidt, William, 156
Gore, Al, 184
Great Man-Made River Project, 166
Green Revolution, 52, 55–56, 58, 81, 196–197
greenhouse effect, 175, 176–177, 178, 179, 180–185
greenhouse gas emissions, 89
groundwater: polluting of, 193–203; scarcity of, 162–169

H

habitat degradation, 18, 19, 110, 124; as threat to Earth's oceans, 170, 172
habitat fragmentation, 110, 114
Hardin, Garrett, 142, 143, 144, 146, 148
Hazardous Substances and New Organisms Act of 1996, 119
health, coal use and, 72, 73–74
health indicators, of the environment, 213
heavy metals, as groundwater pollutant, 201
Heffernan, Bill, 155, 156
hemlock woolly adelgid, 17
herbicides, 204
Holdren, John, 44
Homer-Dixon, Thomas F., 112–113
Homo sapiens, 109, 111, 112, 114, 115
Honduras, predicament in, 8–9
Honey, Martha, 136
horseshoe crabs, 145–146
human-induced bioinvasions, as primary threat to oceans, 117
hunger. *See* malnutrition
hunting, 11; as cause of extinction, 124; of wildlife, 130–139
Hurricane Mitch, 8–9
hurricanes, 180
hydroelectricity, 89
hyrogren experiment, in Iceland, 87–96

I

Icelandic Hydrogen and Fuel Cell company, 92, 93
Independent System Operator (ISO), 100
India: arsenic poisoning in, 198; groundwater scarcity in, 163, 164–165; malnutrition in, 62
industrial monocultures, 12
Industrial Revolution, 79, 89
industrialized countries, 52
infant methemoglobinemia, 197
infectious diseases, 18, 19
inorganic chemicals, 206
insecticide, 54
instituting programs, of restoration, 25
integrated pest management (IPM) 58
International Food Policy Research Institute (IFPRI) model, 53
International Plant Protection Convention of 1951, 119
invasive species. *See* bioinvasions
IPAT identity, 44, 49, 50
Iron Gates Dam, 14, 15
irrigation, groundwater scarcity and, 162–169

J

Jablonski, David, 109, 110, 111, 114, 115
Japan, decline of farmers in, 152
jellyfish, 14, 15
jet skis, 146
junk food, 67

K

Karl, Thomas, 182
Kuhn, Thomas, 31
Kyoto Protocol of 1997, 89–91, 96, 179, 184, 185

L

labor, decline in coal use and, 77
Landless Workers Movement, 160
landscape conversion, 111, 114
Levins, Dick, 157
Libya, 166
life-saving cures, from marine species, 171
Living Machines, 28
low-energy precision application (LEPA) sprinklers, groundwater scarcity and, 167, 168

M

MacMillan Bloedel, 35
macroalgae, 172
Madagascar, 111
malaria, 8, 9
malnutrition: of children, 52; poverty and, 61–69
Malthus, Thomas, 81
"managed invasion," 118–119
mangroves, 21
marine conservation, 12
Marine Mammal Protection Act, 127
marine systems, 170
Maruska, Edward J., 123
mass extinction, 109, 113, 123–129
mass-media educational campaigns, 66
material transformation, 46–47
material use, 34; reinventing and, 162–169, 188, 192
materials efficiency, reinventing material use and, 189–191
matrix, of trouble, 18–19
May, Robert M., 110, 113
McDonough, Bill, 27
methane, 54, 76
methanol, 93, 95
microcredit initiatives, 63
micronutrient deficiencies, 68
micronutrients, lack of, 52
minerals, lack of, 52
Mining Law, 191
Mississippi, 150
Mnemiopsis leidyi, 117
monoculture technologies, 11–12, 22
Montreal Protocol, 12–13
Morse, Phillip M., 81
mosquitoes, 8
Mozambique tilapia, 118
MTBE additive, in gasoline, 197
Myers, Norman 9, 110

N

National Energy Plan, 79
National Environmental Trust (NET), 85
National Research Council, 44, 46
natural capital, 24, 27; investment in, 28–29
nature, ethics on, 11
New Zealand, decline of farmers in, 152
nitroaromatic compounds, 195
nitrogen oxide, 16, 74–75
nitrogen pollution, 14, 17, 18, 19, 20, 21–22; in groundwater, 196–197, 201
Nixon, Richard M., 212
nongovernental organizations (NGO), 37
Non-Indigenous Aquatic Nuisance Act, 127
Normal Accidents: Living with High Risk Technologies (Perrow), 15
Norsk Hydro, 92–93
nutritional education, 63

O

obesity, among people, in the United States, 61, 63
oceans, 170–173
official control program, 121
Ogallala acquifer, groundwater scarcity and the, 164, 165–166, 167, 168
oil companies, 79–80, 82
Olsen, David, 123
Oppenheimer, Michael, 182
Ordovician extinction, 109, 128
Organization of Petroleum Exporting Countries (OPEC), 80
organochlorines, 207, 208
Ostrom, Elinor, 143, 147
overfishing, as: cause of extinction, 124; as threat to Earth's oceans, 170
ozone, 16, 17, 74

P

Pakistan, groundwater scarcity in, 163, 164, 165
paleontologists, 108, 109
Pamplona, Spain, wind energy in, 103
participatory learning and action (PLA), 58, 59
passenger cars, and private fuel cells, 93, 94
pathogens, of globalization, 116–122
peasant farmers, 158
peat bogs, 89
Permian extinction, 109, 128
Perrow, Charles, 15
persistent organic pollutants (POP), 172, 204–211
personal watercraft, 146
pesticide-resistant biotype, 118
pesticides, 54; as groundwater pollutants, 196, 201; used in monocultural farming, 210
Peterson, Roger Tory, 133
petrochemicals, 197–198, 201
petroleum, 89–91
Philippines, decline of farmers in the, 152
photosynthesis, 170, 171, 172
photovoltaic (PV) cells, 79; panels, 83, 84
phthalates, 209
physics, consumption and, 46
phytoplankton, 171, 172
pigeons, 113
Pimm, Stuart L., 110–111, 113
plantation, 119
plants: mass extinction of, 123; as providers of human essentials, 126
plastid's genome, 54
pneumoconiosis, 74
Poland, decline of farmers in, 152
pollution: as a cause of extinction, 124; as a threat to Earth's oceans, 170

polychlorinated biphenyls (PCBs), 205, 206, 208
polyvinyl chloride (PVC), 208, 210–211
Population Bomb, The (Ehrlich), 81
population growth, 18, 19, 34–35, 43, 111–112, 152–153; as cause of environmental degradation, 44–51
Population/Consumption (PC) version, 49, 50
poverty, malnutrition and, 6–69
pregnancy, and nutrition, 63
privatization, of environmental resources, 12
product life, reinventing material use and, 190
profit-maximizing capitalists, 24
Project Feeder Watch, 135
prone-to-obesity myth, 63–67
property rights, the commons and, 143–144
protein, lack of, 52
"public benefits trusts," 84
public health approach, 66–67
public lands, 146

Q

Qaddafi, Muammar, 166
quantitative analysis, of consumption, 47

R

Raup, David M., 128
Raven, Peter H., 126, 128
Reagan, Ronald, 80, 83
recycling, reinventing material use and, 189–191
red tides, 14, 21, 172, 206
regional pollution, 54
Reid, W. V., 110
"Renewable Portfolio Standards" (RPS), 84
Republic of Congo, 11
reservoirs, 162, 163
resource depletion, and consumption, 47–48
resource efficiency, increase of, 25
Resources for the Future, 183
"reverse logistics," 28
Reykjavik, Iceland, hydrogen experiment in, 87–96
"ripple effects," 12
riverbedfellows, 160
Robbins, Mark, 151
Royal Dutch Shell, 35
Ryan, John, 48

S

Safina, Carl, 145
Salmonella, 159
salt, as a natural groundwater pollutant, 199
satisfaction, of consumption, 48, 50
Saudi Arabia, 166
Savory, Allan, 28
scarcity myth, 62
"scenario planning technique," 92
Schilham, Jan, 26
Schor, Juliet, 48
sea trade, 170
sea turtles, 20
seabed mining, 171
seagrass, 21
second-order effect, 10
service providers, reinventing material use and, 188–189
ships, as carrier of species, 117
shrimp farming, 118–119
shrinking, of consumption, 48
Sigfússon, Thorsteinn, 93
simplicity movement, 48
Simon, Julian, 110, 112
Sinking Ark, The (Myers), 110
Skilling, Jeffrey K., 99, 100, 101
smog alerts, 212
snowbirds, 131
sociology, consumption and, 46
Solar Energy Industries Association (SEIA), 83
solar power, 33, 79–82, 83–86
Southwire Corporation, 25
Spain, wind energy in, 103
Spiral/Grace Project, 161
sprinklers, groundwater scarcity and, 167
standards, for peasant farmers, 158
starfish, 20
state-of-the-shelf technologies, 25–26
Stavins, Robert, 184
structural adjustment requirements, for peasant farmers, 158
Structure of Scientific Revolutions, The (Kuhn), 31
Stuff: The Secret Lives of Everyday Things (Ryan & Durning), 48
sublimation, of consumption, 48, 50
sulfur dioxide, 74–75
sunlight, 79, 83–86
"Superfund" program, 195, 200
surface fires, 10
surge valves, groundwater scarcity and, 167
Sweden, decline of farmers in, 152
synergism: of environmental change, 9; as primary threat to oceans, 171
synthetic chemicals, 205–206
systems theory, 15

T

taxes: on food, 67, 69; shifting of, from income to environmental, 35–37
technology, as cause of environmental degradation, 44
Texas, groundwater scarcity and, 162, 167–168, 169
textbook optimization, 26
thresholds, 30–39
Todd, John, 28
tornadoes, 181
Toyota, 190
trade liberalization policies, for peasant farmers, 158
tragedy of the commons, 142–149
transgenic cereals, 57
trawling, 172
tree plantation, 119
Triassic extinction, 109, 128
Turner, Ted, 38

U

ultraviolet (UV) radiation, 18–19, 172
United Nations Framework Convention on Climate Change, 179
United States, decline of farmers in, 152
unnoticed trend, of environmental change, 9
United States, obesity in, 61
urban life, 152
urban transport, 32–34
U.S. Army Corps of Engineers, 150
U.S. Energy Policy Act of 1992, 99

V

vitamin A deficiency, 52, 68
vitamins, lack of, 52
volcanoes, in Iceland, 88

W

Ward, Peter D., 128
washing machines, reinventing material use and, 189
waste, 29; climate concept of, 25, 27
waste taxes, 192
watching, of wildlife, 130–139
water, groundwater scarcity and, 162–169
Wilderness Act of 1964, 127
wildlife-watching tourism, 131
Wilson, E. O., 121, 123, 125–126, 133–134
wind energy, 33, 103–105
women, and hunger, 64
World Energy Modernization Plan, 86
World Health Organization (WHO), 61
world trade, as most dangerous form of environmental decline, 116
World Wildlife Fund (WWF), 123, 124, 125, 127
Wurmfeld, Charles, 80

X

Xerox Corporation, 189

Y

yellow fever, 118

Z

"zero-emissions," 94, 95; fuel cells, 183

Test Your Knowledge Form

We encourage you to photocopy and use this page as a tool to assess how the articles in **Annual Editions** expand on the information in your textbook. By reflecting on the articles you will gain enhanced text information. You can also access this useful form on a product's book support Web site at **http://www.dushkin.com/online/**.

NAME: DATE:

TITLE AND NUMBER OF ARTICLE:

BRIEFLY STATE THE MAIN IDEA OF THIS ARTICLE:

LIST THREE IMPORTANT FACTS THAT THE AUTHOR USES TO SUPPORT THE MAIN IDEA:

WHAT INFORMATION OR IDEAS DISCUSSED IN THIS ARTICLE ARE ALSO DISCUSSED IN YOUR TEXTBOOK OR OTHER READINGS THAT YOU HAVE DONE? LIST THE TEXTBOOK CHAPTERS AND PAGE NUMBERS:

LIST ANY EXAMPLES OF BIAS OR FAULTY REASONING THAT YOU FOUND IN THE ARTICLE:

LIST ANY NEW TERMS/CONCEPTS THAT WERE DISCUSSED IN THE ARTICLE, AND WRITE A SHORT DEFINITION:

ANNUAL EDITIONS revisions depend on two major opinion sources: one is our Advisory Board, listed in the front of this volume, which works with us in scanning the thousands of articles published in the public press each year; the other is you—the person actually using the book. Please help us and the users of the next edition by completing the prepaid article rating form on this page and returning it to us. Thank you for your help!

ANNUAL EDITIONS: Environment 01/02

ARTICLE RATING FORM

Here is an opportunity for you to have direct input into the next revision of this volume. We would like you to rate each of the 28 articles listed below, using the following scale:

1. **Excellent: should definitely be retained**
2. **Above average: should probably be retained**
3. **Below average: should probably be deleted**
4. **Poor: should definitely be deleted**

Your ratings will play a vital part in the next revision. So please mail this prepaid form to us just as soon as you complete it. Thanks for your help!

We Want Your Advice

RATING	ARTICLE
	1. Environmental Surprises: Planning for the Unexpected
	2. The Nemesis Effect
	3. Harnessing Corporate Power to Heal the Planet
	4. Crossing the Threshold: Early Signs of an Environmental Awakening
	5. The Population Surprise
	6. Population and Consumption: What We Know, What We Need to Know
	7. Food for All in the 21st Century
	8. Escaping Hunger, Escaping Excess
	9. King Coal's Weakening Grip on Power
	10. Oil, Profit$, and the Question of Alternative Energy
	11. Here Comes the Sun: Whatever Happened to Solar Energy?
	12. The Hydrogen Experiment
	13. Power Play

RATING	ARTICLE
	14. Bull Market in Wind Energy
	15. Planet of Weeds
	16. Invasive Species: Pathogens of Globalization
	17. Mass Extinction
	18. Watching vs. Taking
	19. The Tragedy of the Commons: 30 Years Later
	20. Where Have All the Farmers Gone?
	21. When the World's Wells Run Dry
	22. Oceans Are on the Critical List
	23. The Human Impact on Climate
	24. Warming Up: The Real Evidence for the Greenhouse Effect
	25. Making Things Last: Reinventing Our Material Culture
	26. Groundwater Shock: The Polluting of the World's Major Freshwater Stores
	27. POPs Culture
	28. It's a Breath of Fresh Air

(Continued on next page)

ANNUAL EDITIONS: ENVIRONMENT 01/02

BUSINESS REPLY MAIL
FIRST-CLASS MAIL PERMIT NO. 84 GUILFORD CT
POSTAGE WILL BE PAID BY ADDRESSEE

McGraw-Hill/Dushkin
530 Old Whitfield Street
Guilford, CT 06437-9989

ABOUT YOU

Name _____ Date _____

Are you a teacher? ☐   A student? ☐
Your school's name _____

Department _____

Address _____ City _____ State _____ Zip _____

School telephone # _____

YOUR COMMENTS ARE IMPORTANT TO US!

Please fill in the following information:
For which course did you use this book? _____

Did you use a text with this *ANNUAL EDITION*? ☐ yes ☐ no
What was the title of the text? _____

What are your general reactions to the *Annual Editions* concept? _____

Have you read any particular articles recently that you think should be included in the next edition? _____

Are there any articles you feel should be replaced in the next edition? Why? _____

Are there any World Wide Web sites you feel should be included in the next edition? Please annotate. _____

May we contact you for editorial input? ☐ yes ☐ no
May we quote your comments? ☐ yes ☐ no